BIOLOGY OF MOLLUSCA

By

Dr. D.R. Khanna
Reader
Reader in Zoology
Gurukul Kangri University
Haridwar (Uttaranchal)

&

Dr. P.R Yadav
Reader
Department of Zoology
D.A.V. College
Muzaffarnagar (U.P.)

DISCOVERY PUBLISHING HOUSE
NEW DELHI-110002

First Published-2004

ISBN 81-7141-898-8

Published by
DISCOVERY PUBLISHING HOUSE
4831/24, Ansari Road, Prahlad Street,
Darya Ganj, New Delhi-110002 (India)
Phone: 23279245 • Fax: 91-11-23253475
E-mail:dphtemp@indiatimes.com

Printed at:
Tarun Offset Printers,
Delhi–110 053

Preface

The present title "Biology of Mollusca" has been carefully organized for undergraduate, post graduate and those involved in Agricultural studies. The text has been designed to approach the classification, morphology, anatomy, physiology and development of all required types in a simple lucid and straight forward manner. The text is not encyclopaedic in scope, but also of introductory coverage, commensurate, concise, comprehensive yet exhaustic presentation. We sincerely feel that this publication will adequately meet the long felt need of the students for a simple and upto date book and hope that the method of presentation will be liked and found beneficial.

It has been the constant endeavour of the authors to furnish maximum substances, keeping in view the limitations of size of the book. Efforts have been made to condense the matter as far as practicable. Attempts have also been made to simplify the language and style of narration to the extent that it may be within the reach of students.

A text book can not be written without the support and professional contributions of many people. This book is no exception. The authors are grateful to those teachers and colleagues whose stimulating discussions have clarified certain vague points for them, but all errors, omissions, and corrections needed are solely their responsibility.

The authors tried hard to be accurate and upto date in statement and realise the impossibility of completely avoiding errors therefore, the authors will greatly appreciate having their attention called to any questionable statement.

The authors are deeply grateful to all their friends and colleagues for their expertise and advice during the preparation of the present title.

The authors express their gratitute to Mr. Wasan and staff of M/s Discovery Publishing House for their whole hearted co-operation in the publication of this book.

Authors

Contents

1

SURVEY OF MOLLUSCS

Members of the phylum *Mollusca* are among the most conspicuous invertebrates and include such familiar forms as *clams*, *oysters*, *squids*, *octopods* and *snails*. During the last 18th and 19th centuries, when natural history, occupied the time of many well to do gentlemen, shell collections were as popular as today. Such collections, often containing species gathered from various parts of the world, have contributed considerably to our knowledge of the phylum. In abundance of species, molluscs comprise the largest invertebrate phylum aside from the Arthropoda.

GENERAL CHARACTERISTICS

1. Next to the arthropods, Mollusca is the largest described invertebrate phylum, numbering perhaps 100,000 or more species. At least 35,000 species of fossils are known. There may be more species existing among the nematodes and protozoans, but as yet not as many species have been described in those groups.
2. The plan of structure involves a body divided into two functionally distinct regions—the upper region (visceropallium) composed of a visceral hump and an overhanging mantle specialized for mucous secretion and ciliary action and the lower region (head-foot complex) composed of the head and a foot whose activity is chiefly muscular.
3. Although molluscs have diverged into numerous evolutionary lines, the body plan of all can usually be traced back to a single ancestral pattern. The plan of structure and function is uniform throughout the phylum, but there is no standard shape. Homologies involving the mantle and gills and functional homologies stressing ciliary

and mucous mechanisms in feeding and other processes are easily established.

4. As a group molluscs have added or stressed a soft extension of the outer layer of body wall (the many functioned *mantle*), the flexible file-like tongue (the *radula*) the highly protective *shell*, the ventral muscular *foot*, an effective respiratory system of *gills* or *lungs*, and efficient *eyes*.
5. Although most molluscs are marine, one group—the gastropods—has been successful also on land and in fresh water. Bivalves are basically filter feeders and so are unknown on land. Cephalopods differ from other mollusks in their very high level of efficiency in locomotion and behaviour. They stand high in invertebrate evolution.
6. Mollusks range from forms that are only a few millimeters in diameter to the largest of all invertebrates many meters long.

Phylogeny and Adaptive Radiation

Little is known with certainty about the relationship of mollusks to other phyla. A possible connection between flatworms (Turbellaria) and mollusks is seen in the shell-less Aplacophora (solenogastrids) which have some of the features of chitons and also certain reproductive tubes similar to those of primitive flat worms. A relationship with annelids is evidenced by a similar trochophore larva found in both phyla. The original ancestral stock of mollucks may have given rise to two series of schizocoelomates; one of these was unsegmented (mollusks) and the other was arranged metamerically (annelids). The discovery in 1952 of *Neopilina* with its somewhat superficial segmentation has strengthened the belief that annelids and mollusks came from a common stem. However, segmentation is absent in moluscan embryos and in larvae of *Nepopilina*.

Many zoologists, moreover, think the segmentation in *Neopilina* is not the same type as that in annelids but represents an aberrant example of convergent evolution. Molluscs have a basic design of two interacting symmetries—a visceropallium composed of viscera and overhanging mantle, and a head-foot complex. The head-foot part has a definite bilateral symmetry, the visceropallium at first had radial symmetry, with a tendency later toward biradial symmetry. This basic design has been modified in various ways by different groups of molluscs. For instance, the head and foot may be fused together (cephalopods), or the head may be absent altogether (pelecypods). The mantle has been one of the most versatile structure in molluscan evolution and is modified to form the mantle cavity, gills or lungs, siphons or apertures, locomotory structures, feeding processes, etc. Torsion of the body and coiling of the shell (gastropods) produce an asymmetric growth and changes in the location of many basic structures.

The shell has many evolutionary adaptations, such as heavy shells for withstanding stress and pressure, boring and burrowing shells for penetration, swimming shells for locomotion, and reduced (or absent) shells for preventing encumbrance. In cephalopods the body is lengthened along the dorsoventral axis, which becomes the functional anteroposterior axis. Whereas the foot is the chief organ of locomotion in gastropods and pelecypods, locomotion in cephalopods involves jet propulsion of water from the mantle cavity through a funnel.

General Survey of Molluscs

The molluscs include such forms as chitons, clams, mussels, oysters, snails, slugs, octopuses, squids, nautiluses and some others. The group has enormous diversity, both in shape and size. Their ecology is also very diversified, for they have been able to live in nearly all kinds of habitats—from the depths of the oceans to some of the highest altitudes recorded for animal life. The shell-bearing forms are restricted somewhat by the calcium content of their environments, but the shell-less molluscs can live even in regions that are more or less devoid of lime. As a consequence, molluscs are found in the sea, brackish water, fresh water, and on land. They share with the arthropods the capacity to adapt themselves to terrestrial life better than most groups in invertebrates. They include some of the slowest and most sluggish of animals and some of the swiftest invertebrates.

In structural and functional adaptations, molluscs rank with the arthropods because some of the most unique invertebrate behaviour patterns are found among them. Nutritionally they range from primitive filter feeders to highly carnivorous predators. Most molluscs live in the sea. Three of their classes—Amphineura, Scaphopoda, and Cephalopoda—are not represented at all on land or in fresh water. Both gastropods and bivalves have colonized fresh water; however, they are fewer in number than their representatives in the sea. The chief physiologic requirements for conquest of the land involve air breathing, temperature regulation, and water control. Bivalves have successfully invaded fresh water but their ciliary feeding method has prevented their colonization of land. Only the gastropods have successfully invaded terrestrial habitats. Although both prosobranch and pulmonate gastropods are found in fresh water, the larger number belong to the pulmonates.

Mulluscs are ceolomate, unsegmented (except class monoplacophora) and bilaterally symmetric animals. The mollusc typically has a head, a visceral mass, a mantle, a shell, and a foot. The head is located at the anterior end, the compact visceral mass is enclosed in a soft integument; the mantle is a modified dorsal part of the integument and hangs down as free folds or flaps around the body to form a mantle cavity; the calcareous shell is secreted by the mantle; and the ventral muscular foot, modified from the integument, is the organ of locomotion. All these structures may be modified in various ways in the different groups of molluscs. A head, shell or mantle may be lacking altogether in some. The shell may be bivalved, univalved, spiral, or cone shaped.

Most molluscs have a rasping organ, the radula. This is a kind of toothed tongue that bears denticles or teeth. The exact status of the monoplacophorans is very much in doubt. Their metamerism is slightly superficial, involving only some of the systems and lacking external segmentation of the body wall altogether. At present they cannot be considered a basic mollusc type because of their irregular metamerism and isolated case. If metamerism were a primitive molluscan feature, to account for its total absence in other molluscs would be difficult. All this may be interpreted to mean that metamerism in the monoplacophorans has been secondarily acquired. More embryologic evidence may shed some light on the problem. Stronger evidence for

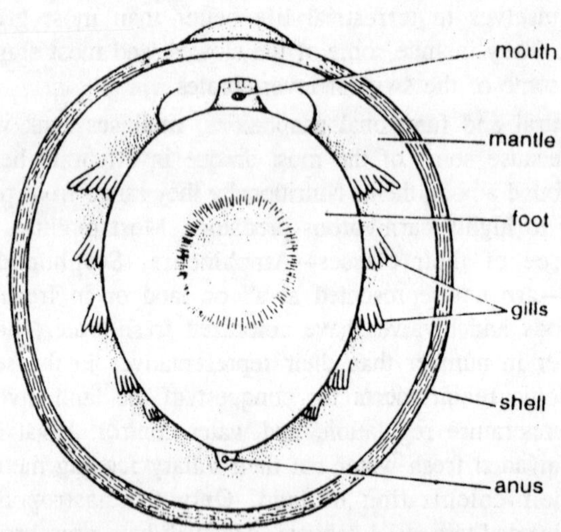

Fig. 1.1. Neopilina galatheae.

the annelid affinity is shown perhaps by the resemblance between the embryology of molluscs and that of polychaetes. A similar spiral cleavage is found in both groups, with the exception of the cephalopods.

The relations between the various molluscan classes are also obscure. Although logical explanations can be made for the modifications of most groups from the hypothetical ancestral type, these throw little light on possible interrelationships. Each class seems to have distinct evolutionary tendencies. The amphineurans have departed least from the primitive plan, although their fossil record is not so old as that of more specialized groups. Convergence of characters may well add to the confusion about relationships. The gastropods and the cephalopods share many characters, although there is a wide gulf in basic plan between the two classes. At present there are few logical reasons for grouping the classes according to relationships.

The discovery (1959) of a bivalved gastropod, *Tamanovalva limax*, off the coast of Japan, and similar ones elsewhere, may suggest aspects of relationship of great interest. This type has the characteristic valves of a bivalve and the sluglike head and foot of a gastropod. The evolutionary appraisal of this and related species awaits further study. Within the phylum two classes, the gastropods and the pelecypods (bivalves), each demonstrate some rather striking phylogenetic relationships and evolutionary differentiations among the members of the class. The most primitive basal gastropods are the Prosobranchia with torsion and a helical shell. Prosobranchs exhibit two grades—those with bipectinate (aspidobranch) gills (2 rows of gill leaflets on the branchial axis) and those with pectinibranch gills (1 row of gill filaments). From the pectinibranchs the evolutionary lines lead to the subclasses—the Opisthobranchia and the Pulmonata.

The evolutionary trends in the Opisthobranchia have led to a reduction or loss of the shell, detorsion loss of streptoneury, and reduction of visceral mass. The pulmonates retain many of the prosobranch characteristics but have lost streptoneury, and some are shell-less. Most evident evolutionary changes in gastropods have been associated with torsion and modifications of the respiratory organs and foot. The fossil record indicates that the prosobranchs date from the Paleozoic era, whereas the Opisthobranchia and the Pulmonata appeared in the Carboniferous period. The bivalves have been a slowly evolving group of organisms. Many date back as far as the early Paleozoic, whereas others have arisen in the Recent epoch. The Protobranchia are considered the most primitive group of existing bivalves, although they have some specialized characters. The most primitive of the living bivalves is the protobranch *Nucula*. *Nucula* has been a pair of equal adductors and two pairs each of protractor and retractor muscles. Its gill structure is not unlike that of the gastropod *Haliotis*. Except in the Septibranchia, there are 2 gills in all bivalves. The primitive form of the gill consists of a vascular axis carrying a series of filaments of opposite sides. This is the type found in the Protobranchia. From this type of gill has emerged the various modifications found in different groups. These will be described later.

Ancestral Form

Students of mollusks have reconstructed a hypothetical primitive mollusc, which is supposed to have existed in Precambrian seas. Such as generalized ancestor was bilaterally symmetric with a dorsally arched body. Its head bore a pair of tentacles with ocelli at their bases, and its mouth was provided with a protrusible radula similar in structure and arrangement to that in modern snails. A dorsal mantle (pallium) extended over the body and secreted the arched, horny shell. Ventrally, a flat, muscular creeping foot was attached by muscles that could draw the foot into the protective shell. Between the mantle and foot at the posterior end was a mantle cavity containing paired gills openings to the nephridia (kidney), and the anus. Within the axis of each gill, blood vessels provided for the exchange of respiratory gases. Water currents were drawn into the cavity by beating cilia on the gill

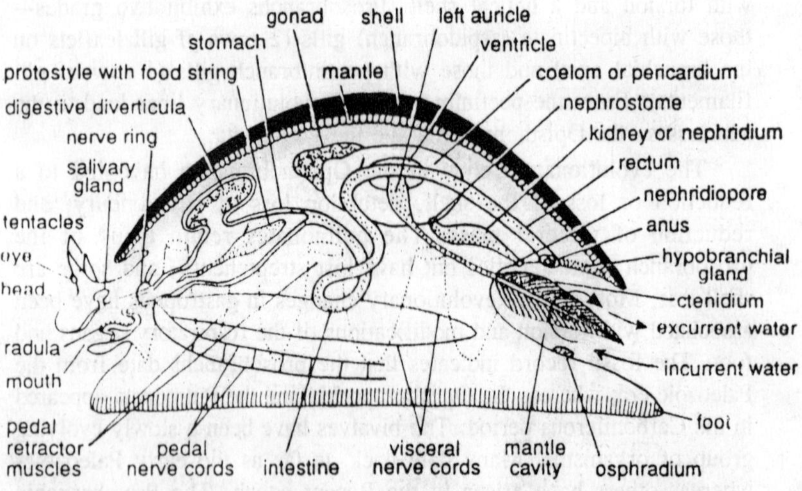

Fig. 1.2. Schematic lateral view of an archimollusc (hypothetical molluscan ancestor).

filaments. A hypobranchial gland between each gill and the rectum helped carry waste particles into the dorsal branchial chamber for rejection. A small sensory patch, the osphradium, at the base of each gill tested the chemical nature of the incoming current. The digestive system was lined with cilia and mucous glands.

The posterior part of the stomach was differentiated into a style sac and a ciliated groove extended along the stomach floor. The food of this early ancestor was probably algae and other material scraped from rocks. Digestion may have been primarily intracellular. The early coelomic cavity was mostly the pericardium, surrounding the three-chambered heart. Blood circulation was similar to that of present-day molluscs. There were a pair of gonads and a pair of metanephridia, which emptied their products into the mantle cavity.

The nervous system was made up a central nerve ring around the esophagus, with two pairs of longitudinal nerve cords, one running to the mantle and viscera and the other to the muscles of the foot. Nerve cells were probably diffused and not gathered into ganglia. Sense organs were represented by the tentacles, ocelli, statocysts in the foot, and osphradia. From such a hypothetical ancestral form the various groups of mollusks could have evolved by adaptive radiation into the diversity demonstrated by both the fossil record and the existing classes. Phylum Mollusca is divided into six classes. Monoplacophora, Amphineura, Gastropoda, Scaphopoda, Pelecypoda and Cephalopoda.

Class–I Monoplacophora

1. Abyssal or fossil, from lower Cambrian to Devonian probably carnivorous.
2. Bilaterally symmetrical and metamerically segmented but segmentation internal.
3. Shell dorsal and consisting of a single piece (hence Monoplacophora).
4. Mantle covering the dorsal surface of body.
5. Foot is ventral with a flat creeping sole.
6. Mouth is anterio-median and the anus is posterio-median.
7. Head is without eyes.
8. 5 or 6 pairs of gills externally and serially arranged.
9. Six pairs of nephridia and two pairs of gonads.
10. Stomach contains a crystelline style and the animal feeds on radiolarians.
11. Nervous system primitive, lacks ganglia.

12. Sexes are separate.
13. Development is indirect with a trochophore larva.

    Example: *Neopalina galatheae.*

Class–II Amphineura

(Gr; *Amphi* = both; *neuron* = nerve)

1. Most primitive molluscs with either elongated and vermi form or dorso-ventrally flattened body.
2. Head indistinct and without eyes and tentacles.
3. Foot broad, flat, sole-like and ventral in position.
4. Shell and mantle covering the body from dorsal and lateral sides.
5. 6-60 pairs of gills situated laterally in the pallial cavity on either side of foot. They may be reduced to a pair.
6. Mouth and anus at the opposite ends of body.
7. Nervous system primitive and ladder-like with or without ganglia.
8. Heart with one or two auricles.
9. A pair of nephridia connected to the pericardium.
10. Sexes separate or united.

Class Amphineura is divided into two orders:

Order-1. Polyplacophora

1. Body elliptical, convex dorsally and flattened ventrally.
2. Shell dorsal and formed of eight calcareous pieces arranged in a longitudinal row.
3. Mantle with spicules disposed along the margins.
4. Gills 6 to 8 pairs, present in the pallial groove on the lateral sides of foot.
5. Head devoid of eyes and tentacles.
6. The whole ventral surface is occupied by the broad and flat foot.
7. Alimentary canal is coiled with well formed redula.
8. Heart with two auricles.
9. Nervous system without definite ganglia.
10. Sexes separate.

    Examples: *Chiton, Chactopleura.*

Order-2. Aplacophora

1. The body is long, narrow, worm like.
2. The shell is absent.
3. Mantle contains tiny calcareous spicules.

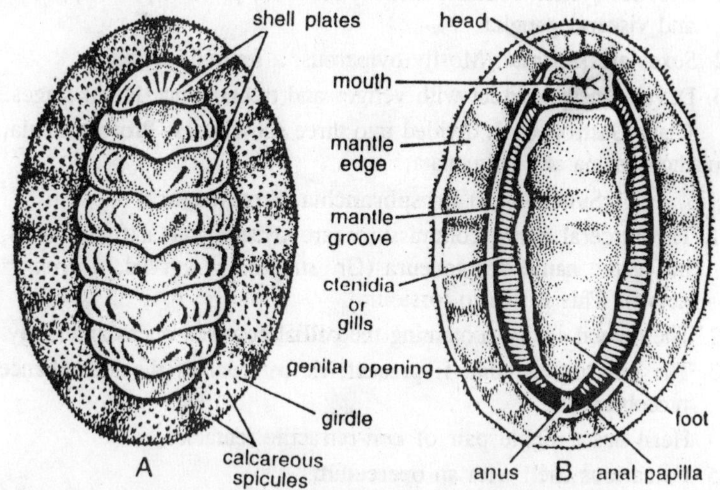

Fig. 1.3. Chiton. A—Dorsal view; B—Ventral view.

4. Foot is reduced or absent.
5. Heart consists of a single auricle and a ventricle.
6. Gills are absent or reduced to a pair located in the cloacal cavity.
7. Alimentary canal is straight without redula.
8. Nervous system with a distinct brain and ganglia.
9. Sexes united.

Examples: *Chaetoderma, Neomenia*

Class–III Gastropoda

(Gr. *Garstros* = stomach; *podos* = foot)

1. Body is asymmetrical with a mantle and shell.
2. The shell is univalved and spirally coiled.
3. Head is distinct bearing tentacles, eyes, and a mouth.
4. Foot is ventral and muscular.
5. Visceral mass usually coiled exhibiting various types of torsions.
6. Anus usually situated anteriosly close to the month.
7. Buccal cavity contains an odontophore with a redula.
8. Respiration is carried on by integument or gills or lungs.
9. Circulatory system open type with a dorsal heart enclosed in pericardium.
10. Excretory system includes one or two kidneys or nephridia.

11. Nervous system contains distinct cerebral, pleural, pedal, parietal and visceral ganglia.
12. Sexes are separate. Mostly oviparous, a few viviparous.
13. Development indirect with veliger and trochophone larval stages.

Class Gastropoda is divided into three sub-classes. Prosobranchia, Opisthobranchia and Pulmonata.

Subclass (A) Prospbranchia (Streptoneura)

1. The visceral nerve commissures are twisted into a figure of 8, hence the name streptoneura (Gr. *streptos* = curved, *neuron* = nerve). This is due to torsion.
2. The mental cavity containing the pallial complex opens anteriorly.
3. The gills or ctenidia, if present, lie in front of the heart hence prosobranchia.
4. Head has a single pair of non-retractile tentacles.
5. Calcarious shell with an operculum.
6. Sexes are separate.
7. Exclusively marine.

Subclass Prosobranchia is divided into three orders: Neogastropoda, Archaeogastropoda and Mesogastropoda.

Order-1. Neogastropoda (Stenopoda)

1. Carnivorous species with an eversible proboscis.
2. Ospheridium is large and bipectinate.
3. Foot operculate.
4. One gill with filaments in one row, heart with one auricle, single nephridium present.
5. Nervous system is lightly concentrated.
6. Shell has a siphonal canal.
7. Embryos are intracapsular. Free swimming veliger suppressed.

Order-2. Archaeogastropoda (Aspidobranchia)

1. Due to incomplete atrophy of organs, the kidneys, auricles nephridia and ctenidia are paired.
2. Ctenidia are bipectinate or plume like.
3. Ospheridium poorly developed.
4. The nervous system is little concentrated.
5. Gonad opens into the right kidney serving as gonoduct.
6. Fertilization external.

The order has been distinguished into the following two suborders:

Surborder (i) Rhipidoglossa

1. Shell spiral.
2. Ctenidia paired.
3. Radula with numerous marginals.

   Examples: *Haliotis, Fissurella, Trochus.*

Surborder (ii) Docoglossa

1. Shell and visceral mass conical.
2. Ctenidium single or absent or replaced by secondary gills.
3. Radula with three marginals on either side.
4. Auricle single.

   Examples : *Patella, Acmaea.*

Order-3. Mesogastrpoda (Pactinibrachia)

1. These are advanced snails with complete atroply of the left side so that usually unpaired auricle, nephridium, ospheridium and ctenidium occur.
2. Ctenidium monopectinate or comb-like, attached to the mantle throughout its entire length.
3. Edge of the shell opening lacks a siphonal notch or canal.
4. Foot may be operculate.
5. Ospheridium pectinate and well differentiated.
6. The nervous system is concentrated.
7. Gonad has its separate duct and opening.
8. Fertilization internal. Development is indirect with a free-swimming veliger larva.

   Examples: *Pila, Triton, Valvata, Natica, Vermetus.*

Subclass (B) Opisthobranchia (Euthneura)

1. Due to detorsion of visceral mass, visceral loop becomes untwisted hence named (Gr. *euthus* = straight, *neuron* = nerve).
2. Mantle cavity displaced posteriorly.
3. Head with two pairs of tentacles.
4. Shell and operculum, if present, reduced
5. Usually one auricle and one nephridium present.
6. Ctenidia are replaced by secondary gills.
7. Hermaphroditism is universal. Larva is veliger.
8. Exclusively marine.

   Subclass Opisthobranchia is divided into eight orders as follows:

Order-1. Acochlidiacca

1. Small without shell and gill.
2. Naked visceral mass projecting behind the foot and covered with spiracles.

   Examples: *Acochlidium.*

Order-2. Saccoglossa

1. These are herbivorous, with a modified redula and suctorial pharynx.
2. They are shelled or naked slug-like, with a gill.

   Examples: *Elysia, Oxynoe.*

Order-3. Cephalaspidea

1. These are burrowing forms having shield like head.
2. Shell and mantle cavity moderately developed.
3. Lateral parapodial lobes are prominent.

   Examples: *Bulla, Acteon.*

Order-4. Thecostomata

1. These are shelled pteropods.
2. Shell is spirally coiled or a non-spirally pseudo-conch.
3. Mantle cavity is well developed.
4. Parapodial fins are large.

   Examples: *Clio, Limacina.*

Order-5. Nudibrandia (Acoela)

1. They are naked bilaterally symmetrical sea-slugs without shell, mantle cavity, ctenida and osphridium.
2. They have secondary gills around anus or surface outgrowths or cerata.

   Examples: *Doris, Aeolis, Tritonia.*

Order-6. Anaspidea

1. These are crawling orburrowing forms having a pair of rhinophores on head.
2. Shell is small and internal.
3. Mantle cavity is reduced on right side.
4. Parapodial lobes are prominent.

   Examples: *Aplysia, Akera.*

Order-7. Gymnostomata

1. These are naked pteropods and planktonic.

2. Shell and mantle cavity absent.
3. Small ventral parapodial fins are present.

   Example: *Pneumoderma.*

Order-8. Notaspidea

1. Shell is external or reduced and internal.
2. Mantle cavity is absent, but a skirt like projection of mantle covers gill on the right side.

   Example: *Pleurobranchus.*

Subclass-(C) Pulmonata

1. The ctenidia are absent and replaced by pulmonary sac or lung.
2. Shell is simple spiral or vestigial or absent and operculum never occur.
3. One or two pairs of tentacles and one pair of eyes are present.
4. The torsion takes place. The nervous system is secondarily symmetrical due to shortening of connectives and concentration of gangila into a circumoesophageal ganglionic complex.
5. The heart with a single auricle lying anterior to the ventricle.
6. The animals are bisexual. Gonad single.
7. Development direct without larval stage.

   The subclass is divided into two orders:

Order-1. Stylommatophora

1. Two pairs of retractile tentacles. Eyes are lodged on the tip of posterior pair of tentacle.
2. Shell is internal, reduced or absent.
3. Mostly terrestrial.
4. Male and female gonopores usually united.

   Examples: *Helix, Limax, Arion.*

Order-2. Basommatophora

1. One pair of tentacles with the eyes at the base.
2. Shell is delicate with a large aperture.
3. Some forms with secondary gills.
4. Male and female gonopores usually separate.

   Examples: *Limnea, Planorbis, Physa.*

CLASS–IV PELECYPODA

1. The pelecypods or bivalvia are marine, fresh-water, bilaterally symmetrical molluscs.

2. Mantle consists of paired right and left lobes, which secrete a bivalved shell.
3. Devoid of head, eyes, tentacles and redula.
4. The muscular foot is usually wedge-shaped or tongue-shaped, which is adapted for burrowing.
5. Presence of two gills, one on either side of mantle cavity, the function of which is to produce respiratory and food-carrying currents of water.
6. Pharynx is absent.
7. Coclom is reduced to a dorsal placed pericardium.
8. Alimentary canal is coiled with large paired digestive glands.
9. Heart is continued with pericardium and comprises two auricles and one ventricle.
10. The renal organs are a pair of coelomic kidneys, which lead into pericardium at one end and to outside at other.
11. The nervous system includes four pair of ganglion, *viz.*, cerebral, pleural, pedal and visceral.
12. Sense organs are statocyst and osphradium.
13. Sexes may be separate or united. Development includes metamorphosis, followed by trochophore stage.

Class-Pelecypoda is divided into following orders:

Order-1. Septobranchiata

1. Gills replaced by horizontal muscular and perforated septa dividing the mantle cavity into upper and lower chambers.
2. Foot long and slender, byssus reduced or absent.
3. Adductor muscles two.
4. True ctenidia are absent.

Examples: *Cupisdaria, Poromya.*

Order-2. Pseudolamellibranchiata

1. The gills are plaited so as to present vertical folds.
2. Interfilamentar junctions may be ciliary or vascular. Interlamellar junctions may be vascular on non-vascular.
3. Shell valves are frequently unequal and single large adductor muscle present.
4. Foot is little developed or rudimentary.
5. Genital organs opens into kidneys or near them.

Examples: *Pecten, Ostrea, Pinna, Lima.*

Order-3. Filibranchiata

1. Single pair of plate-like gills formed of distinct V-shaped filaments.
2. Inter-filamentar junctions are either absent or formed by groups of inter-locking cilia.
3. Inter-lamellar junctions are either absent on non-vascular.
4. Two adductor muscles present, anterior may be redused or absent.
5. Foot small or poorly developed.

   Example: *Mytilus, Arca, Noah.*

Order-4. Eulamellibranchiata

1. Gills are basket like, the gill-filaments are attached with vascular inter-filamentar and inter-lamellar junction.
2. Gills may be smooth or plaited.
3. Two equal adductor muscles.
4. Foot is well developed.
5. Siphon of small of large sized present.
6. Gonads open out by separate openings. They do not open into kidneys.

   Examples: *Unio, Teredo, Anodonta, Venus, Cardium, Mya.*

Order-5. Protobranchiata

1. Presence of paired plume-shaped gills, each bears two rows of filaments.
2. The foot is flattened.
3. Presence of anterior and posterior adductor muscles.
4. Sexes separate, gonads lead into kidneys.

   Examples: *Nucula, Solenomya.*

Class–V Scaphopoda

1. They are aberrant marine animals comprising only of three genera.
2. Body is elongated worm-like, enclosed in a tusk-like shell open at both ends.
3. Eyes, tentacles and gills are absent.
4. Mantle tubular completely enclosing the body.
5. Mouth surrounded by lobular processes or outgrowths.
6. Foot is reduced used for digging.
7. Heat rudimentary.
8. Sexes are separate.
9. Torsion takes place in some forms.

Scaphopoda has two families:

Family (i) Siphonodentalidae

1. Foot is elongated and capable of expansion into a terminal disc.
   Examples: *Pulsellum, Siphonodentalium.*

Family (ii) Dentalidae

1. Foot is trilobed having two epipodial lobes and one hypopodial lobe.
   Examples: *Dentalium, Antalis.*

Class-VI Cephalopoda

1. Exclusively marine.
2. Shell spiral chambered or usually with or without shell embedded in mantle.
3. Body with head and trunk. Head bears eyes and mouth.
4. Trunk consists of symmetrical and uncoiled visceral mass.
5. Mantle encloses posteriorly and ventrally a large mantle cavity.
6. Foot altered into a series of sucker bearing arms of tentacles encircling the mouth.
7. Mouth bears jaws and radula.
8. Two or four pairs of bipactinate gills.
9. Circulatory system closed, heart with two or four auricles.
10. Excretory system comprises two or four pairs of nephridia.
11. Nervous system is highly developed and the principal ganglia are concentrated around the oesophagus.
12. Sexes are separate.
13. Development meroblastic without metamorphosis.
14. Most of the cephalopods have an ink-gland with a duct opening into the rectum.

Cephalopoda is divided into three subclasses:

Subclass-A. Dibranchiata

1. The foot modifies into a circlet of eight to ten arms bearing suckers, around the mouth.
2. The funnel forms a complete tube.
3. The shell is generally internal.
4. Presence of paired gills and kidneys.
5. Presence of characteristic ink-gland and duct.

Dibranchiata is divided into two orders.

Order-1. Decapoda

1. Body elongated and with no lateral fins.
2. Arms ten, eight short and 2 long and known as tentacles.
3. Suckers on the arms sessile and without horny rim.
4. Shell internal and well developed.

Examples: *Sepia, Loligo.*

Order-2. Octopoda

1. Body rounded or oval and with no lateral fins.
2. Arms eight, long arms or tentacles absent.
3. Suckers on the arms sessile and without horny rim.
4. Shell and enidamental glands absent.

Example. *Octopus* (Devil-fish), *Agronauta*

Subclass-B. Nautiloidea or Tetrabranchiata

1. The foot divides into lobes bearing numerous tentacles.
2. The funnel is incomplete.
3. Shell is external which is divided into several compartments.
4. Kidneys and gills are four.
5. Devoid of ink gland.
6. Pericardium communicates to the exterior directly.
7. Eyes simple without crystelline lens.
8. Chrometophore, salivary glands absent.

Example: *Nautilus*

Subclass-C. Ammonoidea

1. All are extinct.
2. Shell variously modified in the form of straight spiral or turreted.
3. Siphon external, siphuncle simple and marginal.
4. Two pairs of ctenidia.

Example: *Ammonites*.

CHAETODERMA

Phylum	—	Mollusca
Class	—	Amphineura
Subclass	—	Aplacophora
Genus	—	*Chaetoderma*

It is a most primitive mollusc. The body is wormlike cylindrical. It is found around North Atlantic. The head is separate off from the body by a constriction. Mantle covers the body completely. Shell is

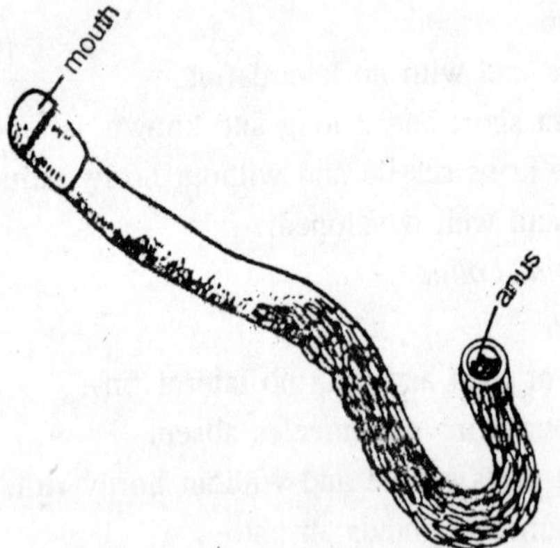

Fig. 1.4. Chaetoderma.

absent but the body is covered by numerous calcareous spicules. Rudimentary foot. Gills reduced to a pair which are situated in the cloacal cavity. Sexes are separated and each has one fused gonad. A cavity is present at the posterior end of the body into which opens the anus.

PATELLA

Phylum	—	Mollusca
Class	—	Gastropoda
Subclass	—	Prosobranchia
Order	—	Aspidobranchia
Suborder	—	Docoglossa
Genus	—	*Patella*.

Patella or limpet is a littorral and is found attached to the rocks. It is found along Pacific coast, Atlantic coast, Europe, U.S.A. Body is oval with convex dorsal surface. Shell is oval and core - like, with a conical projection from its dorsal surface. Operculum is absent. The head is laterally produced into a pair of small, stoat, tectile tentacles. Eyes are simple present at the base of tentacles. Foot ventral and sole-like for creeping. True mantle is restricted to the anterior and secondary mantle cavity is developed between foot and mantle. Redula composed of very few, strong hooked teeth in each row. All around the foot lies a secondary pallial cavity containing a series of secondary branchiae. Heart with single auricle. Herbivorous, feeding upon algae, sea-weeds and diatoms. It is eatern by poor classes in country like France, Italy and Ireland.

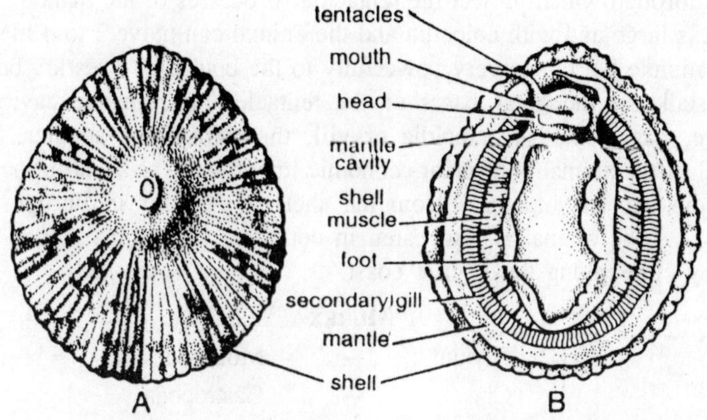

Fig. 1.5. Patella. A—Dorsal view; B—Ventral view.

HALIOTIS

Phylum	—	Mollusca
Class	—	Gastropoda
Subclass	—	Prosobranchia
Order	—	Aspidobranchia
Suborder	—	Rhipidoglossa.
Genus	—	*Haliotis*.

Haliotis or Ear shell is marine and found attached to rocks and feeding on sea-weeds and other marine vegetation. The animals are very common at and below low tide level. *Haliotis* or *Abalone* has ear-shaped shell with small flattened spire. The aperture is very large and without operculum. The shell is perforated by a series of marginal

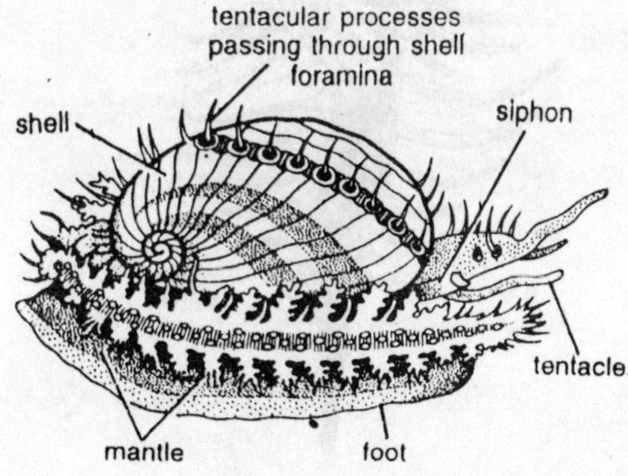

Fig. 1.6. Holiotis.

slits through which project the tentacular processes of the mantle. The foot is large and with epipodia and the animal can move 2 to 4 metres per minute but clings very powerfully to the boulders. Eyes are borne on stalks at the outer bases of thc tentacles. The mantle cavity is large, has bipectinate ctenidia or gill, the right being smaller. The shell of the animal is of great economic importance, as beautiful pearly buttons are manufactures from the shell of *Haliotis* in Europe and U.S.A. The animal is also eaten in countries like China, Japan and many cities along the Pacific coast.

Murex

Phylum	—	Mollusca
Class	—	Gastropoda
Subclass	—	Prosobranchia
Order	—	Pectinibranchia
Genus	—	*Murex*

It is a marine animal found between low tide marks on stones and sea weeds. It is found on the Syrian coast, Italy, India, West Indies and U.S.A. Shell is spirally coiled and with three or more rows of beautiful spines of different lengths. It is carnivorous and highly predator feeding on living animals which it grasps by means of its foot. It is destructive to oyster bed. A very prominent bipectinate osphradium

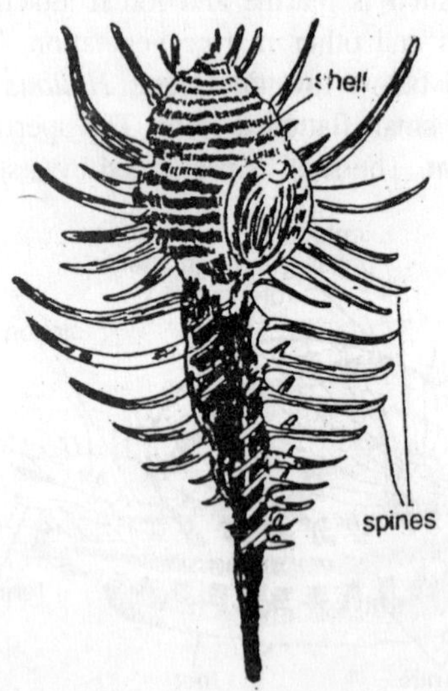

Fig. 1.7. Murex.

persist. Head is prolonged into a long proboscis which can be retracted within the probocis-sheath. Adrectal glands present which secrete a dye the *Jyrain purple* (Yellow in colour). Sexes are separate.

APLYSIA

Phylum	—	Mollusca
Class	—	Gastropoda
Subclass	—	Opisthobranchia
Order	—	Anaspidea
Genus	—	*Aplysia*

It is a marine animal found among sea-weeds. It is found in India, Asia, West Indies and on the Florida coast. It is commonly called 'sea hare.' Body is soft, lumpy with a visceral hump. Head with two pairs of tentacles. The anterior pair is large and ear-like, the posterior pair is olfactory and bear eyes at their base. Shell thin, reduced and transparent, enclosed by mantle. The foot is muscular and elongated pointed posteriorly with lateral out growths, the parapodia. Mantle possesses unicellular ink glands which secrete purple ink used for defence. Hermaphrodite. There is a single generative aperture and a single duct for the sperm and ova but a seminal groove runs from the aperture to the head and reciprocal fertilization is impossible. Nervous system is well developed with perfectly symmetrical visceral loop.

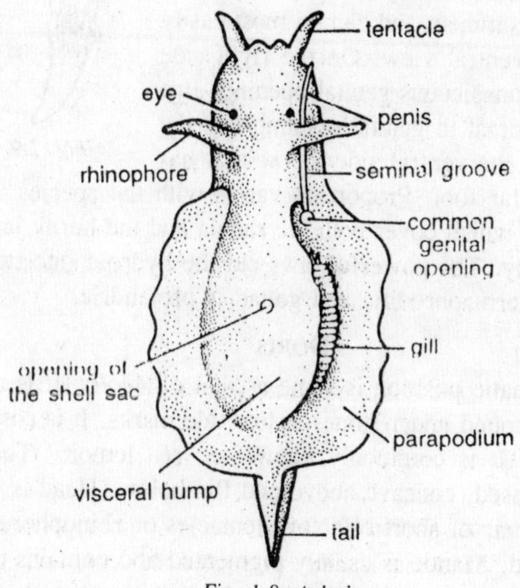

Fig. 1.8. Aplysia.

EOLIS

Phylum	—	Mollusca
Class	—	Gastropoda
Subsclass	—	Opisthobranchia
Order	—	Nudibranchia
Genus	—	*Eolis*

It is a marine slug- like gastropod found in shallow waters crawling on under surface of sea weeds. Found in U.S.A., Europe. *Aeolis* is a small nudibranch 1 or 2 cm in length. The soft body is covered with a transparent integument. The shell, mantle and true gills are absent. The cerata are the respiratory organs. There are cylindrical extensions of the integument, containing hollow extensions of the liver and nematocysts derived from ingested hydroids. The cerata have characteristic arrangement, may occur in many transverse or in few longitudinal rows. The head is small, bears tentacles, usually two pairs, one of which may be retractile. The anterior or oral pair is the longer. The posterior or dorsal pair, known as rhinophores bear simple sense organ which test the surrounding water. At the base of each rhinophore is an eyespot. The mouth is prominent and can be most easily seen from the ventral view. On the right side of the head is conspicuous genital opening. Just posterior and dorsal to genital opening lies the small anus. On the ventral side the animal has a highly muscular foot. Proportion varies with the species. Mouth is guarded by lips which cover both the radula and the horny jaws within the buccal cavity. The powerful jaws cut the hydroid into two pieces. The *Aeolis* is hermaphroditic and gonad is protandric.

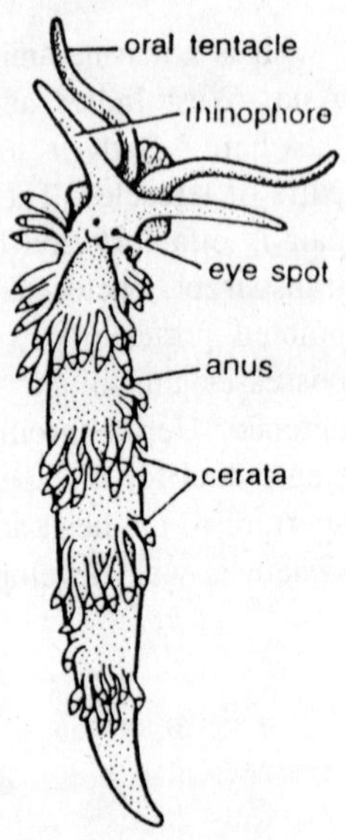

Fig. 1.9. Eolis.

DORIS

The systematic position is same as that of *Aeolis*. It is a sluggish marine animal found under stones at low tide marks. It is cosmopolitan in distribution. It is commonly known as 'sea lemon.' The body is oval and depressed, concave above and flat below. Head is indistinct, but there is a pair of short olfactory tentacles or rhinophores towards the anterior end. Mantle is usually pigmented and contains calcareous spicules or dorsal tubercles. Mouth is present on ventral side. Anus

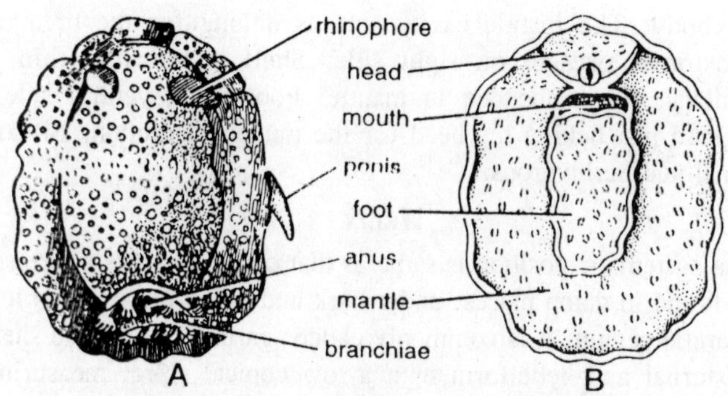

Fig. 1.10. Doris. A—Dorsal view; B—Ventral view.

lies mid-posteriorly and is surrounded by a circlet of feathered retractile secondary branchiae. The broad foot is on ventral side. Hermaphrodite, gonopore is asymmetrically placed on the right side of the body. *Doris* has undergone *detorsion*, therefore, the shell, mantle cavity and primary gills are absent.

Limax

Phylum	—	Mollusca
Class	—	Gastropoda
Subclass	—	Pulmonata
Order	—	Stylommatophora
Genus	—	*Limax*

It is a terrestial gastropod, found in gardens over damp soil. It is cosmopolitan in distribution. It is commonly known a 'gray slug.' Body is elongated and tapering behind and is divisible into head, foot and visceral hump. Two pairs of retractile tentacles are present on head. Eyes are present at the tips of posterior tentacles. The mouth is on the ventral side of enterior end and is bounded by a pair of lateral lips. Mantle forms a shield like area in the dorsal side of the anterior

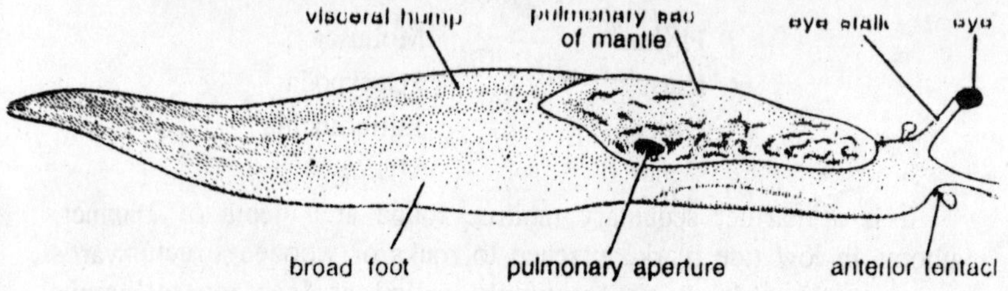

Fig. 1.11. Limax.

part of body. The mantle cavity acts as a lung for the breathing, pneumostome, present on right side. Shell is internal, thin and rudimentary, lies embedded in mantle. Foot is on ventral side. A pedal gland just behind the head for the mucous secretion. Bisexual. Nocturnal and herbivorous.

HELIX

The systematic position is same as that of *Limax*. It is a terrestrial animal found in damp places, under bark and fallen leaves. It is found in palearctic region. It is commonly called 'garden snail.' The shell is thin, external and heliciform with a low conical spire, measuring 4 cm. in length. Head bears two pairs of tentacles, first pair bears organs of small and second pair is large and bears a pair of simple eyes. Foot possesses a flat ventral surface for creeping. Respiration by pulmonary sac. Gills are absent. Hermaphrodite. The genital aperture opens above the right lateral lip. Gonads contain four angled dart and multifid digitiform glands.

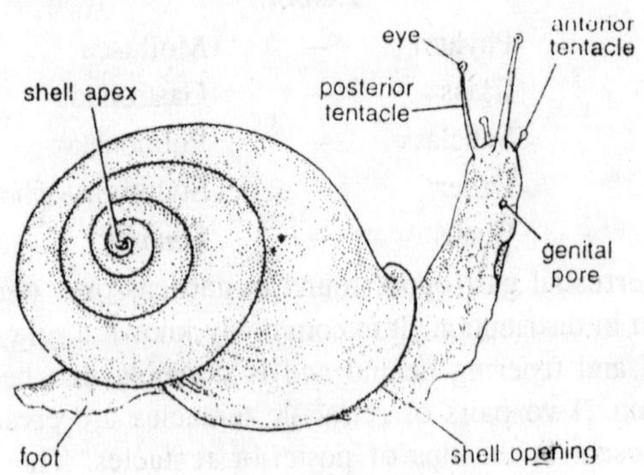

Fig. 1.12. Helix.

MYTILUS

Phylum	—	Mollusca
Class	—	Pelecypoda
Order	—	Fillibranchia
Genus	—	*Mytilus*

It is a marine, sedentary mollusc found at a depth of 2 or 3 fathoms in low tide mark, attached to rocks or wooden structure by its byssus threads. It is commonly called as 'sea mussel' and cosmopolitan in distribution. The shell is elongated, equivalent with

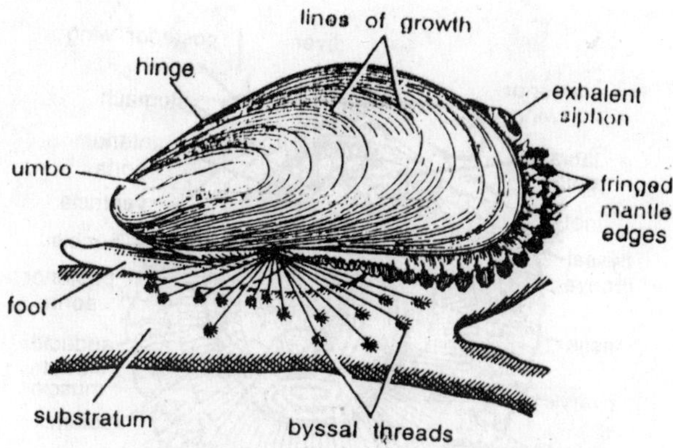

Fig. 1.13. Mytilus.

umbo at or near anterior end. Hinge toothless but may bear cerculations. The foot is tongue like and is modified into a bunch of byssal threads coming out from between the two shell-valves, which are main anchoring organs. The anterior adductor muscle is weaker where as posterior is strongly developed. A pair of gills is present, each gill is provided with gill-filaments. The gills are lamelliform. The eyes are found anterior to the inner gill-lamella. Sexes are separate. Gonads extend into the mantle. It is used as food in Europe.

Pecten

Phylum	—	Mollusca
Class	—	Pelecypoda
Order	—	Pseudolamellibranchia
Genus	—	*Pecten*

It is free swimming marine mollusc lives on the sea bottom of 10 fathoms deep. It is found in India and U.S.A. It is commonly called as 'scallop.' Two shells are unequal. The right shell being larger and more convex and the animal rests on this valve. There is a single large adductor muscle. This is divided into parts and the larger of these served for rapid contraction which cause swimming movements. Foot is very much reduced but in larva very well developed and used for locomotion. Two large and crescentric gills are present. Mantle is *tentaculiferous* and it encloses viscra. Large number of stalked eyes are present at regular intervals along the edges of the mantle. These are hermaphrodite. The ovary has pink colour when eggs are ripe. The testis lies behind and is cream in colour.

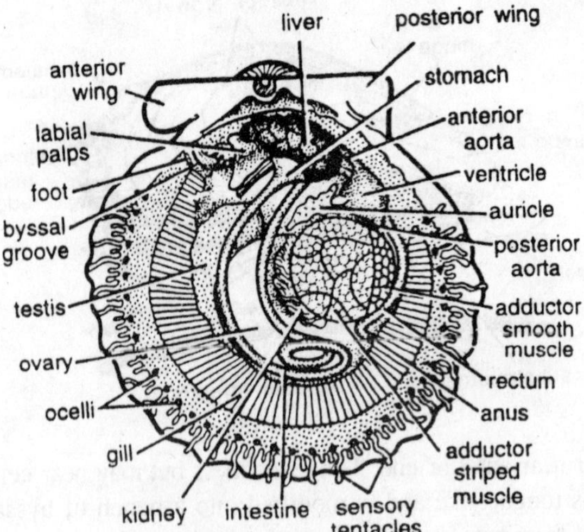

Fig. 1.14. Pecten.

OSTREA

The systematic position is same as that of *Pecten*. It is a sedentary bivalve attached to rocks or other shells and found in shallow and brackish water. It is found in Atlantic and Pacific coasts, India, Gulg of Mexico to Massachusetts. It is commonly called 'edible oyster.' The shell valves are unequal, the left is always larger and permanently attached to the rocks by byssus. Only one adductor muscle i.e. posterior one. It is divided into two parts one is striated and other is non-striated. The foot is absent. Heart with two fused auricles. A pair of curves gills present. It is peculiar in the sense as the individuals function alternately as males and females. Spawning generally takes place at full moon. The eggs develop into vileger larva which metamorphoses into adult.

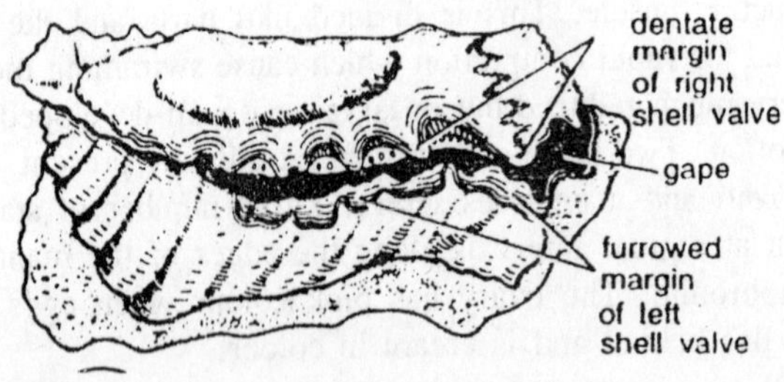

Fig. 1.15. Ostrea.

TEREDO

Phylum	—	Mollusca
Class	—	Pelecypoda
Order	—	Eulamellibranchia
Genus	—	*Teredo*

Marine pelecypod which burrows in wood of the boats, ships and other submerged wood. It is found in India, Europe, Cape Cod to Florida and Massachusetts Bay to Florida. It is commonly called 'shipworm'. *Teredo* has body extremelly elongate and worm-like appearance. Foot reduced. Mantle tubular and opens anteriorly, shell is reduced and covers only the anterior part of the body. The mantle cavity and siphons are long and the latter unites to a large extend and provided with calcareous pallets for protection. A constant current into and out of the mantle cavity is maintained by ciliary action. The saw dust is swallowed by the animal and is digested by cellulose digesting enzymes.

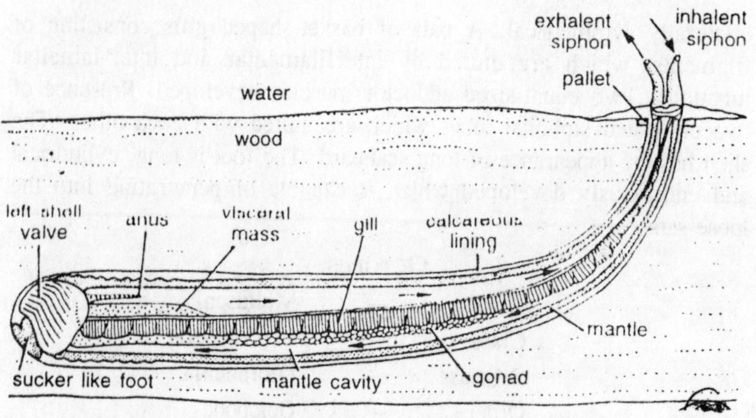

Fig. 1.16. Teredo. Anatomical features (diagrammatic).

SOLENOCURTUS

Phylum	—	Mollusca
Class	—	Pelecypoda
Order	—	Eulamellibranchia
Genus	—	*Solenocurtus*

It is a burrowing pelecypod found in India and U.S.A. It is commonly called 'razor fish' or 'razor shell'. Bivalve shell, mantle divides into right and left lobes. Body is laterally compressed and

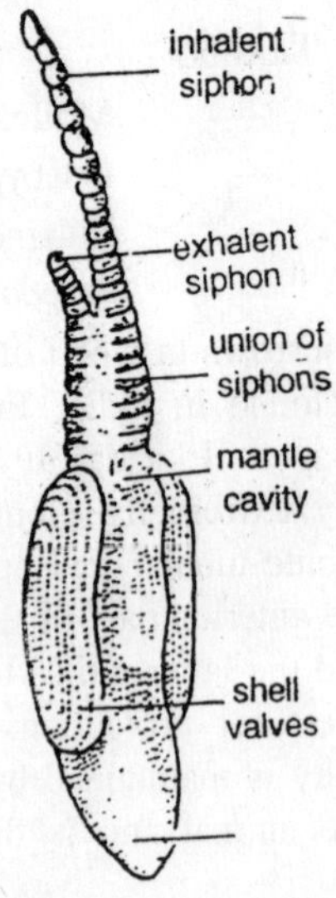

Fig. 1.17. Solenocurius (Razor-fish).

bilaterally symmetrical. A pair of basket-shaped gills consisting of filaments, which are united by interfilamentar and inter-lamellar junctions. Two equal-sized adductor muscles developed. Presence of two elongated siphonal tubes which are united with each other. The shell has the appearance of long scabbard. The foot is long, cylindrical and enormously developed which is capable of penetrating into the loose sand.

Octopus

Phylum	—	Mollusca
Class	—	Cephalopoda
Subclass	—	Dibranchia
Order	—	Octopoda
Genus	—	*Octopus*

It is marine and cosmopolitan in distribution generally found on the Atlantic and Pacific coasts. It is commonly called devil fish. Body surrounded by mantle containing chromatophores. They can change their colour according to surrounding medium. Foot modified into eight equal arms surrounding the head and siphon. The arms are beset with suckers which are sessile and devoid of horny rims. Body is covered with tubercles, short, round aborally and without fins. External shell is absent and body with vestiges of an internal shell. Third arm on the

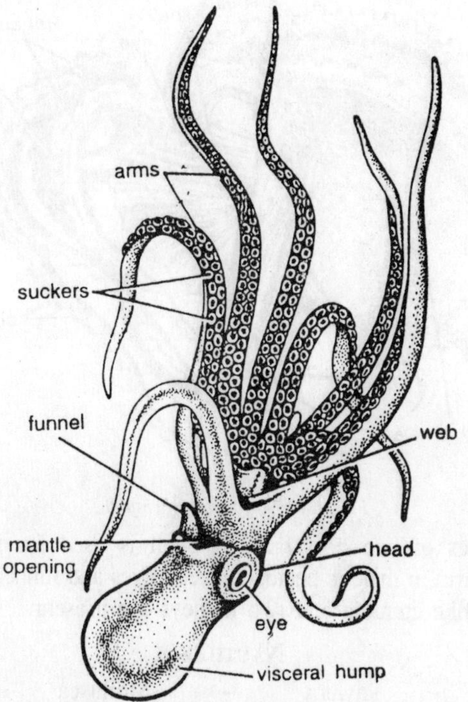

Fig. 1.18. Octopus.

right in male modified to act intermittent organ or hectocotylized. Like *Sepia* and *Loligo*, *octopus* is also capable of producing inky liquids which form a smoky screen water. Sexes are separate and there is no metamorphosis as the development is direct. Female cares the eggs. Mouth with sharp and beak-like horny jaws.

ARGONAUTA

Phylum	—	Mollusca
Class	—	Cephalopoda
Subclass	—	Dibranchia
Order	—	Octopoda
Genus	—	*Argonauta*.

It is a marine animal found in Atlantic and Indian oceans. It is commonly called as 'paper nautilus.' Sexual dimorphism is remarkable and well-marked. Male in 2.5 cm in length and without shell. Female is 20 cm in length and possesses a thin, transparent shell secreted by

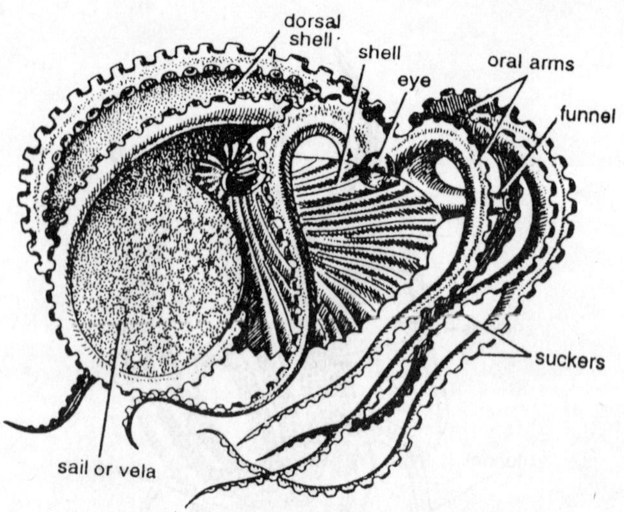

Fig. 1.19. Argonauta (Female).

the extrimities of two dorsal arms. Shell is used for protection of eggs. Third arm in male is hectocotylized. Eyes and funnel and present. Two plume like ctenidia and two kidneys are present.

Nautilus

Phylum	—	Mollusca
Class	—	Cephalopoda
Subclass	—	Tetrabranchia
Genus	—	*Nautilus*.

It a marine cephalopod living in shallow waters and shores and coral reefs in the Indian and south Pacific Oceans. It is commonly known as 'Pearly nautilus.' The shell is well developed, spirally coiled and composed of calcareous matter, which is divided into several compartments by a series of septa. Through the various compartments runs a tube called siphuncle the walls of which are formed of calcium carbonate. A narrow vascular prolongation of visceral region of the animal passes through this tube. Head is well developed bearing the eyes. The anterior part of foot is modifies into several arms provided with tentacles, which are arranged in two series. The posterior portion of the foot is modified into funnel situated below the head. Buccal cavity with a pair of jaws and odontophore bearing several teeth. Two pairs of gills and kidneys are present. They are hyper-polyandrous i.e.

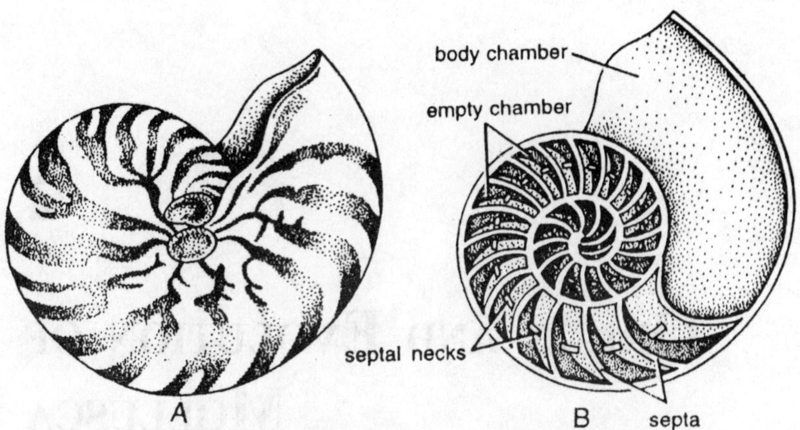

Fig. 1.20. Nautilus. A—Complete shell; B—Shell in section.

male are abundant than female. Presence of 60 tentacles in male and 90 in female. Pericardium leads directly to ecterior by two orifices located near the posterior renal aperture.

2

Origin and Evolution of Mollusca

The position of Phylum Mollusca is in several respects are considerable and positive isolation and any attempt to derive it from one group or other in the view of major evidences to lay any hypothesis for the origin of Mollusca is ill founded. It should be noted with interest that even in such remote time as Cambrian Period, the sponges, echinoderms, Platyhelminths and mollusc have already identify themselves as a separate and distinct group with the same distinction from each other what we observe in the present time. In such circumstances any system of classification or any theory of development which proposes the derivation of mollusca from any of this group is extremely dangerous. It would, therefore, almost certain that the origin of Mollusca is merely a metaspeculation and if at all any attempt is to be made we have to examine the indication of the relationship of the phylum by the consideration of embryonic resemblance.

One of the most important feature of phylum Mollusca is the occurence of a *trochophore larva* in the development. It is interesting to note that such a larval form is also shared by annelids but if the presence of trochophore is in any form to be regarded as important in shading some light from the origin of Mollusca, it should not necessarily be taken as indicative of an annelid ancestor. If we examine the trochophore larva of any mollusc we see a complete gastrulation and the ciliated bands besides a single large cell, which would give rise to the mesoderm in the successive developmental stages, is seen. In addition to the position of an apical organ i.e. *protroch* and larval nephridia its internal organisation also shows similarities with the annelidian trochophore. Such apparent similarities appeal to be discussed and a more careful study would be certain striking differences between the two which can not escape our attention.

We thus say on the dorsal side of advanced molluscan trochophore there is a rudiment of mantle formed by ectoderm. At this stage the earliest sign of the shell can also be seen. The foot has also made its appearance as a ventral prominence. The single large mesoderm cell gives rise to two bands of mesodermal cells in each of which a cavity develop, but by for the most important difference from annelid is afforded by the absence of segmentation in the embryonic mesoderm of mollusc. It would thus appeal that absence of segmentation provides a positive evidence from such group of animals which were devoid of segmentation. From the foregoing account it becomes positively certain that we have to leave our such view for the origin of Mollusca from some annelid type ancestor and possibility of deriving the group from some turbellarian ancestor may help us. Such a connection is proved by the striking similarities that are present between the ladder-like nervous system in certain turbellarians and in placophore, solenogasters and certain aspidobranch gastropods. In either case we find that the ganglia itself—are distributed along the entire length of the main nerve trunk.

The pleuro-visceral and pedal nerve cords should be taken to correspond to the lateral and ventral nerve cords of the turbellarians. Now if such a hypotheical ancestor could suppose to has given rise a primitive mollusc then we should have to postulate following important changes that should have undergone in its organisation for such a purpose:

1. Secretion of dorsal shell which in the beginning might have been in the form of a layer of cuticle containing calcareous particles and would have served as a protective covering.
2. The development of shell would evidenty reduced the general respiratory surface of the body which in turn would make the necessity to develop such specialized respiratory organ as ctenidium as present in molluscs.
3. The ventral musculature which have already been well developed in the turbellarian must have undergone a remarkable development to give rise to such a specialized and efficient molluscan foot.

It is worthy of notice that typical trochophore larva bears close resemblances to certain form of turbellarian larva i.e. *Muller's larva*.

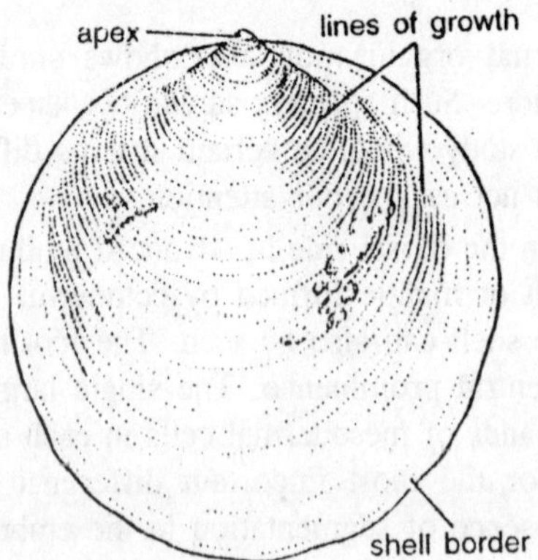

Fig. 2.1. Neopilina galatheae. Dorsal view of shell.

It we have to assume that the primitive molluscs derived from some turbellarian like ancestor we will also have to postulate the conversion of Muller's larva into a molluscan trochophore. Such a postulation would automatically lead us to the contentation that the annelids and mollusc had probably common platyhelminth origin will subsequent extreme divergence into two phyla. If ontogeny repeats phylogeny then a trochophore will be taken to be present a remote ancestor of annelids and molluscs which resembles the ancestors of platyhelminthes represented by Muller's larva. The trochophore larva by certain changes developed into anneliden form. These changes includes:

1. Elongation of body which becomes segmented.
2. Formation of septa.
3. Development of coelom.

On the other hand transformation into molluscan form occurred through the following changes:

1. The apical cilia got transformed into velum which serves not only for locomotion but also for feeding. The cilia of apical plate create a current of water which bring food particles into the mouth.
2. Shell gland got developed which secretes the unique shell.

But in 1952 the whole picture changed when 10 living, specimens of *Neopilina* were found from a depth of 3500 metres off the Paciffic coast of Costa Rica (Maxico). The most unusual feature of *Neopilina* which belongs to the Monoplacophora of Mollusca is its appearance, metameric plane of structure. It displaces 5 pairs of gills, 6 pairs of nephridia and these organs are more or less metamerically arranged, a number of zoologists believe that this metameric plane displaced

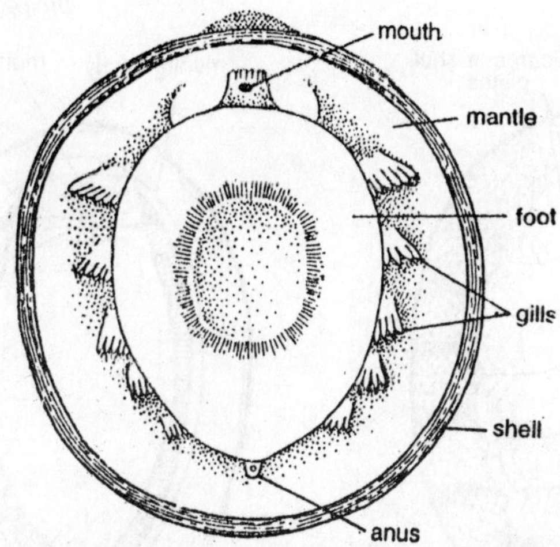

Fig. 2.2. Neopilina galatheae. Ventral view.

from *Neopilina* is an evidence that Mollusa comprise a fundamental metameric phylum and must have evolved from the annelids. The molluscan trochophore is thus believed to be additional evidence of annelid origin of mollusc. In the light of *Neopilina* it may existed to revise our thinking in future, but at present the older thing assuming a common origin of annelid and mollusc seems to possess the greatest weight of evidence.

Evolution

From the primitive archimollusc structure different groups, known today, have diverged and each of the groups, follows the evolution along certain lines of specialization.

Amphineura or Placophora

The Amphineuran are the Chitons, seem to have departed least from the ancestral mollusc. Their body becomes elongated and was protected by eight shell pieces on the dorsal side. On either side of mantle cavity was arranged symmetrically several pairs of bipectinate gills. If we were to assume that the primitive mollusc possesses several ctenidia then it is appearent that the chitons have retained the primitive character. On the other hand if we assume that the primitive mollusc posses a single pair of ctenidia then it naturally employees that the numerous ctenidia of chiton are formed by the repetation of the former. The foot in chiton has retained the primitive from. The nervous system is also primitive in that the ganghonic cells are distributed along the

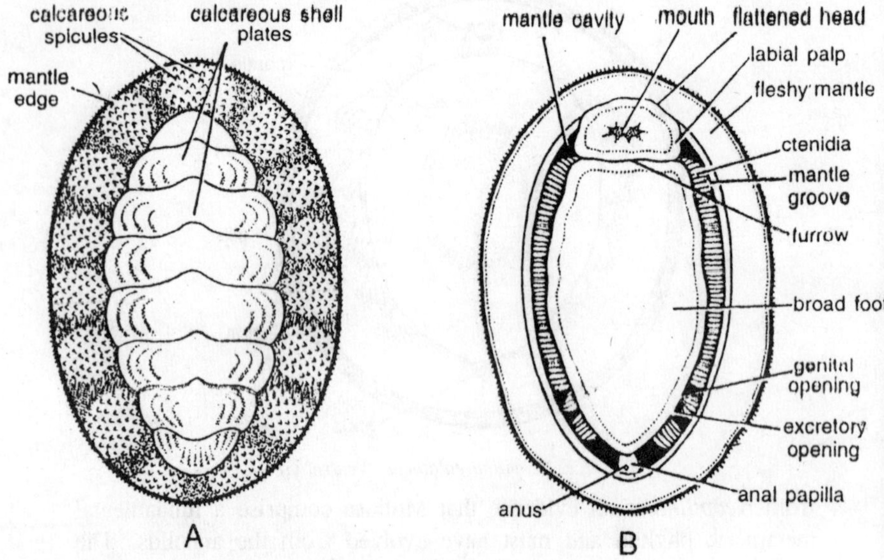

Fig. 2.3. Chiton. A—Dorsal view. B—Ventral view.

whole length of nerve cords. Chitons also possess a redula which is well developed.

Gastropoda

The leading features in the evolution of Gastropoda has been the torsion of visceral mass producing symmetry in body. There is no doubt that the ancester gastropod possessed a body with a straight alimentary canal, leading in a posteriorly placed anus and on either side of anus was placed the ctenidium. We also know well that gastropods have univalved shell which is covering all the soft parts of body and extension of foot on the ventral side tend to remove the pallial complex with and renal openings away from head. The univalved shell, therefore, ment apparently necessary for mantle cavity, ctenidia and excretory aperture to shift at anterior and somewhat in the neighbourhood of opening of shell. The whole of pallial complex was shifted from its primitive posterior position to the anterior position by the torsion.

The large majority of gastropods included in order Prosobranchiata shows torsion in full development. It is also known that each group of Mollusca shows a progressive loss of shell in the course of evolution. It has also been invariably observed in Gastropoda that such a loss of shell is accomplished by a reversion of the phenomenon of torsion i.e. detorsion and consequent rotation of visceral loop back to its posterior

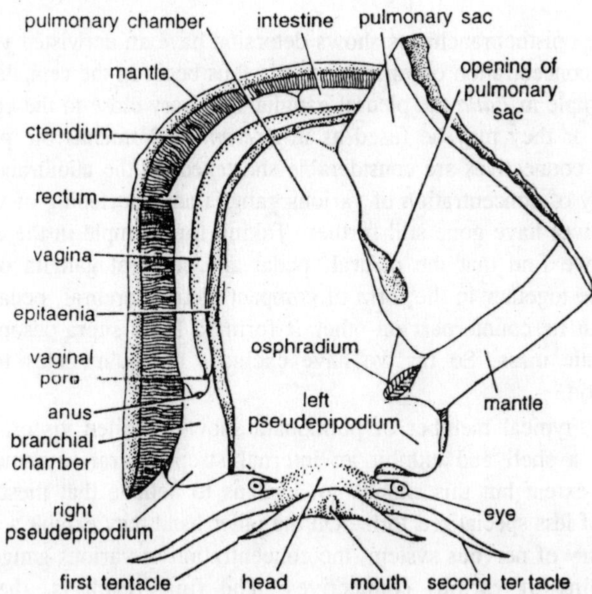

Fig. 2.4. Pila. Mantle cavity and pallial complex.

condition i.e. in opisthobranchiates. The Opisthobranachia are derived from pectinibranchia is proved by the fact that they possess a single ctenidium, single auricle, besides embryological evidences also support such a view. Even those opisthobranchs which are detorted in adult condition the primitive torsion is seen in the course of development. The nudibranches seem to have more step further.

Here the detorsion is complete and have lost the ctenidium as well as the shell. But they have evolved from a fully torted gastropods when we study their embryology. We have tried to show a gradual evolution of different groups of gastropoda on the basis of torsion and its subsequent reverser but nervous system also shows a way to the evolutionary history of Gastropoda. We find the most primitive type of nervous system in this class in exhibited by certain aspidobranchs is that they show a lesser tendency towards the concentration of various ganglia, besides they possess longitudinal pedal cords which can compare to pedal cords of Chiton but the evolution of nervous system in this class has taken along two lines:

1. In the establishment of progressive euthyneury are symmetry.
2. In the concentration of various ganglia and their connectives around the circum-oesophageal nerve ring. The visceral loop remains twisted and in all the prosobranchiats are the result of torsion.

The opisthobranchiates shows detorsion have an untwisted visceral loop, a concentration of various ganglia thus begin in the cephalaspides for example in *Bulla* the pleural ganglia shift very close to the cerebral ganglia or they may be fused as in *Aceton*. In *Notochus* the pleural-visceral connectives are considerable shortened. In the nudibranchs the tendency of concentration of various ganglia and shortening of various connectives have gone still further. Taking for example in the case of *Tethys* we find that the pleural, pedal and cerebral ganglia of each side fuse together in the form of compact pleuro-cerebral, pedal mass and with its counterpart on other it forms a large supra-oesophageal ganglionic mass. So far we have excludes the pulmonates for any discussion.

The typical member of pulmonata shows a coiled visceral mass possess a shell and exhibit an internal streptoneural asymmetry to certain extent but this should not lead us to believe that these are a group of less specialized form. On the other hand they exhibit a perfect symmetry of nervous system, the concentration of various ganglia and shortening of various connectives. The final result is, therefore incorporation of visceral loop together with its ganglia and circum-oesophageal nerve cord. On an examination of respiratory organs one finds a gradual loss of ctenidia as we move from lower groups to the higher ones in class Gastropoda. We thus see that certain aspidobranchs (*Haliotis*) exhibit the most primitive condition as they retain a pair of bipectinate ctenidia.

Other aspidobranchs belong to family Trochiade, Tuberidae and Neritidae have only one ctenidium but that is bipectinate. From this bipectinate type of gills arose monopectinate type of gills of pectinibranchi which have the last leaf-lets of one side of the animals. Certain land prosobranchiates have lost their ctenidia in relation to terrestrial life and consequently respire with the help of pulmonary sac.

In *Pila* while still retained the monopectinate gill has developed a pulmonary cavity and is capable of leading an amphibious life. When a true ctenidium is present among opisthobranchiates it lies posteriorly on right side as a result of detorsion. The nudibranchs have lost the true ctenidium as well as mantle cavity and respiration is carried with the help of secondary gills or outer skin. Finally in pulmonata the ctenidia are totally absent and the respiration is effected with the help of pulmonary sac. So far as the foot in gastropods are connected, in aspidobranchia it is very well developed and is differentiated into propodium, metapodium and epipodium. All of which show remarkable development. In pectinibranchia epipodium is usually absent, in certain prosobranchiates the foot may be reduced in simple discoidal prominence (*Vermetus*). It may be represented by small, ventral appendage in parasitic forms. Passing to Opisthobranchia we find that there is no division of foot of various forms.

The parapodia may be modified into fold-like extensions from the edges of the sole of foot and bent back over the shell. Among certain nudibranchs the foot no longer exists as a differentiated organ while pulmonate show poor development. From what has been said above it can be safely concluded that the most primitive gastropod, aspidobranchia, which in tern gave rise to the pectinibranchia. The pectinibranchia by a series of changes in their internal as well as external organisation gave rise on one end to the Opisthobranchia and on the other Pulmonata and have made specialization among two different lines.

Pelecypoda

As far as bivalvia is connected the fact that bivalved shell at first appearance is univalve and that of the foot of Protobranchiata is of creeping type and the ctenidia are plume or feather-like suggest the derivation of their class from animal resembling a simple gastropod with no redula and undisturbed bilateral symmetry. On the other hand certain zoologists record this proof to have been derived from the hypothetical primitive ancestor with a more or less developed head and redula. Such a primitive mollusc with the process of degeneration on account of leading an inactive sedentary life was modified accordingly either lost or fail to acquire those specialized characters of the organization of highly developed forms. The pelecypods thus arose with absence of any definite region and masticatory apparatus. From the point of view of physiology, the most primitive pelecypods are those in which the foot has a creeping ventral surface and the branchial filaments are free and not reflected those of the protobranchiate. From these are evolved the filibranchiates where branchial filaments are reflected but are still lacking vasscular function. These is their term gave rise to the culamellibranchiates which are most specialized in respect of complication of Ctenidia, Numerous vascular interlammellar and inter filamentar junctions are present.

From eulamellibranchiates are derived septibranchiates in which respiratory organ is represented by branchial sephum. The most primitive condition of nervous system exhibited by Protobranchiata as the cerebral and pleural ganglia are still separated and there is a set of cerebro-pleural, cerebro-pedal and pleuro-pedal connectives on each side. In all other lamellibranchiates, however, the cerebro-pleural ganglia are intimately fuse to one another and there is a single pedal connectives on each side. The two pedal ganglia also lie close to each other and may be reduced in those forms in which the foot is reduced or absent e.g. *Pecten*, *Teredo*, *Ostrea* etc.

Cephalopoda

The cephalopods are bilaterally Symmetrical and have a posterior mantle cavity. The primitive form in this group i.e. tetrabranchiates possess an external shell which may be many chambered besides they possess two pairs of ctenidia, kidneys and auricles in contrast to more specialized form i.e. Dibranchiata which have each of these structure in single pair and have an internal shell which may be reduced or absent. The remarkable features of cephalopods, however is extraordinary modification of foot into arms and tentacles. This class

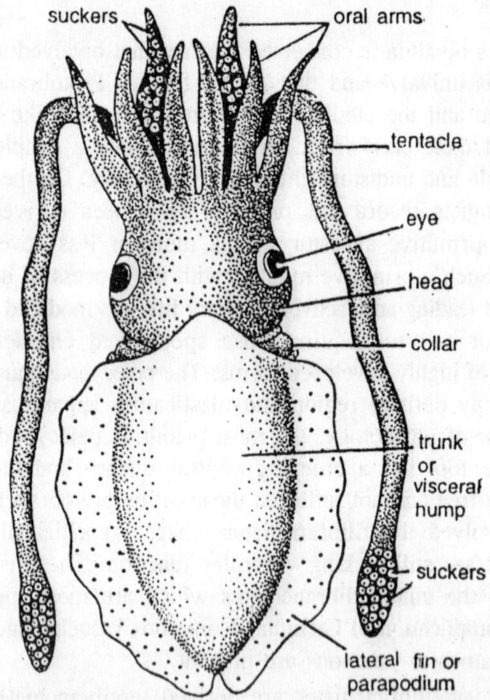

Fig. 2.5. Sepia officinalis. Male in dorsal view.

remains apart from molluscs by its wonderfully high organisation, specialized nervous system and eyes. There is nothing to indicate close relationship with any of lower groups beyond the general resemblance to molluscan organisation and presence of odontophore. The cephalopods form, infact, a singular ray isolated group. Palaeontology has also not given any indication of their origin. Embryology is equally silent. The absence of free-larva and performed modification in the development produced by enormous mass of foot-yolk sharply separates them from all the members of Phylum Mollusca.

3

FIELD STUDY OF MOLLUSCS

Most people including the native fishermen of remote tropical shores, are poor shell hunters. The clam diggers of New England, the oystermen of the Chesapeake Bay and the pearl divers of Japan are all very familiar with their own specialties, but they seldom take note of other molluscs and very rarely collect them. A casual beach stroller or a novice conchologist may see the large and obvious shells cast up by the waves, but the dozens of minute species underfoot go unnoticed. Even a well-experienced malacologist suddenly placed in a new habitat will take five or ten minutes to "turn on his collecting eyes."

It sometimes takes that much time to find the first mollusc in a rock tide pool, but in the next five minutes a dozen kinds and a hundred specimens will be seen. A mud flat, at first seemingly barren will begin to reveal more and more snails, and the sand bars will begin to turn up more clams. Camouflage is a built-in protection for many molluscs. Novice collectors expecting to see the bottom of the sea dotted with bright and colorful shells are bound to be disappointed. The purple-dotted cone is covered with a thick, brown periostracum; the long-spined orange *Spondylus* is hidden by a thick growth of algae, sponges and bryozoans and even the tiny *Nassarius* mud snail is heavily covered with a thick film of mud.

Shyness and a dislike for bright sunlight are two additional mollusc characteristics that make most of them difficult to find. A snail venturing over the sand in broad daylight is a likely target for predators; a clam or mussel nestled in a crevice on the underside of a rock ledge is less likely to be found by a marauding fish or crab. To the novice, the beach is the obvious place to find shells. There are certainly few places as pleasant to stroll when the weather is fine and the clean sand is firm. However, few beaches harbor many molluscs. Dead shells, of course, may be cast up along the tide line, particularly after storms, but usually subsequent tides sweep the strand clean again.

Some beaches, because of the rather sudden drop-off, rarely have any shells. The two great exceptions in the world are Jeffreys Bay in South Africa and the west coast of Sanibel-Captive Island in western Florida. At the latter, the beach becomes strewn with millions of tons of fresh shells once every few years, usually after a chilly three-day blow from the northwest. Minor incursions occur once or twice a year, but for the most part the pickings are poor to bad in the off-seasons. In New Jersey, where the ocean beaches have few shells most of the years; the strand will become deeply littered with surf clams, *Spisula*, and moon snails during February when a combination of near-zero temperatures, heavy offshore winds and unusually low tides kill billions of molluscs.

In the tropics, sand beaches are apt to be more productive, for a number of clams and *Terebra* snails live in the moist slope below the midtide line. As the tide begins to rise and each wave is larger and sweeps farther up the slope, the large *Donax* clams begin to pop up above the surface for a few seconds to feed. At times, the beach may be dotted by small clams wiggling themselves back into the sand as each wave recedes. Higher up on the beach, well beyond the reach of the average wave, there may be a zone shaded by trees and overgrown with vines where moist humus and coconut husks harbor a vast kingdom of salt-loving pulmonate snails, including *Melampus*, *Pythia*, *Ellobium* and *Truncatella*. Collecting on this high land zone is best just after a rain. Beaches are not always sandy. Some are made of boulders and small stones. If the slope is not too steep and if the stones are fairly well anchored, a respectable community of molluscs may exist on the sides of the rocks and in the narrow mud or clay interstices.

In cooler climates, festoons of brown *Focus* weeds add additional protection for mussels, periwinkles and *Nucella* dogwinkles. The sand and mud tidal flats are lower extensions of the beach and harbor vast colonies of molluscs, since that area is underwater for long periods is richer in food and is less subject to the rigors of wave action. Some molluscs roam at low tide, others move about only when there are several inches of water above them. In areas such as this, night is the time that most molluscs come out of the sand and travel in search of food and mates. Although in most parts of the world rocky shores harbor the same general types of molluscs, there are many variations and curious novelties limited to certain geographical regions, to certain types of rocks or to various tidal levels. In England, hard sandstone is bored into by the piddock clams, whereas in the West Indies the coral limestone rocks are penetrated by an entirely different family, the *Lithophaga* date mussels. In Peru, the offshore sedimentary rocks are bored by the elongate ark clam, *Litharca*.

The upper levels of rocky coasts seldom are washed by waves, but during storms sea spray may be carried inland for many yards. The periwinkles, *Littorina* and *Tectarius*, attempt to venture as far inland as possible, moving mainly at night and during rainy periods. Since they must return to the sea or to rocky pools of seawater to lay their eggs, the adults do not completely desert the ocean. A few species, notably the Zebra Nerite, *Puperita*, of the Caribbean, can survive in hot, supersaline pools isolated on the high rocks. Their ability to withstand changing temperatures, from 60° F. to 110° F., and salinities, from very salty to pure rainwater, attests to the adaptability of littoral snails. Rock pools in the midtide zone are usually well-stocked with molluscs, both large and minute.

In cool waters, the pools are lined with colorful algae, sponges and bryozoan growths. Colorful starfish and sea urchins may dot the bottom. Limpets, periwinkles and blue mussels are dominant, but hidden among the more delicate green algae or nestled in the sandy cracks there are apt to be dozens of species of minute snails, some adults being less than four millimeters in size. In the tropics, an even richer fauna inhabits the intertidal pools. Nerites, chitons, *Purpura* rock shells and occasionally a large tun or frog shell from deeper water are found in these deep, rockbound pools.

In the Caribbean, pelagic species of the open ocean wafted shoreward by winds are sometimes trapped in these pools, but lack of oxygenated, cool oceanic water usually results in their death. Among the inhabitants of the large floating mats of brown sargassum weed are the little brown snail *Litiopa* and the brown-and-white nudibranch *Scyllaea*. In surface waters nearby are usually found the violet snails, or purple sea snails, *Janthina*, that keep afloat by bubble rafts they create with their feet. A whole world of molluscs is found living as commensals or symbionts with other marine organisms.

The phylum of starfish, sea puddings and sea urchins, known as the Echinodermata, plays host to dozens of species of univalves and bivalves. There are snails, such as *Thyca*, that live like limpets enwrapped about the spines of sea urchins. Deep inside the urchins' pencil like spines live the tiny, white *Stylifer*. Shaped like an old-fashioned electric light bulb, the *Stylifer* creates for itself in the spine a hollow cyst that has a small opening for taking in fresh seawater. A parasitic clam attaches to the underside of urchins. Even the lumbering sea cucumber, *Holothuria*, has tiny white *Melanella* snails attached to its anal end, and deep within the intestine is sometimes found another shell-less family of parasitic snails. Almost everywhere one looks, one is apt to find molluscs—clams boring in waterlogged planks, manila rope, ship bottoms and floating coconut husks. Snails live on the leaves and roots of mangroves. Upwards of seventy kinds of minute molluscs have been found living on a Red Abalone.

How to Watch Molluscs

All too often the real enjoyment and profit of shell collecting is lost by the hasty and greedy impulse of grabbing up a living mollusc, throwing it into a dark bag and not looking at it again until it is time to drop it into boiling water or some deadly pre-servative. In today's world where man is rapidly brining thousands of species to extinction, it seems little enough to ask of seashore visitors that they spend more time watching than collecting. Overcollecting is occurring in more and more places around the world, although it is not this activity, but rather massive pollution from cities, manufacturing plants and ships, that is endangering coastal marine molluscs.

Locally, some kinds of molluscs are becoming very scarce because of commercial collecting. In some states, even the constant revisiting of biology classes, week after week, has decimated certain small, once-productive spots. Shell watching and limited shell collecting is becoming more popular, if not a necessity, for the conchologist of the future. It is surprising how much information may be obtained by simply observing undisturbed living molluscs for a few minutes. Trails in the sand tell a tale, and if you follow a furrow in the surface of a Florida tidal flat, you may find a hump in the sand caused by an olive shell, a venus clam or a small *Busycon* whelk. If whelk and clam happen to meet, you will see the snail envelop the bivalve in its black foot, press the edge of its shell between the clam's valves and slowly but surely pry open its victim.

Equally fascinating, but seldom seen except by snorkelers or scuba swimmers, is the process of egg laying by a whelk. The capsules are formed in a pore in the sole of the foot and passed up by the oviduct, where several eggs are injected into each wafer-shaped case. When filled, the capsules, strung together like the segments of a rattlesnake's tail, are pushed up out of the sand. Many profitable and interesting hours can be spent observing living molluscs in a homemade aquarium. In fact, the strange habits and life cycles of such molluscs as the fish-eating cone and the sea-anemone-sucking wentletraps were first discovered and documented in saltwater aquariums. The tank need not be elaborate and space-consuming. A clear gallon jar, with a large, round opening, will keep most small molluscs overnight. The tricks are not to overcrowd and not to put natural enemies together. The best aquariums are ordinary square plastic boxes, no larger than cigar boxes, available in kitchenware departments. Keep only four or five of the smaller molluscs, such as periwinkles and nerites, in one miniaquarium.

Mechanical aeration can be set up by attaching a tube to an inexpensive aquarium pump and branching it out to several miniaquariums. Where the lid of the box overlaps the bottom section, file a round hole to accommodate the air tube. For very tiny molluscs, aeration is not necessary, but a lid should be kept on to prevent evaporation of the water. Be sure that the porous stones and algae-covered rocks remain moist. Young snails that have crawled up the dry sides should be brushed down at least once a day. When replenishing evaporated seawater, add only fresh water or the salinity will rise too much. The following genera have been kept in small unaerated miniaquariums for from one to five years: *Cerithium*, *Batillaria*, *Planaxis*, *Littorina*, *Neritina*, *Nerita* and *Marginella*. When feeding carnivores, say once a week, in aerated tanks, remove pieces of uneaten shrimp, clams and mussels. Many important habits and morphological details of small molluscs may be observed in small watch glasses of new seawater. Although a hand lens of seven-power magnification will reveal a great deal, for comfort and better vision, a binocular dissecting microscope is recommended.

How Shells Mate

It is believed, on the basis of biological studies made on today's molluscs, that all primitive univalves and bivalves had separate sexes, that there were no accessory sexual organs and that the ova and sperm were directly and freely discharged into the seawater. This simple and rather chancy way of fertilization has been retained in many modern bivalves, the tusk shells and a few marine gastropods, such as the limpets and some top shells. Spewing sperm and eggs loosely into the ocean is fraught with danger. They may be eaten by sea creatures or destroyed by adverse temperatures and currents or be unable to find each other. To offset these disadvantages, the molluscs produce enormous numbers of egg cells and sperm cells, or gametes, all in as short a period as possible and all at the same time. *Patella* limpets will breed only when the wind and waves have been extremely strong for several days. Northern oysters will shed eggs only when the water temperature goes above 60°F.

Bivalves that cling to mangrove trees in estuaries will breed only on the first very high tide after a full moon. This assures the young of being spread about over as wide an area as is possible. The sperm of all oysters contain a strong hormone that causes the muscle of the female to relax and at the same time increases the pumping rate of water. This allows the entry of more sperm. Some hermaphroditic wood-boring clams, *Xylophaga*, have a special sac for storing live sperm. Later, when the eggs have become mature, the sperm are released to fertilize them. This ensures a lone female adrift at sea in a log a supply of male gametes.

Sex changes in an individual is not uncommon in some species. Some bivalves begin life as males and after a year change to functioning females. The edible oyster alternates its sex—male one year, female the next. The common slipper shell, *Crepidula*, lives in clusters with one shell clinging on top of its neighbor. The four or five larger individuals at the bottom of the pile are all females. The younger and smaller ones at the top are males. Very often the middle ones are in the process of changing sex from male to female. Sexuality is controlled by a hormone that is constantly being produced by the female and exuded into the water.

As long as the level of this hormone is high enough, it will prevent males from turning into females. Should the females die or cease to function because of old age, the hormone disappears and the snails at the top of the pile begin to lose their testes, the penis shrinks away and ovaries begin to develop. In a week, the males have become females. Hermaphroditism is limited to certain groups of molluscs. The duck clam family, Anatinacea; most air-breathing land snails and the sea hares and nudibranch sea slugs are hermaphrodites.

The cephalopods have separate sexes; in fact, some octopuses go through a period of courtship. During this time, certain color changes are displayed by the male, and he may at the same time arouse the female by stroking her. The sperm, which occur in packets, are transferred into the female's oviduct by the tip of one of the male's arms.

Egg Capsules and Brooding

Ensuring protection for their eggs is one of the principal problems for the molluscs. Even the marine forms that shed their eggs directly into the sea supply a basic protective membrane around the ovum. A few bivalves lay their eggs in gelatinous capsules, while others deposit them with a sticky mucus to ensure their remaining safely on the bottom. A more advanced egg capsule, somewhat resembling a flying saucer, is produced by the Common European Periwinkle. Less than one-fourth the area of the head of pin, the clear, pill-shaped capsules, each containing an egg, float through the water for a few weeks before hatching. Although molluscs generally do not take care of their eggs or young, a few do. The female cowries, for instance, sit on top of their egg clusters to protect them from enemies. To keep off silt and parasites and help the flow of oxygenated water, the female octopus strokes the festoons of eggs she has deposited in her rocky lair. Brooding is done in various manners among both univalves and bivalves. Many clams hatch their eggs within the mantle chamber and allow the veligers to grow into young clams within the protective gill curtains.

Some boring clams permit their young to remain within the burrow. Then they later begin their own independent tunnels in the wood. The *Clanculus* snail and some sundials permit the young to crawl inside the umbilicus of the adult shell, and sometimes a mucus sheath is laid over the numerous infant snails that have lodged themselves in the spiral furrows of the mother's outer shell. Many slipper shells attach their leathery capsules to the substrate and then clamp down on them so that the mantle cavity serves as a protective hood. The *Trivia* cowries hollow out holes in ascidians—lowly marine chordates—and deposit their eggs deep below the surface of their hosts.

There are many dozens of kinds of capsules laid by marine univalves. In some cases, the eggs hatch and immediately escape as free-swimming veligers. However, in many kinds the veligers remain inside the capsule untii they have metamorphosed into miniature shails. The young may eat their way out or as in the case of some whelks, they eat their young brothers and sisters until only one large individual is left. The *Colus* whelks, some volutes and the *Phalium* helmet shells lay towers of interlocking capsules.

The *Busycon* whelk of the eastern United states lay water-shaped, leathery, capsules attached by a cord in a snakelike row. Each capsule may contain from 50 to 120 miniature snails. The *Natica* moon snails that live on sandy bottoms lay a stiff, gelatinous collar with the inside containing microscopic eggs and the outside densely packed with sand grains. A number of snails give birth to live young. The eggs are retained and hatch inside either the swollen oviduct or a special brood chamber located on the back of the head. Smaller young and deformed eggs serve as food for the survivors. Among the live bearers is a *Turritella* snail, which has been found in fossiliferous rocks many millions of years old, showing the tiny snails closely packed inside the mother's shell.

How Shells are Built

The hard shells of molluscs are made of a limy mineral, calcium carbonate. The degree of hardness and the microscopic structure of the shell material vary, depending upon the type of crystallization. The six-sided crystals may be prismlike, becoming calcite, or they may be very flat and thin, becoming aragonite. The latter is a heavier form of lime that gives a nacreous, or mother-of-pearl, effect. The blood of the mollusk is rich in a liquid form of calcium. Most outer parts of the soft visceral organs–in particular, the fleshy cape known as the mantle–concentrate the calcium in areas where it precipitates into additional crystals.

Most of the shell growth takes place at either the edge or the mouth of the shell, although a great deal of thickening continues farther back inside the shell. Solid crystals form either next to old crystals or on a matrix of conchiolin. The latter is a soft, brownish material composed of proteins and polysaccharides that serves as a microscopic latticework for new crystals. This same material is also used in the formation of the outer coating, or periostracum, that covers the entire shell. The mantle is the main architect of the shell, producing the various ribs, cords and spines. Production of new shell material is controlled and influenced by a number of factors, including sexual hormones, intrinsic rhythmical patterns, diet, acidity of the water and temperature. The size of the crystals is dependent upon the rate at which they are formed, and this, in turn, is determined by the temperature of the surrounding water.

In some molluscs, the inner layers of the shell, next to the viscera, may be reabsorbed to make room for growing gonads and to be reused to build a larger lip on the shell. A cross-section of a large, adult cone will show how the old whorls, once thick-walled, are now paper-thin. The calcium carbonate has been changed back into liquid form and carried by wandering amoebocyte cells to the mantle at the anterior end of the snail. There the mantle has secreted the calcium back into the edge of the growing shell. In the event of a major injury to a shell, this mode of calcium transport is used to mend cracks, to fill holes bored by sponges and snails and to repair a broken apex or damaged outer lip. Old spines that are beyond the reach of the mantle cannot be repaired. Certain areas of the foot of gastropods may be endowed with glands that produce either soft conchiolin or hard shelly material. Such a region is located on the dorsal side near the back end of the foot. Here the operculum, or trapdoor, is secreted.

The last part to be withdrawn into the mouth of the shell, it serves as a protective barrier. The diameter and thickness of the operculum are increased on the underside where it is attached to the foot. The sides of the foot in the olive shells contribute to the glossy outer layers of the shell, whereas the posterior end of the foot of the Angular Volute from Brazil produces a long, shelly spike at the apex of the shell. The glossy and thickened backs of cowries, however, are the product of an enveloping extension of the mantle. The largest prisms of calcite produced among the molluscs are found in the bivalve pen shells. The large crystals are readily seen with the aid of a hand lens or magnifying glass. The conchiolin separating the prisms contains a high percentage of water, thus giving the matrix and the entire shell an almost flexible quality. When recently collected pen shells are allowed to dry out, they become brittle.

4

Body Wall and Mantle

The whole body is covered by a single layer of epithelium, which, in parts not protected by the shell, may be more or less ciliated. This layer is very rich in glands which are almost exclusively unicellular; some of these lie in the epithelium itself, while some have sunk into the subjacent tissue, their ducts, however, passing between the epithelial cells. The layer immediately beneath this body epithelium is called the corium, and consists of connective tissue and muscle fibres. It is, however, not distinctly marked off from the tissues beneath it. The pigment is almost always found in the cells of the subepithelial connective tissue.

Placophora

The *Chitoa* is provided dorsally with eight consecutive shell-plates which overlap in such a manner that the posterior edge of each plate covers the anterior edge of the next. These plates are bilaminar. The outer and upper layer which forms the dorsal surface is called the *tegmentum*, the lower hidden layer the *articulamentum*. As a rule, the tegmentum of the anterior plate only is as large as the articulamentum beneath it; in the other plates, the latter is the larger and projects beyond the former laterally and anteriorly. These projecting parts of the articulamentum called *apophyses*, slide under the plate next in order anteriorly. Between these two layers tissue is found, which is a continuation of the dorsal integument. The tegmentum is penetrated by canals of various sizes, which open at its surface through characteristically arranged pores.

The tegmentum consists of a horny or chitinous substances, which may be considered as a cuticular formation, impregnated with calcareous salts. The articulamentum is compact and free from canals; it contains little organic substance, and much calcareous salt. It alone answers to the shell of other Molluscs, while the tegmentum must be considered as a calcified cuticle covering the true shells (the articulamenta) as a continuation of the cuticle of the zone which encircles the eight shell-plates. This zone carries chitinous or calcareous spines, setae, scales, granules, etc., varying in number and arrangement according to the genus and species. Each spine, as a rule, arises as a globular vesicle within an epithelial papilla and above a very large formative cell. As it grows, it is pushed upwards by the newly forming euticular layers.

The formative cell at its base persists, but remains connected with the epithelial papilla only by a protoplasmie process which continually lengthens, and may surround itself with a nucleated sheath. In fully-developed spines, the remains of this cell are still found as a small terminal swelling (Endkölehen). There are, however, spines are specially flat-scale-or plate like calcareous formations in the integument which do not arise from single large formative cells, but are probably produced by several cells in the base of an epithelial papilla. Just as we have recognised the tegmentum covering the articulamenta to be merely a special portion of the general cuticle, so we may further recognise in the articulamenta the homologues of the calcareous spines, scales, etc. which are developed in the integument of the mantle. The articulamenta would thus be nothing more than very large and expanded calcareous scales. This view, finally leads to the conclusion that the shell (if it may hare be so called) of the Molluses originally consisted of isolated calcarous spicules or spines, which were enclosed in a thick cuticle, and projected from the same as in the *Proncomcnia*, *Nemenia*, etc..

The *Cryptochiton* the shell is internal, *i.e.* it is entirely covered by a fold of the integument, which grows over it from all sides. It consists exclusively of the articulamentum, since the whole dorsal integument is covered by an even cuticle, which therefore forms no tegmentum. The only part of a *Chiton* which can be called the mantle fold is the marginal zone of the body, the ventral side of which encircles the head and foot and forms the lateral boundary of the branchial groove or furrow. Just as the dorsal side of this mantle, which is called the zone, carries large spines, setae, or scales, so may the under surface be covered with small closely crowded spines. The rest of the integument is bare, being merely covered with a simple epithelium.

The genus *Chitonellus* is of great importance in comparing the outer organisation of the Placophora with that of the *Solenogastres*. The body is not dorso-ventrally flattened, as in the *Chiton*, but nearly cylindrical; the ventral surface, however, is flattened, and has a median longitudinal groove. The foot is not externally visible, but can be discovered, much reduced, in the base of the median groove, itself possessing a ventral median groove representing a narrow contracted sole. The flat ventral surface is therefore the mantle. In the narrow cleft on each side, between mantle and foot, in the posterior half of the body, lie the gills. The lateral margin of the body in *Chiton* is represented in *Chitonellus* by a mere blunted ridge, which is almost exclusively caused, as may be seen in transverse sections, by a great thickening of the cuticle.

Solenogastres

In the Solenogastres (Aplacophora), whose outer organisation has already been sufficiently described, the shell is altogether wanting, but the cuticle secreted by the epithelium over the whole body is usually exceedingly thick. It contains caleareous spicules, which sometimes project above the surface. These, like the spines of the *Polyplacophora*, rise from cellular cups, which are connected with the basal epithelium of the cuticle by nucleated stalks. There can be no doubt that the spicules are formed by these cups and nourished by them during growth. The foot, as we have seen, is reduced to a narrow ciliated longitudinal ridge, which rises from the base of the medio-ventral groove.

The term mantle is here inapplicable, except perhaps to the integument which forms the lateral boundary of this groove. In *Chactoderma* the foot finally atrophies, and the medio-ventral groove also disappears. The long series of undoubtedly primitive characteristics in these two groups—the *Placophora* and Solenogastres—obliges us to place them, as we shall have repeatedly to point out, near the root of the Molluscan phylum. In some points the Solenogastres are perhaps more primitive than the *Polyplacophora*, and the vermiform body, the slight development of the mantle, the foot and the gills have been thought to be primitive characteristics.

More recently, however, it has been maintained, as the present writer thinks, with justice, that these conditions are rather the result of secondary adaptation to a limicolous habit of the life (most Solenogastres inhabiting mud). The shell, mantle gills and foot are such essential characteristics of the Mollusca that we must assume their existence in the racial form. The series *Chiton*, *Chitonellus*, *Neomenia*, *Choctoderma* does not, therefore, illustrate for us the rise and development of typical Molluscan characteristics, but rather their progressive obscuration and disappearance.

Gastropoda

Integument

The free edge of the mantle, which takes the chief part in the formation and growth of the shell, is particularly rich in *mucous*, *pigment*, and *calcareous glands*. The epithelium is ciliated over areas of varying extent, especially in aquatic Gastropods. In many of the shell-less *Opisthobranchia* the whole surface of the body is ciliated. The remarkable marking and colouring of the integument especially seen in the *Nudibranchia* are caused by pigment cells, which are more often found in the cutis than in the epithelium. Where there is no firm shell, calcareous granules or spicules may be found scattered throughout the cutis. In several *Nudibranchia stinging cells* have been discovered in the integument.

Mantle, Visceral Dome

The mantle fold is, as a rule, well developed in Gastropods, and covers a spacious pallial cavity. Whenever the fold is small or altogether wanting, the condition is secondary rather than primitive.

Prosobranchia

In the Prosobranchia, the mantle fold develops on the anterior side of visceral dome, and there covers a spacious cavity. It further usually extends like a narrow collar right round the base of the visceral dome. In the symmetrical *Fissurellidoe*, the mantle cavity is short, and opens outwardly by means of a dorsal aperture through the mantle fold, which corresponds with the perforation at the apex of the shell. A circular fold, provided with a highly sensitive fringe, is formed by the mantle around the aperture, and projects for a short distance beyond the perforation in the shell. The water needed for respiration passes into the pallial cavity through the slit-like aperture at the free edge of the mantle fold, over the nuchal region, and flows out through the apical aperture just described. This aperture also serves for the ejection of excretory matter from the rectum, which lies immediately behind it.

In *Rimula*, the apical apertures in shell and mantle have moved somewhat forward, and lie anteriorly between the apex and edge of the shell. In *Emarginula*, the mantle has an anterior cleft, the edges of which, in the living animal, are folded in such a way as to form a

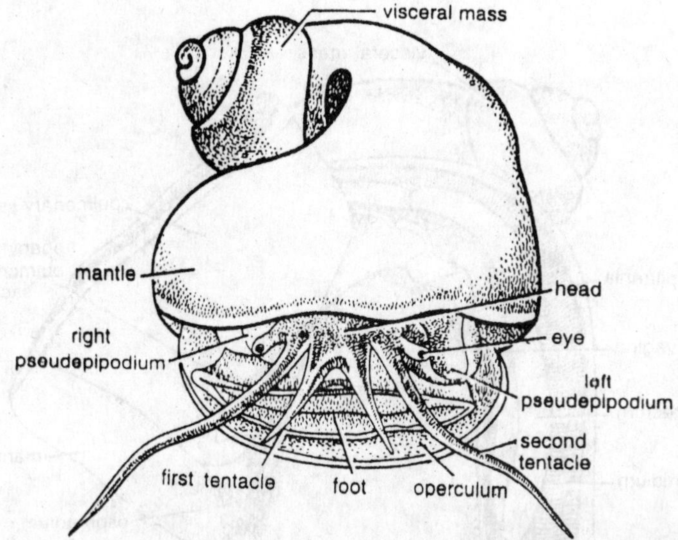

Fig. 4.1. Pila globosa. Front view of the animal after removal of the shell.

tubular siphon, which can be protruded through the marginal cleft of the shell. In *Parmophorus* there is no second opening into the mantle cavity, but the lateral edges of the mantle are very much widened, and bent back dorsally over the outer surface of the shell in such a way as to cover the greater part of it. In *Haliotis*, the enormous development of the columellar muscle on the right side confines the mantle cavity to the left. The mantle fold has a long slit reaching from its edge to the base of the pallial cavity. This slit lies under a row of perforations in the shell which are characteristic of *Haliotis*, and through these the respiratory water is expelled.

In the spaces between the consecutive perforations, the edges of the mantle cleft are apposed, merely separating beneath each aperture to allow of free communication between the cavity and the exterior. The edges carry three tentacular processes, which can be thrust outward through the perforations. The anus is always found under the posterior perforation. The edge of the mantle surrounding the body splits into two narrow lamelliae, which bend round to form a groove for the reception of the edge of the shell. The *Trochidoe*, *Turbinidoe*, *Neritidoe*, and nearly all *Monotocardia* have no second aperture and no mantle cleft. In *Docoglossa* (*Patella*, etc.) the mantle forms a circular fold round the visceral dome, which is the form of a blunt cone. It covers the edge of the almost circular broad-soled foot. The mantle is broadest

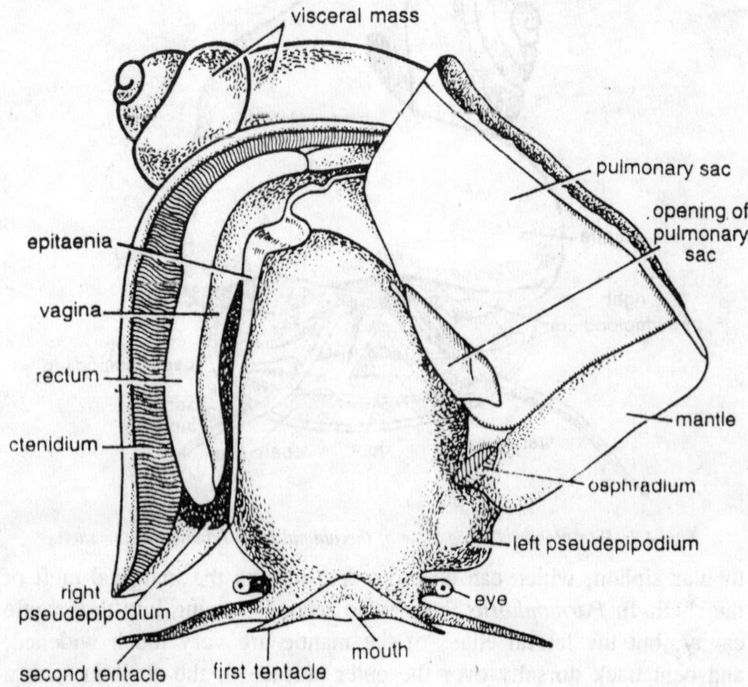

Fig. 4.2. Pila globosa. A female individual dissected to show the organs of the pallial cavity.

anteriorly, where it covers the head and neck, *i.e.* the pallial cavity or groove is here deepest. The visceral dome, in the *Monotocardia*, is almost always distinctly constricted at the base, and spirally coiled. The pallial cavity occupies its typical position. In many *Monotocardia*, the free edge of the mantle fold is prolonged on the left side, projecting forward, sometimes to a great extent; the lower edges of this projecting fold bend round towards each other to form a tube or semi-cylindrical channel, which is called the siphon. Through this siphon, the water needed for respiration flows into the mantle cavity.

It can generally be told, by the shape of the shell, if there is a siphon or not, since most *Monotocardia* which possess one have either a notch in the edge of the shell at the columella, or a process called the canal or beak at this some point, which endless this siphon. The length of this latter canal need not, however, correspond with that of the siphon. The *Monotocardia* have been grouped, according to the presence or absence of a siphon, into the *Siphoniata* or *Siphonostomata*, and the *Asiphoniata* or *Holostomata*; but this classification is artificial, since siphons are sometimes present and sometimes absent in forms which are undoubtedly nearly related.

In most *Monotocardia*, the shell is not outwardly covered by the mantle, but in some groups, the edges of the mantle bend back over the shell, and finally grow over it to such an extent as to unite above it. The external shell in such cases becomes an internal shell. In the *Harpidoe* among the *Rhachiglossa*, the mantle bends back over the columellar lip of the shell. In the *Marginellide*, it covers a large part of the outer surface, and the same is the case in *Pyrula* among the *Taenioglossa*, in most *Cyproeidoe* and in the *Lamellaridoe*. In *Lamellaria*, the shell is completely grown over by the mantle. In *Stilifer* among the *Eulimidoe* also, the shell is more or less covered by the mantle. The edge of the mantle may be fringed on notched, or (*cypracidoe*) provided with wart-like, tentacular, or branched appendages.

Pulmonata

In the Pulmonata, the arrangements of the mantle fold and visceral dome and of the shell, which is intimately connected with them, are of great interest. We have, on the one hand, forms such as *Helix*, with large protruding, spirally-coiled visceral dome and large mantle fold enclosing a spacious cavity; on the other forms such as *Oncidium*, without distinct visceral dome or mantle fold and without shell. Between these two extremes there are numerous transition forms; indeed, complete series of such forms may be found even within some of the natural divisions of the Pulmonata. The following are a few characteristic types.

Helix

The visceral dome is large and spirally coiled, and is covered by spiral shell sufficiently large to shelter with case the whole body. The mantle fold covers a cavity lying anteriorly to the visceral dome (pulmonary cavity). Its free thickened glandular edge unites with the nuchal integument near it in a way characteristic of the Pulmonata, leaving only one aperture, the respiratory aperture—on the right. The apertures of the hind-gut and excretory organ are close to the respiratory aperture, through which their excreta have to pass out. In many species of the genus *Vitrina*, the shell cannot contain the whole animal. The mantle fold projects in front of the shell, and has a process which is bent back over the shell, and is used for cleansing it. In *Daudebarida* (*Helicophanta*) the visceral dome and shell are in comparison with the rest of the body, much smaller than in *Vitrina*. The animal cannot be

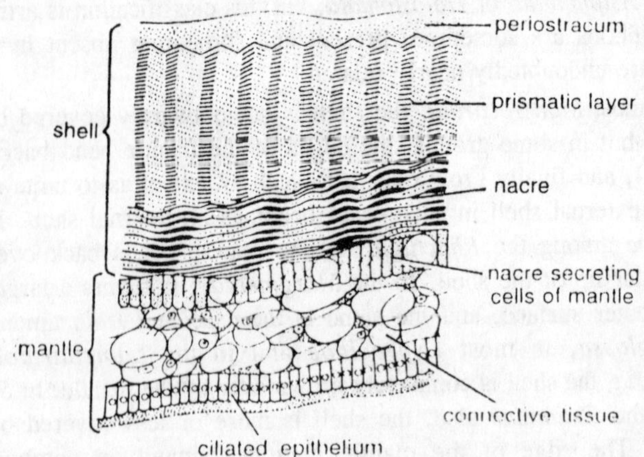

Fig. 4.3. Unio. Cross section of shell and mantle.

sheltered by the shell. The visceral dome begins to be levelled down to a certain, extent, disappearing into the dorsal surface of the foot. It lies far back on the body, the respiratory aperture being on its right side. A somewhat similar arrangement is found in the genus *Homalonyx*, in which the low visceral dome lies on the centre of the back. The respiratory aperture lies to the right at the edge of the mantle. The edge of the flat ear-shaped shell is fixed into the mantle fold. *Daudebardia* and *Homalonyx* begin to look like slugs.

The *Testacella* a visceral dome hardly exists. The only remains of its is a small mantle at the dorso-posterior end of the body, which is covered by an ear-shaped shell. Beneath the mantle lies a reduced respiratory cavity. The respiratory aperture lies to the right posteriorly, beneath the edge of the shell. The viscera lie dorsally on the foot. The common terrestrial snails *Limax* and *Arion* resemble *Testacella* in the reduction of the visceral dome, but in them the mantle or so-called shield which takes its place lies place lies anteriorly behind the head. At its right edge lies the respiratory aperture.

In *Limax* there is a small round rudimentary shell which is internal, *i.e.* it is entirely enveloped in or overgrown by the mantle fold. In *Arion* this shell is represented by isolated calcareous granules. In *Oachidium* and *Vaginulus* there is no trace of a visceral dome, nor, in the adult, of the shell. The visceral dome has to a certain extent spread out over the whole dorsal surface of the foot and his disappeared. There is, further, no outwardly recognisable mantle fold distinct from the rest of the dorsal integument. A longitudinal furrow still divides the dorsal part of the body from the foot. The respiratory aperture with the anus lie posteriorly the median line.

In the genus *Physa*, the edge of the mantle takes the form of lobe-like or finger-shaped processes, which bend back over the shell, and can be applied to its outer surface. In *Anaphipeplea*, the mantle is much widened and, when bent back over the shell, covers all but an oval spot on the dorsal side of the last coil. The dorsal integument of the *Onchidia* has wart-like protuberances of (in *Peronia*) branched appendages. These are richly supplied with blood-vessels, and serve for respiration. In *Peronia* there are besides these also dorsal prominences which carry eyes. The dorsal integument projects all round the body above the foot, and thus forms as in *Chiton*, a peripheral zone, which is ventrally separated from the foot by a groove. In *Oncidiella* the edge of this zone, *i.e.* the lateral edge of the body is dentate or fringed.

OPISTHOBRANCHIA

The typical outer organisation of the Gastropoda here suffers even more varied and thorough modification then in the *Pulmonata*. We have, on the one hand, forms with head, foot, visceral dome, shell, mantle and gill; on the other, forms which possess none of these organs and nevertheless are both Gastropods and Opisthobranchia. In one principal division of this order, the *Palliata* or *Tectibranchia*, the mantle fold is retained on the right side of the body, and partially covers a typical Molluscan etenidium; in other divisions both mantle and etenidia are wanting. We do not here apply the term mantle to the fold or edge of the dorsal integument which surrounds the body at the part where the head and foot take their rise; such an edge is more or less developed in most Opisthobranchia and distinctly marks off the foot and head from the rest of the body or back. The mantle here means only the broader fold which covers the mantle cavity, in which lies a typical molluscan gill. The edge of the mentle never forms a distinct siphon in the Opisthobranchia, though there is an approach to such a structure in the *Ringiculidoe*.

Tectibranchia

Reptantia

In this division we have, on the one hand, forms which still have a distinctly projecting visceral dome, whose integument secretes a coiled shell, into which the whole body can be withdrawn. On the other hand, forms occur in which the flattened visceral dome has spread out over the whole dorsal surface of the foot, the shell being rudimentary and internal. Examples of the former are found in the *Cephalaspidoe*, e.g. the *Aetaeouidoe*, *Tornatinidoe*, and some *Seaphandridoe* (*Atys*, *cylichnat*, *Amphisphyra*), a few *Bullidoe* (*Bulla*), and the *Ringiculidoe*. In *Seaphander* among the *Seaphandridoe*, and *Acera* among the *Bullidoe*, the body cannot be completely withdrawn into the shell. In the *Cephalaspidoe*, to which so far reference has been made, the shell is external. In *Gastropteron*, the mantle is rudimentary, and is provided posteriorly with a filiform appendage. It covers a delicate membranous internal shell, into which the body cannot be withdrawn.

The same is the case in *Philine* and *Dorulium*, where there is also a delicate internal shelly covering only a small portion of the viscera; this shell, in *Doridium*, is produced in the form of two lobes, the one to the left ending in a filiform process. The visceral dome in the *Anaspidoe* is small as compared with the size of the animal, but rises distinctly above the rest of the body, and is covered by a thin inconspicuous shell. The mantle and shell often only partially cover the gill.

In *Aplysia*, the shell is internal, *i.e.,* it is entirely overgrown by the mantle; in *Dolabella*, this enveloping overgrowth is not quite complete, as a circular median dorsal aperture is left, through which the dorsal surface of the shell is visible. The mantle in *Dolabella* forms a small anal siphon posteriorly. *Notarchus* has a microscopically minute shell. In certain species of this genus, the integument forms protuberances or delicately branched appendages. In the *Oxynodiea*, the shell is only partially covered by the mantle, and is, further, much too small to shelter the body. Among the *Notaspidoe*, the *Umbrellidoe* have a small flattened cap-like visceral dome lying upon the massive foot. The visceral dome is surrounded by a mantle fold, which, on the right side, covers the gill. The integument of the dome and mantle is covered by a flattened disc-shaped shell.

In *Pleurobranchia*, the visceral dome is relatively large. The right and left margins project as short mantle folds, but there are no such folds to the front and back, so that at these latter parts the flattened visceral dome is not distinct from the rest of the body. In *Pleurobranchia*, the integument of the flattened visceral dome broadens out into a large, flashy dise which projects on all sides beyond the large board soled foot; its margin (mantle fold) is separated from the foot by a deep continuous groove running right round the body; in this groove, to the right, lies the large gill, while in *Pleurobranchus* a small flat internal shell, thin and membranous, is still found; in related forms this may be wanting. The dorsal integument is often strengthened by a layer of calcareous granules.

Natantia

Pteropoda thecosomata. The *Limacinidoe* have a well-developed visceral dome and corresponding shell, with sinistral twist; the shell can be closed by means a typical operculum. The mantle fold covers a cavity which lies anteriorly to the visceral dome. The anus is to the right. The animal can withdraw into its shell. In the *Cavoliniidoe* the dome and shell are bilaterally symmetrical not twisted, and the body can be entirely hidden within the shell. The mantle cavity her lies posteriorly to the visceral dome, on what is usually called its lower side. The symmetrical shell of the *Cymbuliidoe* does not correspond with the shell of other Thecosomata; it is a cartilaginous "pseudoconch" covered with body epithelium. In the *Cymbuliidce* the mantle cavity also lies posteriorly. We shall return later to the varying position of the cavity among the Thecosomata.

The mantle, in the genus *Cavodinia*, shows peculiarities which can best be described in connection with the shell. In the latter, two surfaces are distinguished, a slightly arched anterior surface (usually described as the upper), and an arched posterior surface. The anterior surface projects forwards and downwards beyond the posterior for a third of its length. The shell has three slit-like apertures, one anterior and ventral, through which the fin-like processes of the foot can be protruded, and two lateral apertures stretching far up, so that the shell appears almost bivalve. At these lateral slits, which admit water to the mantle cavity, the mantle bends round on to the outer surface of the shell, covering the greater part of it; and, at the upper angles of the slits, has two freely projecting processes.

Pteropoda gymnosomata. In these, the long outwardly symmetrical body is naked and without a mantle, and the foot, which is much reduced, is found on the ventral side of the most anterior part of the body.

Ascoglossa and Nudibranchia

In mature Ascoglossa and Nudibranchia, with the single exception of the *Steganobranchia*, a shell is always wanting, as also a distinctly demarcated visceral dome. The latter, indeed, spreads, out over the whole dorsal surface. The dorsal integument, nevertheless, forms a circular fold (mantle fold) separated from the foot by a groove sometimes deep sometimes shallow; but except in the *Phyllidiidoe*, no gills lie in this groove. Where this groove has nearly disappeared, the animals strongly resemble Planaria.

Phyllidiidae. In these, the mantle fold is distinct, and carries on its lower surface, to the right and left, a row of branchial leaves, herein recalling *Patella* and *Chiton*. The genus *Dcrmatobranchus*, which judged by its organisation, belongs here, has, however, no gills.

Dorididae. The dorsal integument (notaeum), which here covers the body like a shield, being generally distinctly demarcated from thc foot and the head, contains numerous calcareous particles, which give it a firmer consistency. Anteriorly, there are two feeler-like processes, the rhinophores, which can generally be withdrawn into special sheaths or pits; these are not to be confounded with the tentacles. The anus lies in the median line, generally behind the middle of the body, and is surrounded by an ornamental circlet of pinnate gills. The notaenm is often covered with prominences, and in some genera the margin carries variously shaped processes.

Cladohepatica. Here there are no anal gills. The dorsal integument has variously formed and variously arranged appendages; these may be conical, club or finger-shaped, lobate or branched; they are, for the most part, very striking in colour and appearance. Sacs of nematocysts are generally found at their tips, and caeca of the intestinal canal (branches of the digestive gland) penetrate them. These dorsal, appendages, which, like the rest of the body, and ciliated have, at least partly, a respiratory function. In many forms they easily fail off, and are later regenerated. Many Cladohepatica have a certain external likeness to Planaria with dorsal papillae (Thysanozoon), but this likeness is still more marked in the following family:

Ascoglossa. Anal gills and also, as a rule, dorsal appendages are here wanting. The whole body is naked and ciliated. The back is indistinctly demarcated from the head.

Phyllirhoe. This Nudibranchia genus, of all Opisthobranchia shows least of typical external organisation of the Mollusca. The body here is naked and laterally compressed with sharp dorsal and ventral edges. It has neither foot nor gills.

Scaphopoda and Lamellibranchia

From each side of the body there typically hangs a large leaf-like mantle-fold of the same shape as the shell-valve formed by it. These mantle folds project beyond the body in front, below, and behind, and enclose a mantle cavity which everywhere, except dorsally opens outward by means of the silt left between the edges of the folds. This large single cleft serves for the admission of nourishment and water into the mantle cavity, and for the expulsion of the excreta, genital products, and respired water; through it also the foot is protruded. Such a primitive mantle is thus completely open, its simple edges (*i.e.* without folds, papillae, tentacles, or eyes) are quite free coalescing nowhere. The above serves for a description of the mantle of *Nucula*—one of the *Protobranchia*—and must be considered as the primitive arrangement. In most Lamellibranchia, however, special differentiations of the margin of the mantle occur; these take the form of folds, thickenings, protuberances, papillae, tentacles, glands, eyes, etc., and this is the case both in forms which have an open mantle and in those in which the mantle is partially closed. The partial closing of the mantle is brought about by the concrescence at one or more points of the free edges of the mantle folds.

(A) *A completely open mantle, i.e.* one single large cleft entirely separating edges of the mantle, is found:

(i) Among the *Protobranchia* in *Nucula*.

(ii) Among the *Filibranchia* in the *Anomiidoe*, *Arcidoe*, *Trigoniidoe*, and a few *Mytilidoe* (*Pinna*).

(iii) In all *Pseudolamellibranchia* except *Meleagrina*.

(iv) Among the *Eulamellibranchia*, only in a few species of *Crassatella*.

(B) *The mantle folds of the two sides grow together at one point*. In this case this point of concrescence almost always lies high up posteriorly; and marks off a small aperture from the originally simple cleft. This aperture, occurring on a level with the anus, forms the *exhalent* or anal aperture of the mantle. Its edge may be more or less prolonged posteriorly to form an anal siphon, which can be protruded beyond the valves of the shell. At a point a little below this exhalent aperture, the mantle edges usually become applied to one another, *although no concrescence takes place*.

Above this point, between it and the anal siphon, they separate to form an *inhalent* or branchial aperture. The edges of this aperture also may be produced posteriorly into a branchial siphon, which, however, in this case, has a cleft extending along the whole of its lower side, which is a continuation of the large cleft of the mantle. A branchial siphon formed in this way, by mere apposition of the man the edges, is found in the genus *Malletia*, among the *Protobranchia*. An anal aperture, separated by a joint of conerescence from the large mantle cleft, is found in the following Lamellibranchia:

(i) Among the *Protrobanchia* in *Malletia*.

(ii) Among the *Filibranchia* in most *Mytilidw*.

(iii) Among the *Pseudolamellibrancnhia* in the *Aviculiloe* (genus *Meleagrina*).

(iv) Among the *Eulamellibranchia*, in the *Carditidoe* (*Vencrieardia*, *Cardita Milneria*), the *Astartidw*, and most *Crassatellidce*; among the *Cyrendce*, in the genus *Pisidium*; among the *Unionidce*, in the *Unionince* (*Unio*, *Anodonta*); and among the *Lucinacea*, in *Cryptodon Moseleyi*.

In *Solenomya*, among the *Protobranchia*, the two mantle edges grow together only at one point, but to such an extent as to close the whole posterior half of the ventral mantle opening. In this way the mantle cleft is divided into two; the anterior aperture serves for the protrusion of the foot, while the posterior serves at the same time as inhalent (branchial) and exhalent (anal) aperture. *Solenomya* is the only bivalve in which this arrangement is found.

(C) *The mantle folds grow together at two points, thus forming three apertures*. This condition arises in consequence of the complete separation (through concrescence) of the branchial aperture from the rest of the large anterior mantle cleft. The anal and branchial apertures may remain as slits, or may be produced into longer or shorter anal and branchial siphons. The large anterior and ventral mantle cleft serves for the protrusion of the foot, and is called the pedal cleft. These two points of concrescence are found:

(i) Among the *Protobranchia*, in *Yoldia* and *Leda*.

(ii) In most *Eulamellibranchia*, viz in most *Lucinidoe*, most *Cyrenidoe*; among the *Unionidce*, in the *Mutelince*, in the *Donacidce*, *Psammtobiidce*, *Tellinidce*, *Scrobiculariidce*; among the *Veneracea*, in the *Veneridce*, in the *Cardiidce*, the *Mactridce*, *Mesodesmatidce*, and the *Solenidce* (excepting *Solea* and *Lutearia*).

(iii) In all *Septibranchia* (*Poromyia*, *Cuspidaria*).

In the above forms the mantle is still wide open, *i.e.* the points of conerescence are small and local. But these points may become lines of concrescence of considerable length. In the *Chamacca*, for example, and especially in the *Tridacnidoe* among the *Eulamellibranchia*, the three apertures of the mantle are found at considerable distances from one another, being divided by long intervals where the edges have grown together. In some groups of Lamellibranchia, the concrescence between the anal and branchial apertures or siphons remains short, *i.e.* the one aperture lies directly below the other, but in such cases the edges anterior to the branchial aperture grow together to a greater extent, so that the pedal aperture becomes reduced to a small anterior fissure.

In this condition the mantle is *closed*. Such a mantle is found: Among the *Eulamelibranchia* in the *Modiolarea*, *Decisseusia*, *Petricola*, all *Pholadidoe*, (*Pholas*, *Pholadidea*, *Jouannetia*, in which the pedal aperture is said to close entirely in old animals, *Xylophaya*, *Martesia*); in the *Teredinidce*, and among the *Pandoridce*, *Pandora*, the *Verticordiidec* and *Lyonsiidce* (*Anatinacea*).

(D) There are some Lamellibranchia with closed mantle, in which a fourth aperture is added to the three found in the above groups, *the mantle thus having three points of concrescence*. The fourth aperture is always small, and is found between the pedal and branchial apertures; it probably corresponds with a rudimentary fissure for the byssus. This arrangement is found in the *Eulamellibeanchia*; among the *Solenidoe*, in *Soleca* and *Lutraria*; among the *Pandoridce*, in *Myochama*; in *Glyeymeris*; among the *Anatinacea*, in the *Teredinidce*, in the genus *Thracia*; in the *Pholadomydae* and the *Clavagellidoe* (*Clacagella* and *Breehites* [*Aspergillum*]); and finally, in *Lyonisa norvegica*. The anal aperture is often and the branchial aperture nearly always fringed, or in various ways edged with protuberances, papillae, or tentacles, and this is the case whether these apertures are found on the edge of the mantle or at the ends of (longer or shorter) siphons. The siphons can be contracted and extended, and either wholly or partly withdrawn into the shell, by means of special muscles.

These muscles are attached on the inner surface of the shell-valves to the right and left posteriorly, and their line of attachment forms the pallial, which will be described later on. The length of the siphons varies greatly. Specially long siphons are found in the *Mactrudie*, *Donacidoe*, *Psammobiidoe*, *Tellinidoe*, *Serobieulariidoe*, many *Veneracea* and *Cardiidoe*, the *Mesodesmatidoe*, *Lutraria*, the *Pholadidoe*, *Teredinidoe*, *Anatinidoe* and *Clavagellidoe*. The siphons may be separated throughout their whole, length and often diverge one from the other (e.g. *Galatea* among the *Cyrenidoe*, the *Donacidoe*, *Psammobiidoe*, *Tellinidce*, *Scrobicularidoe*, *Mesodesmatidoe*, *Plarus*, etc.). In other forms they coalesce along their entire length; they may even look like a single tube, w hich is, however, always internally divided into an upper (anal) and a lower (branchial) channel. This common siphon is sometimes protected by a special sheath of epidermis, particularly in these forms in which it cannot be withdrawn into the shell. Siphons united throughout their whole length are found in the *Maetridoe*, a few *Veneracea*, *Lutraria*, *Solenoeurtus*, *Solen*, the *Pholadidoe* many *Anatinidoe*, and the *Clavagellidoe*.

In some cases, siphons which are united for some distance at the base, separate near their ends and even diverge, *e.g.* in *Petricola* among the *Vencracca*, *Teredo*, etc. The two siphons are often of unequal length. In the *Modiolaria* (*Mytilidoe*) only one, the anal, is developed, while the branchial aperture remains unseparated from the large mantle cleft. The reverse is the case in *Dreissensia* and *Scrobicularia*, where the branchial siphon is much longer than the anal. The siphons are sometimes provided with valves; these occur more often in the anal than in the branchial siphon.

Significance of the development of the Anal and Branchial Apertures and Siphons

Most Lamellibranchia inhabit mud or sand, into which they sink the anterior part of the body, burrowing by means of the protrusible foot. The water necessary for bathing the gills and for respiration can only be received into and expelled from the mantle cavity through the cleft at the posterior end of the body which projects above the mud. The foeeal mass from the anus near this point must also here be ejected from the cavity. The development of localised inhalent and exhalent apertures is explained by the fact that a constant regulated stream of water into and out of the mantle cavity is necessary both for respiration and for conducting particles of food to the mouth. The most advantageous point for the exhalent aperture is obviously directly behind the anus. Siphons attain development in consequence of the habit of the life of many bivalves, which bury themselves deep in mud, sand, wood, and even rock.

By means of their siphons they can still remain connected with the water which bathes the surface of their place of concealment, and as long as the animals remain undistributed, a constant current enters the mantle cavity through the branchial and leaves it through the anal siphon. Where the mantle folds have grown together to a large extent (closed mantle) the siphons are always well developed. Such closing of the mantle is found principally among bivalves which bore into wood, clay, rock, etc., and in which the foot of the adult is weakly developed, or altogether rudimentary. The degeneration of the foot leads to the shortening of the pedal aperture which originally served for its protrusion.

The mantle is found completely open with only slightly developed anal and branchial apertures or none at all, in bivalves which do not burrow, but live surrounded by water, either attached to the bottom or lying freely on it. In such animals the surrounding water can circulate through the usually open mantle cleft and the mantle cavity. We here find protuberances, papillae, tentacles, etc., carrying sensory organs, all along the free edges of the mantle, whereas, in bivalves which inhabit mud or bore into wood, rock, etc. such organs are mostly found massed together round the edges of the branchial and anal apertures.

Edge of the Mantle

The edge of the mantle often forms a number of diverging folds, which in transverse section look like finger shaped processes. The outermost fold is always applied to the shell. The edge of the mantle may also be beset with one or more rows of protubrances, papillae, or tentacles, and often contains uncellular or multicellular glands, mucous glands, and others which have been considered to be poison glands for protective purposes. Tactile sensory cells are very common. Eyes are rarely developed on the edge (*cf.* section on the Sensory Organs). In the *Pectinidoe*, *Spondyidoe*, and *Limidoe*, the inner fold of the mantle has a somewhat broad border, which, when the shell is open, projects from the mantle towards the median line of the body. The free opposite edges of these folds (flaps, or curtains) springing from right and left, may meet in such a way as to shut off the central part of the mantle cavity even while the shell is open, apertures only remaining anteriorly and posteriorly.

CEPHALOPODA

Integument

The integument of the Cephalopoda consists of an outer cylindrical epithelium, and a subjacent cutis in the form of thick connective tissue. In this cutis, not far removed from the epidermis, and above a layer a connective tissue plates (which are refractive and often shimmer like silver), there are large pigment cells or chromatophores which, by their alternate contraction and expansion, bring about the well-known changes of colour in these animals. These chromatophores are single cells containings yellow, brown, black, violet or carmine pigment, either as fluid or in small granules. The layers containing them are either single or double; in the latter ease, the colour of the pigment in the one layer of chromatophores differs from that of the chromatophores in the other. Radial fibres, arising from the surrounding connective tissue, are attached to each chromatophore, round that equator which lies parallel to the integument.

The chromatophores are enveloped in a special, possibly elastic, membrane, and when contracted are almost globular; the pigment corpuscles are then crowded together. The chromatophores expand equatorially diminishing the distance between their poles, *i.e.*, they become much flattened. In this condition, they may, further throw out fine branches, the pigment granules being thus spread out over a large surface. It was formerly believed that the expansion of the chromatophores was caused by the contraction of the radial fibres, which were thought to be muscular, but more recent investigations have shown the fibres to be of the nature of connective tissue. The changes of colour, which are of great physiological and biological interest and which are partially under the control of the animal, are brought about by the alternate contraction and expansion of these variously coloured chromatophores.

Mantle, Visceral Dome

Some of the most important points connected with the mantle and visceral dome have already been mentioned. In *Nautilus*, the body is attached right and left to the inner surface of the shell of the last or inhabited chamber by powerful muscles, which may make a slight impression on the shell. Between the points of attachment of these lateral muscles, the integument of the visceral dome coalesces with the inner surface of the shell of the inhabited chamber in a narrow circular zone, so that the gas enclosed in the upper chambers of the shell cannot escape. While the integument and mantle beneath this zone of concrescence (*i.e.* towards the free aperture of the last chamber) are rough, fleshy, and muscular, the integument of that portions of the visceral dome which lies above the zone and is applied to the last septum is delicate and soft. The siphuncle, which arises at the dorsal end of the visceral dome and passes through all the septa, and is membraneous and hollow and filled with blood. It is said to communicate with the pericardium.

In the female, *Nautilus*, the nidamental gland (see Genital Organs) lies in the free mantle fold, near the point at which it separates from the visceral dome. We thus have parts which usually lie in the visceral

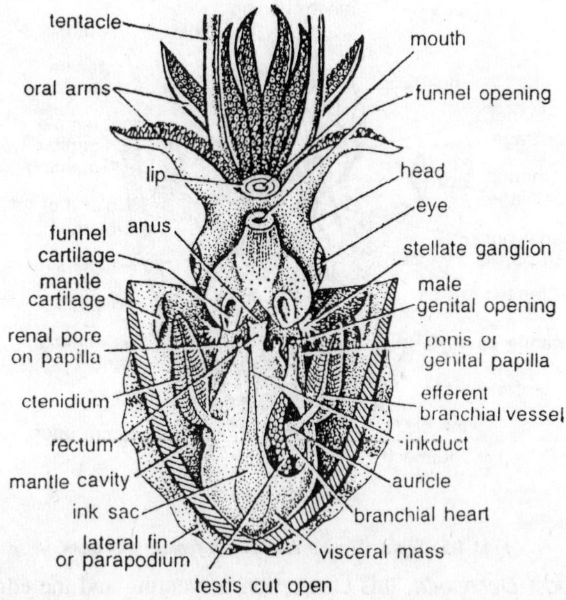

Fig. 4.4. Sepia. A male specimen with mantle cavity laid open.

dome wandering into the mantle fold. Among the *Dibranchia*, which are good swimmers, fins are found. In the *Octopoda*, which are distinguished by the round, compact form of the visceral dome, these are wanting, except in the remarkable genus *Cirrhoteuthis*. Fins are universal among the *Decapoda*, and very much in form, size and arrangement. In *Sepia* and *Sepioteuthis*, the fins are inserted on the lateral edges of the body, along the whole height (length) of the visceral dome, forming the boundary between the anterior and posterior (physiologically the dorsal and ventral) surfaces of the latter.

In *Rossia*, *Sepiola*, and *Sepioloidea* they are almost semicircular, and are like distinct appendages situated on the anterior surface of the dome, about half-way up it. This is also the case in *Cirrhoteuthis*, where the more or less circular fin-lobes rise from the body on stalk like bases. The triangular or semicircular fins of *Cranchia*, *Histioteuthis*, *Onyehoteuthis*, *Loligo*, *Loligopsis*, *Ommastrephes*, etc., are found at the dorsal end of the visceral dome, on its anterior side. In many *Dibranchia* there is a concrescence of the free edge of the mantle fold with the integument of the "head" (Kopffuss), which lies below it. This connection is effected by means of a muscular band, which pass over the neck (nuchal band).

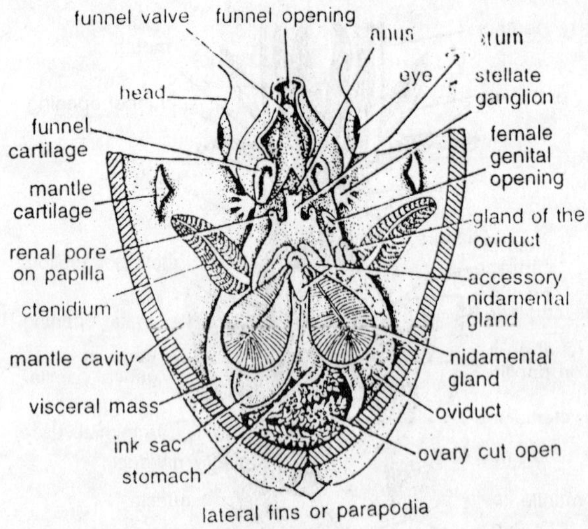

Fig. 4.5. Sepia. Mantle cavity of a female specimen.

In most *Decapoda*, this connection is wanting and the edge of the mantle is free all round the body; the exceptions are the genera *Sepiola*, *Cranchia*, and *Loligopsis*, which have a narrow connection of this sort. All *Octopoda* have this concrescence, commencing with the *Argonauta*; it lengthens in *Philonexis* and *Octopus*, till in *Cirrhoteuthis* it spreads to the posterior surface (physiologically the ventral surface), so that the edge of the mantle remains free only at the aperture through which the funnel or siphon is protruded. Arrangements for fastening the mantle fold to the adjacent body wall are very common. Such attachment is either temporary or permanent. In the former case, there are prominences with corresponding depressions for locking the mantle (*apareil de résistance*); in the latter case, dermal or muscular fusions take place between the mantle and body wall.

Apparatus for locking the mantle. These are paired or unpaired. The former are to be found on the posterior side of the body, in the mantle cavity, near its lower end; they lie to the right and left at the base of funnel, and on the corresponding points of the inner surface of the mantle fold. The unpaired, on the contrary, are found on the anterior surface of the neck. Since all the arrangements serve the purpose of cutting off the mantle cavity from the external medium, it is easy to see that their development is in inverse ratio to the extent of the concrescence of the edge of the mantle round the neck before mentioned. Where no concrescence is found, as in *Sepia*, the arrangements for locking the mantle are highly developed; while, where the line of conerescence is very long, as in *Octopus*, the locking apparatus may be altogether wanting.

The locking apparatus consists, in general, of cartilaginous prominences (often accompanied by depressions) on the inner surface of the mantle fold, *i.e.* the surface turned towards the mantle cavity, which exactly fit corresponding cartilaginous depressions accompanied, as the case may be, by prominences, on the opposite body wall. The special forms of the mantle and nuchal locking cartilages are of importance in classifications. The cartilaginous arrangements for locking which are almost always found in the *Decapoda* (they are wanting only in *Owenia* and *Cranchia*), are still retained in a few *Octopoda* in the form of more or less modified fleshy processes (*Argonauta*, *Tremoctopus*). The nuchal locking apparatus is the first to disappear on the rise of the pallio-nuchal concrescence. It has disappeared among the *Decapoda* in the genus *Sepiola*, when the mantle is firmly attached to the neck.

Permanent connections between the mantle fold and the adjacent body wall traversing the mantle cavity are found only in these Cephalopods which have no locking apparatus. Thus in *Octopus* and *Eledone* the mantle is attached to the body wall by means of a median muscle above the funnel. This muscle consists of two closely-applied lamellae, have the anus between them. In *Cronchia* the free dorsal edge of the funnel (at its so-called base) has become united by an integumental band on the right and left with the mantle fold, and a similar arrangement is found in *Loligopsis*.

Water pores. Near the mouth, or at the bases of the arms, or laterally on the head, in many Cephalopods, there are apertures leading to integumental pouches of the varying size. The function of these organs is unknown.

5

Molluscan Exoskeleton

All the various forms of shell found in the Mollusca are deducible from a cup or plate like shell covering the dorsal region. Such a shell affords sufficient protection for animals such as *Fissurella*, *Patella* etc. which can firmly and almost immovably attach themselves to a hard surface by the sucker-like foot. The soft body is in this case protected on one side by the shell, and on the other by the surface of attachment. Free moving Mollusca, however, show a tendency to protect the whole body exclusively by means of their shells and this object is attained in various ways: In the *Chitonidae*, for instance, the shell is made up of consecutive joints, overlapping in such a way as to the movable one upon the other. This segmented shell can protect the whole body, since it allows the *Chiton* to roll up like an Armadillo or a Wood-louse. In the *Lamellibranchia*, the protection of the whole of the soft body is provided for the development of a bivalve shell, from which the foot can be protruded, and which, by the closing of its two valves, completely envelops the soft body as well as the retracted foot. In the *Gastropoda*, *Scaphopoda*, and *Cephalopoda*, the most complete protection on all sides of the body by means of the shell is attained on another plan. The shell becomes much elongated and turret-like, and is thus so capacious that not only the visceral dome but the head and foot also can find place in it.

Even the only remaining unprotected aperture, the one week point of this fortification, can very often be completely closed by a hard operculum. A long, turret-like shell is an inconvenient burden for a freely moving animal, being, in consequence of its large surface, a hindrance to locomotion. A reduction of the surface is brought about in the *Gastropoda* and *Cephalopoda* by the coiling of the shell, either

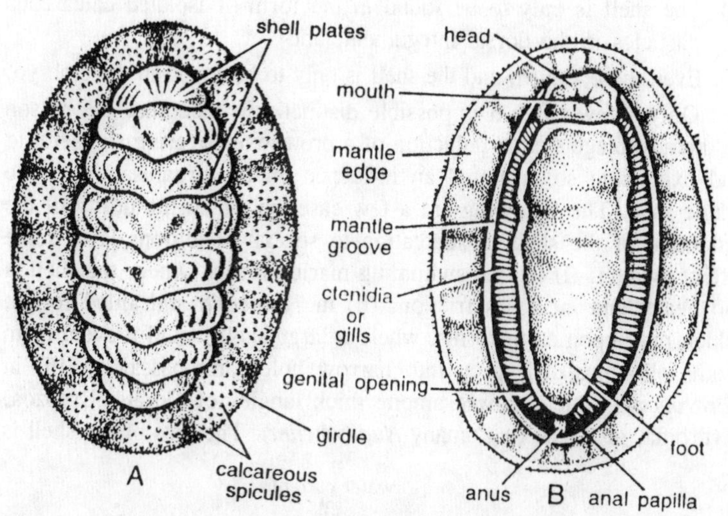

Fig. 5.1. Chiton. A—Dorsal view. B—Ventral view.

in one plane or in a conical spiral. In the latter case the spiral twist is almost always right-handed or dextral. In order to decide the direction of the twist, the shell should be held in such a manner that the point of the spiral is uppermost, while the aperture is directed downwards and towards the observe. If, in this position, the aperture lies on the right of the axis, the shell has a dextral twist if to the left its twist is left-handed or sinistral.

We have a striking and in most cases unexplained phenomenon in the reduction and even complete disappearance of the shell, which takes place not only in nearly all the classes, but even within some of the smaller groups of Molluscs, *e.g.* the *Solenogastres* among the *Amphineura* a few *Heteropoda* and *Titiscania* among the *Prosobranchia*, many *Pulmonata*, very many *Opisthobranchia*, and most extent *Cephalopoda*. In almost all cases the forms in which the shell is rudimentary or wanting can be shown to be derived from forms in which it is well developed. All shell-less snails (slugs) have shells in the early stages of their development. The process of the gradual reduction of the shell to a rudiment, which will be more fully described later on, is often as follows:

1. The shell becomes internal.
2. It decreases in size, so that it no longer can cover the body.

3. The visceral dome disappears.
4. The shell is only to be found in the form of isolated calcareous particles in the dorsal integument; and
5. Even these vanish, and the shell is only to be found in the embryo.

Only a few cases it is possible distinctly to recognise the reason or the advantage of this reduction of a protective covering so useful to and exercising so profound an influence on the organisation of the whole race. The following are a few cases in which the utility of the reduction of the shell in adaptation to special conditions is to some extent evident: (i) In free-swimming marine forms, where the shell is too heavy and increases friction; (ii) in *Testacella* and allied forms, which prey upon earthworms, where a large shell would prevent them from following their prey into narrow holes and passages; (iii) in *Gastropods*, which browse among thick tangles of Corals, Bryozoa, Hydroida, or Algae (*e.g.* many *Nudibanchia*). The loss of the shell is

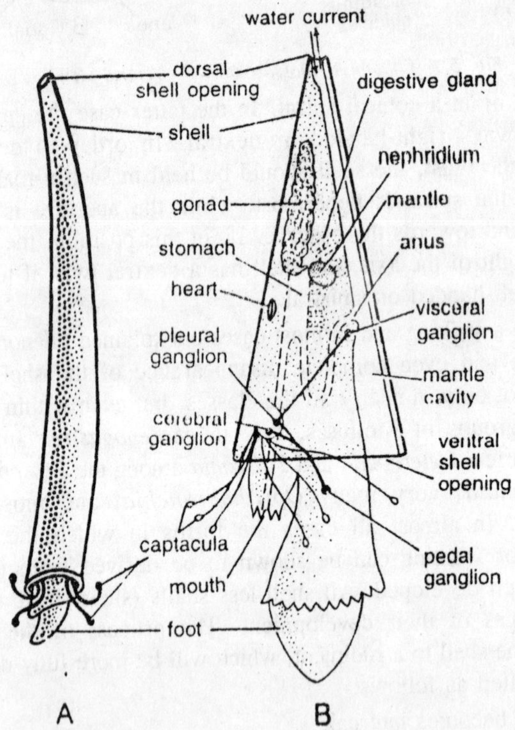

Fig. 5.2. Dentalium. A—External features. B—Internal (anatomical) features.

generally followed by compensatory adaptations for protection, such as great capacity for regeneration, especially of the easily detachable appendages, voluntary amputation of portions of the body, stinging cells, and colouring which may be protective in various ways.

The carnivorous *Cephalopods* are protected (1) by their extraordinary swimming powers, which are in keeping with their highly developed organisation; (2) by their well-developed sight; (3) by the great muscular strength; (4) by strong jaws; (5) by the discharge of the secretion of the ink-bag; (6) by their mimetic changes colour, etc. Certain pecularities of organisation, which can only be understood as remains of a shelled condition (*e.g.* the lateral position of the genital apertures and also to some extent of the anus in the *Nudibranchia*), always persist after the shell has disappeared.

Chemical Composition of the Shell

The shell of the Mollusca consists principally of carbonate of lime, with traces of phosphate of lime and of an organic substance related to chitin—conehyolin. Besides these various colouring matters may occur.

Structure of the Shell

The shell of the *Lamellibranchia* consists of three layers of the innermost layer being applied to the surface of the mantle. The shell is to be regarded as a cuticular structure. The outer layer (shell-integument, epidermis, cuticle, periosteum) is so, far as its physical constitution is concerned, horny and wanting in lime. It generally disappears off the older portions of the shell. The middle layer

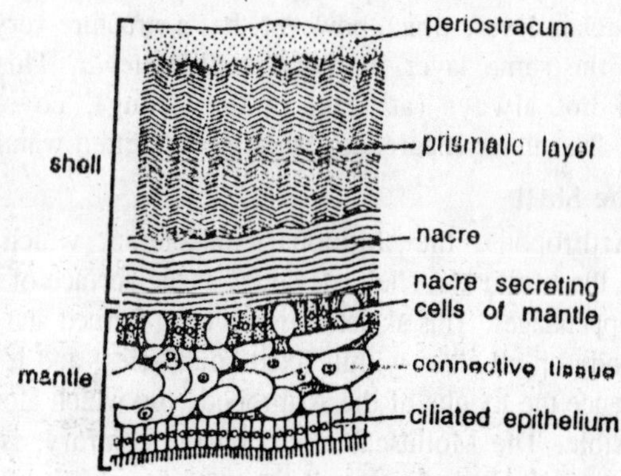

Fig. 5.3. Unio. A part of shell and mantle in T.S.

(columnar, prismatic, or porcelanous layer) consists of slender prisms of carbonate of lime, usually perpendicular to the surface of the shell and closely crowded together. The inner (nacreous) layer has a finely lamellated structure. The very delicate transparent laminae of which it is composed are thrown into slight waves; these cause the wavy lines on that surface of the shell which lies on the mantle, which by interference, produce the characteristic nacreous lustre. The pearls of the pearl oyster are formed of the same substance as this layer.

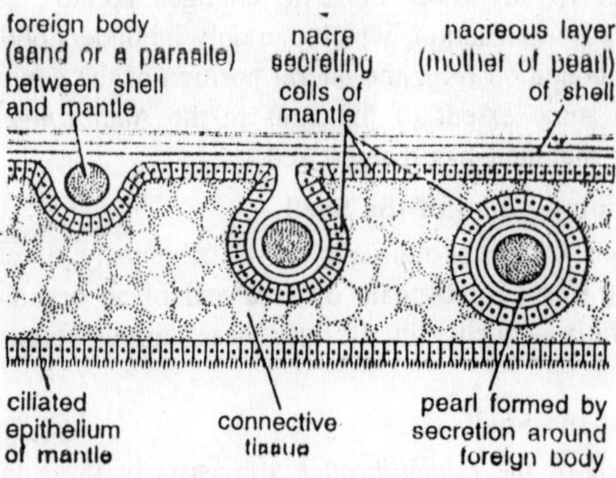

Fig. 5.4. Unio. Pearl formation.

The constitution of these three layers varies greatly in details both in the *Lamellibranchia* and in other Mollusca. The outer and middle layers are formed at the free margin of the mantle, the inner layer is yielded by the epithelium of its whole outer surface. The shell in the *Gastropoda* and *Cephalopoda* consists principally of the middle or procelain layer, which however, has a structure very different from that of the same layer in the *Lamellibranchia*. This layer is generally, if not always (at least in the young), covered by a periostraeum. The inner (nacreous) layer is very often wanting.

Growth of the Shell

In the Arthropoda, the chitinous exoskeleton, which we may compare with the Molluscan shell, develops at the surface of the whole body and its appendages. This skeleton, when once formed and hardened, encases the body on all sides within fixed boundaries, and is incapable of growth. Hence the moults of the Arthropoda, by which alone growth becomes possible. The Molluscan shell, on the contrary, is open. In the *Gastropoda* and *Cephalopoda*, it assumes the shape of a conical mantle, coiled round a single axis and open at the base of the cone.

By continual additions at the edge of its aperture, it grows with the growth of the animal, without materially altering its form. The lines of the surface of the shell of the adult snail register its phases of growth. During growth, the oldest, uppermost coils or whorls of the shell either continue to the filled by the apex of the visceral dome (in many *Gastropods*), or are deserted by the animal which, as the shell grows, withdraws farther and farther from its tip. These whorls may remain empty, or may be partially or completely filled with shell substance. In the latter case, they may be successively thrown off.

The *Nautilus* and allied forms, during growth, periodically form transverse septa, so that the forsaken parts of the shell become chambered and filled with gas, the animal occupying the largest and last-formed chamber, which opens externally. In the *Lamellibranchia*, the growth of the shell keeps pace with the growth of the body is exactly the same manner, the free edge of the shell valve continually receiving additions of the substance from the edge of the mantle to form the periostracum and the prismatic layers, while the whole external surface of the mantle yields an additional nacreous layer. The consecutive phases of growth are here also registered by the concentric markings on the surface of the shell.

Special

Gastropoda

A few details concering the shell of the Gastropods must here be added. As a rule, the shell is coiled spirally round an axis. This spiral is, in rare instances, so flattened that the coils come to lie almost in one plane, giving rise to a nearly symmetrical shell (e.g. *Planorbis*). There are, however, among the Gastropoda, uncoiled shells which are symmetrical, and these require special attention. The most important are the cup-shaped or more or less bluntly conical shells of the *Patellidae* and *Fissurellidae*. Since (1) we derive the Gastropoda from bilaterally-symmetrical ancestors with symmetrical shells; and since (2) the *Fissurellidae* undoubtedly possess the most primitive organisation of all the Gastropoda and thus stand nearest to the racial form, and are moreover (3) strikingly symmetrical in their organisation, it seems at first sight, natural to consider this symmetry a primitive feature.

Certain pecularities of the nervous system, however, especially the crossing of the pleuro-visceral connective, taken in connection with other conditions explained more fully elsewhere, make it certain that the cup-shaped shell of *Fissurella* is only *secondarily* symmetrical, *i.e.*

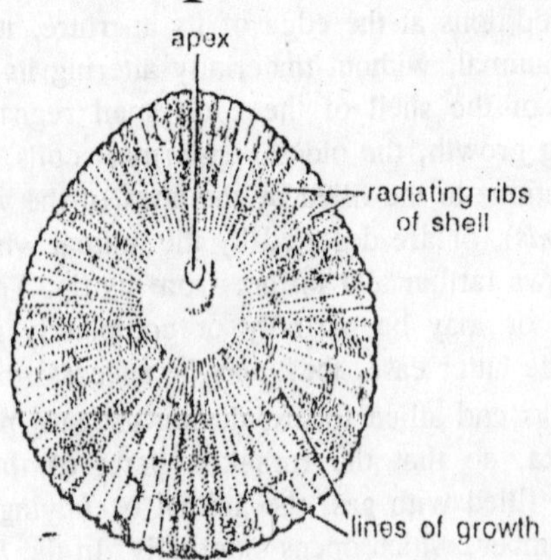

Fig. 5.5. Patella—Dorsal view.

that *Fissurella* is descended from forms which possessed a spirally coiled shell. The same is the case with the *Patellidae*. The following important facts are in harmony with this conclusion; (1) the young shell of *Fissurella* is asymmetrical and coiled, and it only gradually assumes the symmetrical form; (2) the apparently symmetrical shape of some forms nearly related to *Fissurella* and *Patella* prove on closer inspection to be somewhat asymmetrical, the apex especially being more or less excentric; (3) other forms nearly allied to *Fissurella*, such as *Haliotis*, *Seissurella*, and *Pleurotomaria*, have spirally coiled shells.

In the *Fissurellidae*, many *Pleurotomariidae*, and the *Holiotidae*, *i.e.* in the most primitive Gastropods, peculiar and noteworthy perforations of the shell occur, such as are occasionally found in other divisions. These perforations lie above the silt in the mantle which is characteristics of this order and they everywhere establish communication between the mantle cavity and the exterior, especially needed when the mouth or edge of the shell is closely applied to the object on which the creature crawls. In *Scissurella*, *Pleurotomaria*, and *Emarginula*, there is a median indentation in the anterior edge of the shell, which corresponds with an incision in the mantle edge. This is the case in the young *Fissurella*, but during further development, the edge of the shell grows across the incision, so that in the adult animal the aperture lies near the apex. Beneath it is the anus, placed

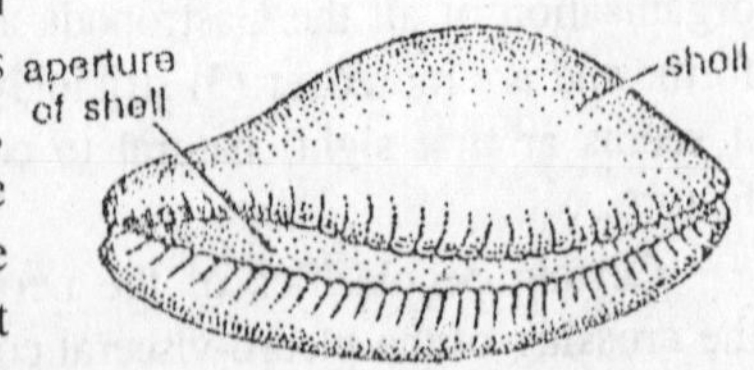

Fig. 5.6. Shell of cowrie.

high up in the mantle cavity. If such a cleft were to arise at both the anterior and posterior edges, and to become very deep, a double shell would result comparable with the bivalve shell of the *Lamellibranchia*. It is in fact probable that this notching of the shell edge is of great phylogenetic significance.

In *Haliotis* we have a row of perforations of the shell, the process of formation of the perforation in *Fissurella* being often repeated; the older apertures are always, however, closed by shell substance, and the younger only remain open as long as they lie immediately over the respiratory cavity.

In very many *Prosobranchia* (the *Siphoniata* of earlier writers), there is, at the columnar edge or lip of the shell, a notch which gives passage to a channel-like fold of the mantle margin. This channel keeps up communication between the mantle cavity and exterior, even when the shell is closed by the operculum. Instead of a notch, a more or less long process or beak may enclose a corresponding process of the mantle, the siphon. The latter may become a tube by the apposition of its edges. It has already been mentioned that the shells of most Gastropods are dextrally twisted. There are, however, a few families, genera, or species in which the shell has a sinistral twist; and in some species where the twist is dextral, a few individuals with sinistral twist may occur, and *vice versa*.

It is a curious fact that some species in which, the shell has a sinistral twist, show the asymmetry of the dextral twist in the soft body, whereas, in others, the asymmetry of the soft body corresponds with the twist of the shell. We shall return to this point. For details as to the growth of the shell, and the capacity of the animal to dissolve the shell already formed, both of which are points full of interest, we must refer to special works on Conchology as also for detailed descriptions of forms of the shell and opercula, and differences due to age. Progressive reduction of the shell occurs in each of the three divisions of the Gastropoda. In the *Prosobranchia* this has only been observed in marine, free-swimming *Heteropods* and in *Titiscania*; in the *Pulmonata*, it is much more common; and in the *Opisthobranchia*, so frequent that nearly all the members of this division have more or less rudimentary shells.

Many adult *Opisthobranchia* have even lost every trace of a shell (*Pteropoda gynnosomata*, *Nudibranchia*, and most *Ascoglossa*), although, in their earliest stage at least, they possessed a coiled shell, for the closing of which there may even be an operculum, secreted by the foot, as in the *Prosobranchia*. The following are some of the principal stages and concomitant phenomena of the reduction of the shell:

(*a*) The well-developed shell ceases to be large enough to shelter the whole body.

(*b*) The shell, which becomes thinner and smaller, is dorsally overgrown, partially or altogether, by extensions of the mantle.

(*c*) As the shell (which is then either cup-shield or ear-shaped) becomes continually smaller, the visceral dome begins to be levelled down, till it no longer rises above the rest of the body, its contents spreading out to a certain extent over thc dorsal surface of the foot.

(*d*) The external asymmetry of the body passes by degrees into symmetry, whereas the internal asymmetry never entirely disappears.

(*e*) The shell is reduced to a number of isolated caleareous particles in the integument of the flattened visceral dome.

(*f*) There is at last no trace of the distinct visceral dome; calcareous particles are to be found in the dorsal integument of the long and now naked Gastropod.

(*g*) Even these particles finally disappear.

In connection with the reduction of the shell in *Opisthobranchia* and *Pulmonata*, compare the section on the mantle. The Heteropoda present the following interesting series:—

Atlanta. The shell is very light and thin, but large and spirally coiled (with an incision at its aperture); the animal can entirely withdraw into it, and close it by means of an operculum developed on the distinct metapodium.

Carinaria. The shell is thin, light and delicate; it is cup-shaped, and covers the large stalked visceral dome, but it incapable of sheltering the long and thick cylindrical body and foot. There is no operculum.

Pterotrachea. The visceral dome is small, and there is no shell and operculum.

Lamellibranchia

The two lateral valves of the Lamellibranch shell are connected, at their dorsal edges, by means of a *hinge* and a *ligament*. The ligament counteracts the muscles of the shell, which will be described later on, and which, by their contraction close the shell. It is usually composed of two layers, the inner layer being elastic, while the outer is not. The outer non-elastic layer passes into the epidermis or periostracum of the shell. The inner layer of the ligament is elastic and calcareous, and is often called cartilage, but this is histologically incorrect. The ligament lies either externally, distinctly seen dorsally between the prominences of the hinge edges of the valves, or internally, stretched between the apposed edges of the hinge, which are furnished with depressions of its reception. These depressions can easily be distinguished from those belonging to the hinge itself, since the former are alike on the two valves, whereas the furrows and other depressions belonging to the one face of the hinge correspond with teeth, ridges, etc. on the opposite face.

When the elastic "cartilage" of the ligament is at rest, as in a dead bivalve, or when the abductor muscles of the living animal are relaxed, the valve open. When the adductors contract, the "cartilage" is apparently in all cases—compressed. On the other hand, when the adductors are relaxed, the elasticity of the "cartilage" forces the shell open again. The continuity established between the two valves, by means of this dorsal ligament, causes the Lamellibranch shell to appear to consist, strictly speaking, of one dorsal piece, developed to the right and left ventrally into two valves. The constitution of the ligament and hinge are of importance in classification. We must refer the reader to systematic zoological works for the special forms taken by the shell, and content ourselves with the following remarks:—

The Lamellibranch shell is originally symmetrical, that is to say, the two valves, apart from the almost invariable asymmetry of the hinge, are exactly alike (equivalve). This is the case in most of the Lamellibranchia. The two valves may, however, become unlike, *i.e.* the shell (and to a much lesser extent and only in unimportant details,

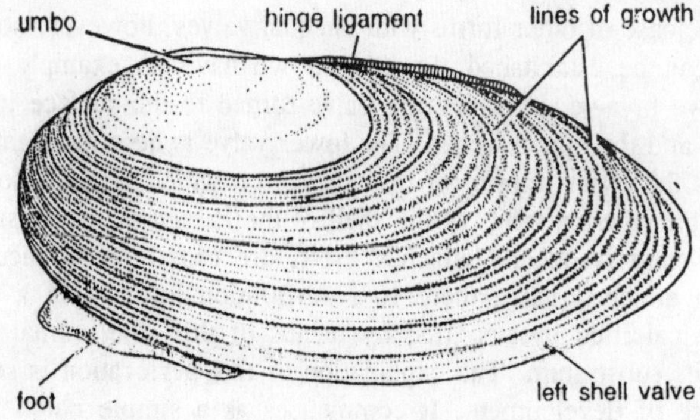

Fig. 5.7. Unio. External features (lateral view).

the soft body also) may become asymmetrical. As far as we can at present judge, this asymmetry is caused by adaptation to an attached manner of life. The left valve of the Oyster is firmly cemented to the surface on which it rests. This valve is thicker, more convex and specious, and forms a sort of basin in which the soft body lies, while the right valve acts rather as a lid, and is thinner and flatter. We have thus an 'upper' (the right) and a "lower" (the left) valves, but it is hardly necessary to point out that this use of terms upper and lower has as little morphological significance as in the Pleuronectidae among the fishes. The attached valve is sometimes the right sometimes thc left, and this variation may occur within one and the same genus (*Chama*), or even species (*Aetheria*). Besides the above-named, the following bivalves are also attached, and have dissimilar valves: *Spondylus*, *Gryphoea* p.p., *Exogyra* p.p., and especially the fossil *Hippurites* (*Rudistes*), in which the right valve assumes the form of a high cone attached by its point, while the left looks like a lid.

The conical valve has, however, no corresponding internal cavity, but is almost entirely filled up with shell substance, so that, in spite of form of the shell, the space occupied by the animal between the two valves is very limited. This same condition is found in certain fossil *Chamacea*. In *Requienia*, the left valve is produced spirally and is attached by its point, while the spirally-coiled flattened right valve covers it like a lid, so that the whole shell closely resembles a Gastropod shell closed by its operculum. There are also free, unattached bivalves with unequal valves, *e.g.* many *Pectinidae*. In these animals, however, many pecularities of organisation, such as the rudimentary foot, the constitution of the mantle edge, and the absence of siphons, indicate descent from sedentary forms.

In the case of other forms with unequal valves, however, no such descent can be established. In *Anomia* we have an example of an inequivalve bivalve, in which the valve turned to the surface it rests on is flat and the upper arched. The lower valve is here the right one, and takes the exact imprint of the surface on which it rests, so that, for example, the markings of the shell of the *Pecten* or the Oyster, to which *Anemia* frequently attaches itself, are exactly reproduced. In this right attached valve there is a perforation into which a shelly plug, the calcified byssus, fits; by means of this, the animal fixes, itself to its substratum. The explanation of this perforation is seen in the course of development. It commences as a simple notch at the edge of the shell, as found also in other bivalves, for the passage of the byssus. By the further growth of the shell, this notch to a certain extent is grown round, and thus apparently travels away from the edge of the shell, with which, however, it is still really connected. In related forms (*Caeolia*) this aperture becomes quite filled up by a homogeneous calcareous mass.

Impressions on the inner surfaces of the shell

Various organs of the Mollusc, attached to or adjacent to the inner surface of the shell, leave more or less distinct impressions on this surface, which are visible when it is empty. These impressions are of great importance, especially to the palaeonotologist, for by their means fairly safe conclusions may be arrived at as to certain points in the organisations of the soft body which has disappeared.

1. The most distinct impressions, are these caused by the adductor muscles. Where there are two powerful adductor muscles, one anterior and the other posterior (Dimyaria), there are two impressions in the corresponding parts of the inner surface of the shell. In cases where the anterior muscle is rudimentary, while the posterior is usually powerful, and has moved anteriorly towards the middle of the shell (Monomyaria), there is only one large impression. Thus anus is always to be found close to the posterior (which in the Monomyaria is the only) adductor.
2. Parallel to the edge of the shell, and more or less removed from it, we find on the inner surface of the shell the so-called *pallial line*, caused by the muscle fibres which attach the edge of the mantle to the valves.

   The course taken by this line undergoes characteristic modification in such Lamellibranchs as have siphons; at the posterior part of the shell it suddenly bends forward and upward, and then again passes backward and upward towards the lower edge of the posterior adductor. The pallial line, in this case forms an indentation, leaving a sinus or bay opening posteriorly, the *pallial sinus*, which has been utilised for systematic purposes (*Sinupalliata*, *Integripalliata*). The sinus marks the line of attachment of the sipho-retractor muscles: the stronger these retractors and the better developed the siphons the larger and clearer is the sinus.
3. The forgoing impressions are the most distinct and constant, but others may occur as well, caused by the protractors and retractors of the foot, by the muscles or ligaments which attach the visceral dome to the shell, etc.; but these cannot be further described.

In most Lamellibranchia, when the shell is closed, the edges of the two valves meet exactly, so that the soft body can be entirely enclosed and cut off from the exterior (*closed*) shell. There are, however, shells in which, in the closed condition the valves gape posteriorly, or, more frequently, both posteriorly and anteriorly (e.g., *Myidae*, *Glyeymeridae*, *Solenidae*). This is accounted for by the great development of the siphons and of the foot, which can only partially (*Myidae*, *Solemocurtus*) or with difficulty be withdrawn into the shell. Such *gaping* shells are found in most boring bivalves, whose shell formation is specially interesting owing to the development of accessory valves or caleareous tubes. In this respect *Pholas*, *Pholadidea* and *Jouannetia* represent the most important stages in the remarkable series.

The shell of *Pholas* is elongated longitudinally, and gapes anteriorly and ventrally for the passage of the short club-shaped foot, and posteriorly for that of the strongly developed siphons. As many as three accessory valves are developed dorsally (prosoplax, mesoplax, metaplax). The shell of *Pholadidea* somewhat resembles that of *Pholas*. In the young animal it gapes anteriorly, as in *Pholas*, for the passage of the foot. Posteriorly, each valve is produced into a horny process, which is succeeded by an accessory piece (siphonoplax), hollowed out like a trough. The siphonoplax of the one valve often fuses with that of the other to form a single tubes for the reception of the siphons. There are two pieces of prosoplax, while the meso and metaplax are rudimentary. In the adult the boring activity is suspended, and the anterior opening becomes entirely closed by the secretion of an accessory piece, the callum (hypoplax). The functionless foot atrophies, and the animal can move no farther in the substance into which it has bored.

The shell of the adult *Jonanuetia* is much shortened longitudinally, and is globular, and the animal cannot move in the round hole it has bored for itself in a block of coral. Any alteration in its position in the hole, which might be fatal to the animal, is avoided by means of a posterior tongue-like process of the shell, which, however, only belong to the right valve. The shell is completely closed anteriorly, and a foot is wanting. The adult condition of *Jouannetia* is explained by its developmental history.

The shell of the young animal is like the segment of a sphere, whose greatest height is hardly half of the radius. It covers the dorsal upper portion of the body, its free edges thus bounding a very wide aperture, which corresponds with the anterior pedal gape of *Pholas*. In this *Pholas*-stage, in fact, *Jouannetia* really possesses a foot. Twisting the body about and rasping the stone with the anterior edge of the shell, the animal excavates a hole, which is spherical in consequence of the shape of its shell. When this hole is made, new accessory shell material is secreted at the free edge of the shell; this forms the "callum," and as the edge of the mantle follows the lines of excavation, the form of the accessory shell is here (as in *Teredo*) determined by the form of the hole, and the sphere of which the original shell was but a segment is completed.

Setting aside a few related forms (*Martesia*, *Teredina*, *Xylophaga*, *Gastrochaena*, and *Fistulana*), in which the conditions are somewhat similar, we come to the shipworm *Teredo*. This animal has a long tubular mantle which is produced posteriorly in a two long-siphons. The body lies at the anterior end of the mantle. *Teredo* bores cylindrical passages in wood. The valves of the shell are very small in comparison with the body; they take the form of tri-lobate, pieces, which encircle the anterior end of the mantle. This rudimentary shell gapes anteriorly for the passage of the pestle-shaped foot, and very widely posteriorly. The mantle further secretes over its whole surface a calcareous tube which lines its burrow, but which does not fuse with the shell valves. Two small accessory shell-pieces, the so-called "palettes," lie at the place where the siphons separate. If the anterior portion of the animal reaches (*i.e.* if it bores through to) the water, the calcareous tube is rounded off and closed.

Aspergillum (*Brechites*) and *Clacagella* show similar conditions. In the club-shaped shell, which inserts its anterior thicker end into rock, shell, coral or sand, we can distinguish a true and a false shell. The false shell forms by far the larger portion of the tube, and corresponds with the secreted tube of *Teredo*, and with a callum like that of *Pholas*. The true shell is very small and lies anteriorly. The two valves of this true but rudimentary shell are, in *Aspergillum*, placed saddle-like over the anterior end of the tube, with which they are firmly fused. Were they isolated, their gape would be unusually wide, not only anteriorly and posteriorly, but ventrally. The shell-tube is open posteriorly, over the apertures of the siphons; anteriorly, however, it is closed (in the adult) by means of a disc perforated like the rose of a watering-can, which corresponds in position with the callum of the *Pholadidae*.

The perforations at the edge of the disc, or even over its whole surface, are sometimes produced into calcareous, and at times dichotomously branched tubules. In the middle of the disc there is sometimes found a narrow slit-like aperture corresponding with the pedal aperture in the mantle beneath, but this is often wanting. Less frequently, we find another aperture is the ventral middle line, corresponding with the fourth mantle aperture above described. *Asergillum* buries its anterior end in mud or sand, but its whole organisation, and especially its shell arrangement, point to a former burning mode of life. *Clavagella*, which is nearly related to *Aspergillum*, bores into rock or the calcareous shells of various other animals. The arrangement of its shell differs from that of *Aspergillum* chiefly in the somewhat greater size of its true valves, and in the fusion of only the left valve with the calcareous tube, the right lying free within that tube.

In the *Pholadidae*, the ligament, which is still found at the hinge, no longer acts for opening the shell. In consequence of a peculiar arrangement of the anterior adductor, the opening of the shell, such as it is, is brought about by the muscles. The anterior and upper edges of the valves are bent outward, and to these edges the anterior muscle is attached. We thus have external instead of internal points of attachment, and the whole shell may be compared to a double armed lever acting along the longitudinal axis of the body, its fulerum being at the point where, in other bivalves, the hinge is found. When the anterior muscle contracts, the shell opens posteriorly and ventrally; when the posterior adductor contracts, the shell closes.

Cephalopoda

The Cephalopoda are all to be derived from an ancient fossil form which possessed a chambered shell, in the last and largest portion of which the animal lived, leaving the rest of the shell empty, or rather fulled with gas (or water) are traversed by the siphon or siphuncle. Such a shell is now found only in the sole living representative of the *Tetrabranchia*, the *Nautilas*, an animal of great importance to the comparative anatomist. Many fossil forms allied to the *Nautilus*, and grouped in the order *Nautiloidea*, possessed such a shell, as did also the *Ammontoidea*, with their enormous wealth of forms which, rightly or not, are generally considered to be nearly related to the *Nautiloidea*, i.e. to belong to the *Tetrabranchia*.

In nearly all these animals the shell, when coiled at all, is, unlike the *Gostropod* shell, coiled anteriorly or exogastrically. One group of the *Nautilodea*, the *Eudoceratidoe*, which includes only very old forms (Cambrian and Lower Silurian), is distinguished by the fact that the chambers of its straight shell, which were filled with gas (or water),

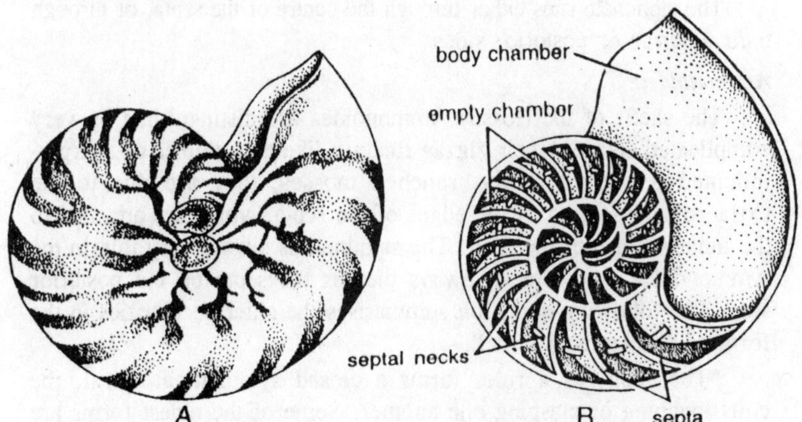

Fig. 5.8. Nautilus. A—Complete shell; B—Shell in section.

lay at the side of and not behind the inhabited chamber. There was no real siphuncle, but the upper end of the visceral dome, much narrowed by the air chambers, stretched as far as for the apex of the shell.

In other *Nautilodea*, the air chambers always lie, as in *Nautilus* above the occupied chamber, and are traversed by a thin membraneous siphuncle, which now ever, in old forms, is much thicker, and represented the narrow prolonged portion of the visceral dome. Some forms of *Nantilaidea* have shells coiled endogastrically; this is never the ease, however, when the shell forms a completely spiral. The sutures, which correspond with the lines of insertion of the septa, are simple in the *Nautilodea*, as compared with those in the *Ammonodea*, in which they are folded in a complicated manner.

Nautiloidea

In the following table we have the chief forms of the shell among the *Nautiloidea*:—

(*a*) *Orthoeeras* group— Shell straight or slightly bent. Silurian—Trias.

(*b*) *Cyrtoceras* group— Shell curved like a horn, but not regularly spirally coiled. Cambrian–Permian.

(*c*) *Gyroceras* group— Shell regularly spirally coiled, the coils, however, not touching each other. Silurian—Permian.

(*d*) *Nautilus* group— Shell regularly spirally coiled, the coils touching, or the outer clasping the inner. Silurian—recent.

(*e*) *Lituites*— Shell at first regularly spirally coiled, straightening later. Silurian.

The siphuncle runs either through the centre of the septa; or through their anterior or posterior sides.

Ammonoidea

The shells of the (fossil) Ammonoidea are distinguished by very complicated sutures, their zigzag line are like the outlines of sharply-indented leaves or richly-branched mosses, they are due to the extraordinary folding of the edges of the septa, which are attached to the inner surface of the shell. The siphuncle is always very thin in the Ammonoidea, and almost always pierces the septa on the posterior side. The following quotation summarises the chief pecularities in the form of the Ammonite shell:—

"The shell, as a rule, forms a closed symmetrical spiral, the coils touching or clasping one another. Some of the oldest forms are straight, on in youth incompletely coiled. In certain groups of the *Ammonoidea* we find a tendency repeated at different times (Trias, Jurassic, Chalk) to depart from the close symmetrical spiral, and to adopt what are called accessory forms. The first step in this process of change is in most cases the detachment of the occupied chamber from the next inner whorl; then, little by little, the inner whorls also separate, through they still remain in the same plane—the *Crioceras* stage. Sometimes the shell grows straight for a time, then becomes hooked—the *Aneyloceeras* and *Hamites* stages, and, if only the occupied chamber separates from the coiled part—the *Scaphites* stage. Finally, entirely straight shells arise in the *Baculites* stage. Rarely, the coils leave the symmetrical plane and assume the shape of a snail's shell; in this case, the shells may be either closely or loosely coiled—the *Turrilites* stage."

Dibranchia

The shells of all known *Dibranchia*, extinct or recent, and more or less rudimentary, since they are never capable of sheltering more than a small portion of the animal. Further, they are always internal, on the anterior side of the visceral dome, and are overgrown by a fold of the integument. In *Spirula* alone, the shell is not completely overgrown, a portion at the apex of the visceral dome remaining uncovered. The shell of the (fossil) Belemnites is straight, conical, and chambered; the septa are near one another, and are pierced on the posterior or ventral side by the thread-like siphuncle, which is enclosed in short, calcareous sheaths. The apex of the shell (*phragmocone*) is protected by a conical calcareous sheath (*rostrum* or guard), the only part usually preserved. The anterior wall of the last chamber is produced downwards into a broad thin process, the *pro-ostracum*.

In *Spirulirostra*, the phargmocone begins to bend posteriorly (endogastrically). The rostrum is triangular and pointed at the top. In *Spirula*, the shell is coiled spirally and endogastrically. The siphuncle is thick, are is surrounded along its whole length by septal envelops. The rostrum is rudimentary, and there is no pro-ostracum. Starting again from the *Belemnites*, the modification of the shell may take another direction. The phragmocone may become smaller and shorter is comparison with the continually lengthening pro-ostracum (e.g. *Ostracotenthis*). The rostrum also may become thinner and smaller. Finally, the shell may be reduced to a very small hollow cone at the end of a long narrow horny lamella which corresponds with the pro-ostracum, and is called in the extant *Decapoda*, the gladius or calamus (or pan) (*Loligo*, *Ommastrephes*, *Onychotcuthis*).

In *Dosidicus*, this terminal cone is almost solid, and in *Loligopsis*, it is nothing more than a thickening at the upper end of the gladius; in other *Decapoda*, there is no trace of it on the gladius. It is *Octopoda*, the shell has completely disappeared. Again starting from the *Belemnite*, the shell may develop in a third direction to form the *Sepia* shell. The transition form is found in *Belosepia* (Eocene), that is, if this interpretation is correct, This shell is somewhat bent, the septa are crowded together and slope downwards anteriorly. They are penetrated posteriorly by an extremely thick siphon, which is enclosed throughout its whole length in an envelop with a very thick anterior wall.

The completely enclosed siphuncular space is thus a wide funnel running through the chambers of the shell on its posterior side. The phragmocone is enclosed in a thick, strongly developed rostrum, and its anterior and lateral walls are produced downwards into a broad, posteriorly concave shell (pro-ostracum?). These arrangements seem to have eulminated in the extent *Sepia*. The siphuncular space fits over the visceral dome likes a mould. The anterior portions of the septa slope downward much more obliquely from behind anteriorly, so that, in a back view of the shell, the whole area of the last septum is visible at the surface.

The septa are thin calcareous lamellae, closely superimposed one upon the other, with very narrow air chambers between them; and these latter are traversed by perpendicular trabeculae. The shell is thus very light, its specific gravity is less than that of water. Behind the siphuncle, on the posterior very much shortened side of the shell, the short septa are closely contiguous, without any intervening air spaces. The dorsal end of the shell is enclosed in a small pointed rostrum. The whole anterior surface is covered by a thin lamella of conchyolin, which projects laterally beyond the edge of the shell, and is itself covered by a calcareous layer which is an anterior and ventral extension of the rostrum.

The female *Argonaut*, is the single exception to the rule stated above, in the Octopoda the shell has entirely disappeared. This animal has a light, thin external shell coiled anteriorly or exogastrically, which is not firmly attached to thc body at any point and serves more for receiving the eggs than for protecting the body. This shell is surrounded and secured by lobate processes of the anterior pair of arms. It has no nacreous layer, but it porcelanous, and is apparently produced from the integument of the visceral dome and the mantle. The dorsal pair of arms is said only to deposit the so-called black layer on its surface. It is usually considered that this *Argonant* shell is not the homologue of the shell of other Cephalopods, but is a formation peculiar to the *Argonant* female. An opposite view has, however recently been very ably advanced—that the *Argonaut* shell is an *Ammonite* shell which has lost its septa and siphuncle and also its naereous layer. Should this view prove correct, the Cephalopods would have to be differently classified.

The division into *Tetrabranchia* and *Dibranchia* would have to disappear, as we cannot tell whether the fossil *Ammonoidea* were Tetrabranchia, and are also ignorant as to when the Dibranchia developed from the Tetrabranchia. The Cephalopods would then have to be divided into (1) *Nautiloidea* with the extant genus *Nautilus*; (2) *Ammonoibdea* with the still living *Octapoda*; and (3) *Belemnoidea* with the extant *Deeapoda*. Bivalve shelly plates called aptyehi have been found sometimes in the last chamber of the *Ammonoidea*, sometimes isolated. These have been proved to belong to the bodies of certain species of *Ammonoidea*, and have been considered by some to be protectives for the nidamental gland, by others as opercula, and by other again as the analogues or homologues of the infundibular cartilage of the *Decapoda*. No one of these three views has as yet been generally accepted.

6

Coelom

The Mollusca are said to have a primary and a secondary body cavity. The former is the system of lacunae and sinuses, into which the arteries open, and out of which the veins, where these are present, draw their blood. *It has no epithelial walls of its own,* its boundaries are formed by connective, nerve, or muscle tissues, or by epithelia, which, however, belong to other organs, such as the intestine, the kidneys, or the body wall. The so-called *secondary body cavity* or *coelom* is, in most Mollusca, very much reduced, usually consisting of only two small cavities, *the pericardium and the cavity of the gonads* (testes, ovaries, or hermaphrodite glands).

The coelom is always lined by an epithelium of its own, the coelomic epithelium, and corresponds with the true coelom of the Annelida, which, also possesses such as epithelium. Like the latter, it is connected, by means of the nephridial funnel, with the nephridia, which lead to the exterior, and in Molluscs are usually found only in one pair. A probe can therefore be introduced through the kidney into the coelom, *i.e.* into that part of it which, containing the heart, is called the pericardium.

The geminal layers must be considered as proliferations of the coelomic endothelium. The epithelium of the pericardium is, in very many Molluscs differentiated into glands, called the pericardial glands; these probably may be classed together with the kidney as excretory. We should be justified in assuming, a *priori*, that the lumen of the genital glands of the Mollusca is part of a true coelom, and that the germinal layers themselves, *i.e.* that complex of cells which yields the eggs and spermatozoa, are outgrowths of the endothelial wall of

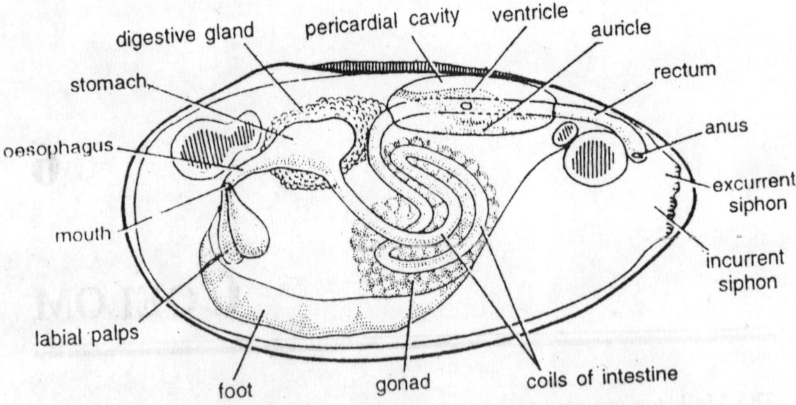

Fig. 6.1. Unio. Alimentary canal and digestive gland.

this coelom. Direct support is, however, given to this assumption by the fact that in the *Solenogastres*, *Sepia*, and *Nautilus*, the sac of the genital glands is in open communication with the rest of the coelom, forming, in fact, an only partly distinct division of the same.

In the *Solenogastres* (e.g. *Proneomenia*), the hermaphrodite gland lies above the mid-gut as a long tube, which in transverse section appears or kidney shaped, as its lower part bulges out on each side. Its shape in determined by the fact that the mid-gut forms dorsally a narrow but deep furrow, which cuts into this glandular tube from below. The tubular gland is divided into two lateral spaces by a partition, whose endothelial wall is the place of formation of the eggs; these lateral chambers may again be traversed by septa, on which the genital products develop. This division is specially distinct at the posterior part of the tube, the two chambers being there completely isolated, and entering the pericardium separately as genital ducts. If the secondary body cavity of *Proneomenia* is completed with that of an Annelid, we find the following differences:

In *Proneomenia*, the dorsal vessel is wanting in the region of the mid-gut. The coelom is much less spacious, and instead of surrounding the intestine lies only on its dorsal side. It is developed merely as a hermaphrodite glandular sac, its endothelial wall yielding the genital products. In the region of the hind-gut the vessel lying in the dorsal mesentery is developed as a heart, the coelom being here represented by the pericardium. The pericardium is connected with the cloaca by two canals; these may be considered as the morphological equivalents of nephridia. As the genital glands have been recognised as part of the

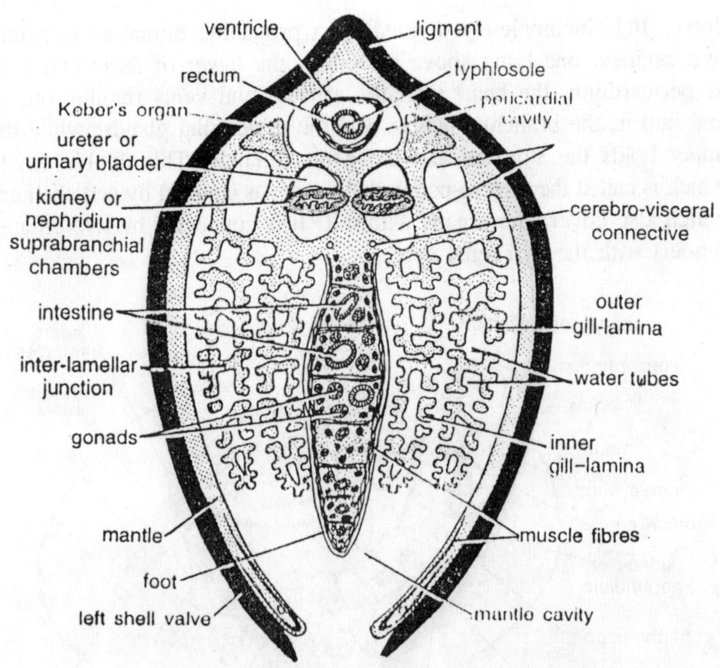

Fig. 6.2. Unio. T.S. body through middle region of gills.

coelom in the *Solenogastres*, *Nautilus*, and *Sepia*, they must necessarily fall under the same category in all other Molluses, even when no longer in direct connection or in open communication with the same.

In the *Chitonidae*, the coelom is large, and falls into three distinct divisions. One contains the intestine and digestive gland (liver), which are accordingly outwardly (*i.e.* on the side turned to the coelom) covered with an endothelium. The mesenteries, however, which originally attached the intestine to the body wall, and along which the parietal endothelium passed into the visceral endothelium of the intestine and liver, have disappeared, with the exception of portions retained on the hind-gut. The two other divisions of the coelom are: (1) the pericardium, and (2) the genital gland. Certain bands, by means of which the three divisions are connected together, have been regarded as the constricted remains of communications between the three divisions of the originally single coelom.

The *Cephalopoda* may with advantage be considered in connection with the *Amphincura*. In *Nautilus* and the *Decapoda* (e.g. *Sepia*) a spacious secondary body cavity is found in the dorsal part of the visceral dome. It is incompletely divided by a projecting dorsal septum into two cavities, one lying above the other; the lower of these contains, as pericardium, the heart with the arteries and veins running out of and into it, the branchial hearts, and the pericardial glands; while the upper holds the stomach and the genital glands. This double cavity which is called the viscero-pericardial cavity, is covered by endothelium, which also covers the organs within it. It is connected by two ciliated funnels with the two renal sacs.

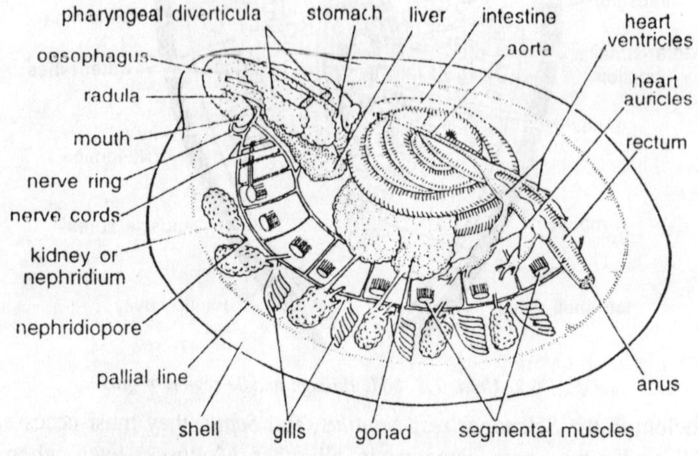

Fig. 6.3. Neopilina. Internal anatomy in side view.

In *Nautilus* it also opens direct into the mantle cavity by two canals, whose apertures lie close to the renal apertures. While the coelom in *Nautilus* and the *Decapoda* is very spacious, in the *Octopoda*, on the contrary, it is very much reduced. It consists merely of a narrow system of canals, which, however, have thick walls; this was formerly called the water vascular system. The organs, which in *Nautilus* and the *Decapoda* lie in the coelom, viz. the arterial heart with its afferent and efferent vessels, the branchial hearts and the stomach, are no longer found within the body cavity, but outside of it, and are therefore no longer covered with endothelium. Nevertheless this canal system of the *Octopoda* shows the same morphologically important characteristics as the coelom of the *Decapoda*.

There are, for instance, on each side three canals which open together, one entering the renal sac, the second widening round the pericardial gland (appendage of the branchial heart) to form a flask-shaped capsule, and the third running to the genital gland to the continued into its wall. In so far as in the *Octopoda* the heart is excluded from the coelom, which has been reduced to the "water canal system," the reduction of this cavity has gone further in these Mollusca than in any others, which all retain at least the heart in one portion of the coelom, the pericardium.

In the *Lamellibranchia* and *Gastropoda*, the only part of the coelom retained, besides the genital glands, is the pericardium. The pericardium and the gonad are, however, entirely separated from one another. In *Lamellibranchs*, there is in the pericardium, besides the heart, a part of the hind- gut which transverses, it in the *Gastropoda* (except in those *Diotocardia* in which the hind-gut penetrates the heart), only this latter organ. Rarely (e.g. *Phylliroe*) the auricle also is excluded from the pericardium. The *pericardial gland* is found in most Mollusca. It is a glandular differentiation of the endothelial wall of the pericardium, and perhaps, as already suggested, share the excretory functions of the kidney. Its position in the pericardium varies, but it seems in all cases shut off from the blood vascular system, with which it is, however functionally connected. Its secretions or excretions must be discharged into the pericardium, and thence outwards through the kidney.

Among the *Prosobranchia*, in the *Diotocardia*, the pericardial gland is found on the auricle, its walls forming dendriform branched outgrowths into the pericardial cavity, these being covered with pericardial endothelium. Where pericardial glands are found in the *Monotocardia*, they lie on the wall of the pericardium itself. Similar lobate formations occur among the *Opisthobranchia*, in *Aplysia*, and *Notarchus*, on the anterior aorta which runs along the pericardial wall; in *Pleurobranchus* and *Pleurobranchaea* on the lower, in *Doridopsis* and *Phyllidaea* on the dorsal pericardial wall. The lateral furrows of the pericardium of *Doris* form niches, which may again have accessory niches. These enlargements of the surface of the pericardial epithelium have also been considered as pericardial glands. Pericardial glands are much more common among the *Lamellibranchia* than among the *Gastropoda*, but are wanting in the most primitive forms (*Nucula*, *Solenomya*, *Anomia*).

The gland is usually of a rusty red colour, and occurs in two forms, consisting either of glandular protrusions of the endothelial wall of the auricles into the pericardial cavity, or the galndular tubes protruding from the anterior corner of the pericardium into the mantle (*Keber's organ*, *red-brown organ*). The first form is found specially strongly developed in *Mytilus*, *Lithodomus*, and *Saxicava*, more or less developed in *Dreissena*, *Uni*, *Anodonta*, *Venus*, *Cardium*, *Scrobieularia*,

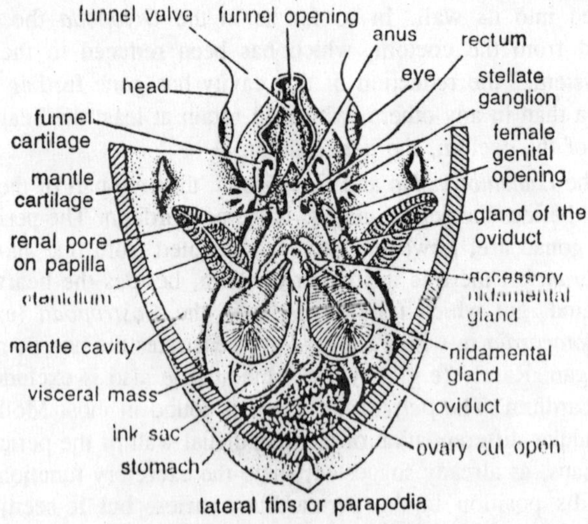

Fig. 6.4. Sepia. Mantle cavity of a female specimen.

Solen, *Pholes*, and *Teredo*, and more or less rudimentary in *Pecten*, *Spondylus*, *Lima*, *Ostrea*. The second form has been observed in *Unio*, *Anodonta*, *Venus*, *Cardium*, *Serobicularia*, *Solen*, *Pholas*, *Montacuta*, and *Dreissensia*.

Pericardial glands may also occur singly in other parts of the pericardium, as in *Meleagrina* (as a projecting ruff in the posterior base of the pericardium, and in *Chama* on the ventricle, etc.) The pericardial gland of the *Cephalopoda* is the so-called appendage of the branchial hearts. This is a structure connected with the branchial heart, and covered with peritoneal endothelium, which projects into the viscero-pericardial cavity, or, in the *Octopoda*, into a flask-like widening of the water-canal system (which has been recognised as a division of the coelom). In *Sepia* this appendage is conical. A deep furrow on the surface which projects into the viscero-pericardial cavity leads into a richly-branched system of canals, the glandular epithelium of which is a continuation of the peritoneal epithelium. Blood sinuses from the branchial heart penetrate in between the canals of this system.

In other *Cephalopoda*, the pericardial gland varies in form and structure; details of these variations cannot, however, be here given. *Nautilus* possesses two pairs of pericardial glands; this fact is connected with its possession of two pairs of gills, with their two pairs of afferent vessels, and on these the two pairs of pericardial glands occupy positions corresponding with those of the branchial hearts.

7

LOCOMOTORY ORGANS

The ventral side of the body in the Mollusca is characterised by the pronounced development of its musculature, which enables the animal to creep, a fleshy foot, provided with a flat sole suited for creeping, distinct from the rest of the body and especially from the head, being developed. This strong ventral musculature must be considered as the remains of the dermo-muscular tube of the racial form, which attained greater development on the ventral side in adaptation to a creeping manner of life, while it degenerated on the dorsal side, being rendered functionless and useless by the hard shell. The flat form of the foot with a sole for creeping must be considered the primitive form. Such a foot is found in the *Chitonidoe* among the *Amphineura*, in most *Gastropoda*, and in certain *Lamellibranchia*, especially in the *Protobranchia*, which for other reasons also must be considered the most primitive form of *Lamellibranchia*.

The musculature of the foot and of all parts which become differentiated from it are innervated from the pedal ganglia or pedal nerve cords. The foot may become much modified in adaptation to various methods of life and of locomotion, in fact, it may entirely lose all resemblance to the primitive organ. It may, be constriction or by the formation of lobes or folds, fall into several parts, of which the following are the most important:—

1. Proceeding from before backward we have the *propodium*, an anterior portion distinct from the rest, and the *metapodium* behind the former and seldom very distinct, which carries the operculum when this is present.
2. From below upward there are the *parapodia*, lobe-like extensions of the edge of the ventral sole, and the *epipodium*, a projecting

ridge or fold round the base, *i.e.,* round the upper portion of the foot. Tentacul: r processes are often developed on this ridge.

Taking the different groups in order, the following variations of the foot and the pedal glands (mucous glands and byssus gland) are to be noted.

AMPHINEURA

The foot is here not divided into separate consecutive portions, and there are no parapodia or epipodia.

GASTROPODA

Prosobranchia

With rare exceptions, which will be described later, the foot, which is well developed in this order, has a simple (undivided) flat sole for creeping.

Propodium

In a few cases, however, the anterior portion of the foot forms a propodium well marked off from the rest of the organ. This is especially the case in the *Monotocardia* (*Olividoe*, *Harpidoe*, certain species of *Pyrulidoe*, *Strombidoe*, *Strombus*, *Pteroccra*, *Terebellum*, *Rostellacia*, *Nenophoridoe*, *Naricidoe*, *Naticidoe*). Among the propodium is particularly well developed in *Oliva*, separated front the rest of the foot by a transverse furrow and forming a semicircular disc. In the

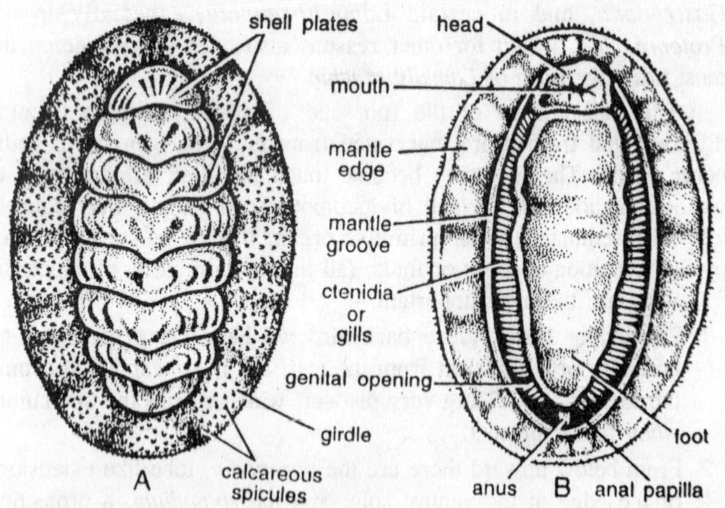

Fig. 7.1. Chiton. A—Dorsal view; B—Ventral view.

large foot of *Natica*, the propodium is also very distinct. It has an anterior lobe which bends back over the shell, and so covers the head.

Sometimes the propodium forms a sort of siphon on the left side, and in other cases the lobe which bends back over the shell shows a bulging. Both these arrangements serve to conduct water to the respiratory cavity. The metapodium also, which, when swollen and expanded, spreads out widely, carries on its dorsal side a lobe which bends forward over the shell, and carries the operculum on the side nearest the shell. In most Prosobranchia the metapodium carries, on the dorsal side, a horny or caleareous operculum which serves to close the shell.

Epipodium

The epipodium is very commonly present in the *Diotocardia*. It is most strongly developed in *Ilaliotis*, where it surrounds the base of the foot in the form of the large integumental fold. This fold, which may aptly be called the ruff, has fringed or digitate appendages as well as long contractile tentacular processes. The tentacles here, as in other Prosobranchia, are organs of touch, and may be provided at their bases with so called lateral organs. In the *Fissurellidoe* this epipodial ruff is replaced by a row of numerous tentacles or papillae, rising on each side from the base of the groove between the base of the foot and the visceral dome.

Among the other *Diotocardia* also, the epipodium is well developed as a simple or fringed, border, which carries a few tentacles (usually

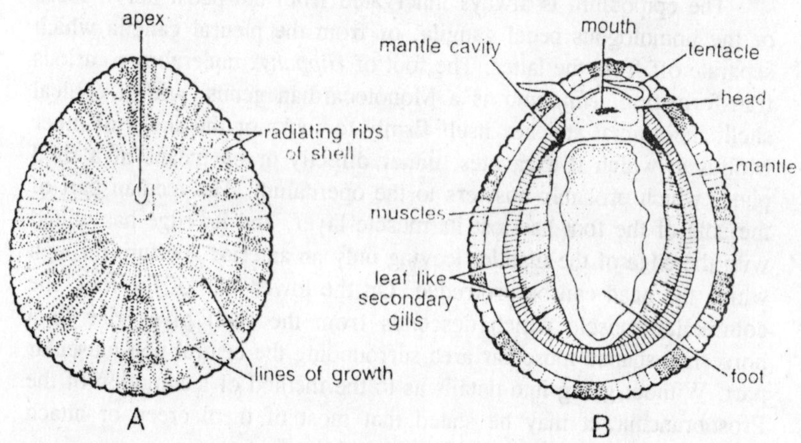

Fig. 7.2. Patella. A—Dorsal view; B—Ventral view.

four on each side) of varying length. At the base of each tentacle there is a lateral organ. Eyes are said to occur at the bases of the epipodial tentacles in *Eumargerita* and *Scissurella*. The epipodium is, as a rule, wanting in *Docoglossa*, but one is found beset with papillae in the genus *Helcion*, and in *Patinella* and *Nacella* it is fringed; these epipodia correspond in position with those of other *Diotocardia*. A well-developed epipodium rarely occurs among the *Monotocardia*, but *Ianthina* has a typical epipodial border, and the *Litiopidoe* and many *Rissodoe* have an epipodium with several (1.5) tentacles on each side. Many other *Monotocardia* have retained either the anterior or posterior portions of the epipodium.

(*a*) Anterior vestiges of the epipodium are found in *Vermetus* in two anterior pedal tentacles, and in *Paludina* and *Ampullaria* in two nuchal, lobes, which must not be confounded with true cephalic tentacles. In *Paludina*, the right nuchal lobe, and the *Ampullaria* the left, forming a longitudinal groove, becomes a sort of siphon. *Calyptroea* possesses on each side under the neck a semicircular epipodial fold.

(*b*) Posterior vestiges of the epipodium are found in *Lateuna* in the form of an epipodial fold with a process on each side above the foot. *Narioa* has, above the metapodium on each side, a wing-like epipodial lobe.

(*c*) Median and posterior vestiges of the epipodium are found in *Choristes*, where there is a median papilla on each side, and posteriorly a pair of tentacles below the operculum.

The epipodium is always innervated from the pedal nerve cords or the liomologous pedal gangila, or from the pleural gangila which separate off from the latter. The foot of *Hipponyx* undergoes a curious transformation. *Hipponyx* is a Monotocardian genus, with a conical shell: the animal attaches itself firmly to rocks or the shells of other Molluscs, which it excavates, either directly or by means of a shell plate, which probably answers to the operculum. The median part of the sole of the foot has lost its muscle layer, and its edge has united with the edge of the mantle, leaving only an anterior aperture through which the head can be protruded. On the lower side of the foot, the columellar muscle which descends from the shell gives rise to a horseshoe-shaped muscular area surrounding the central non-muscular part. Without going into details as to the method of locomotion of the Prosobranchia, it may be stated that most of them creep or attach themselves by means of the flat sole of the foot.

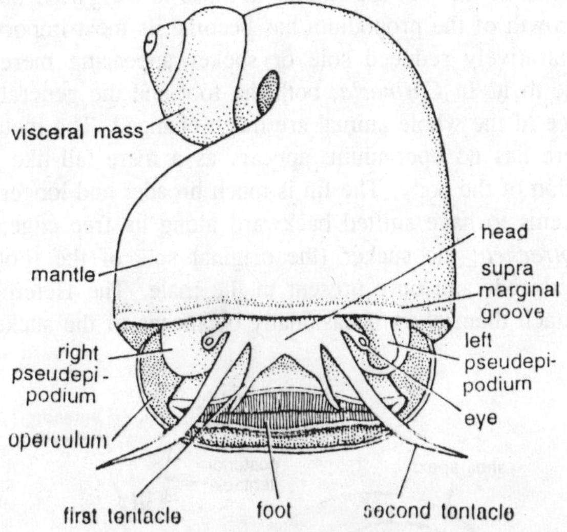

Fig. 7.3. Pila. Body after removing the shell.

Heteropoda

The Heteropoda are pelagic Prosobranchia (*Monotocardia*), which have exchanged the creeping for the swimming manner of life. The foot has in them become peculiarly adapted to this new method of locomotion. The propodium has become changed into a narrow vertical rowing fin (carinate foot), which when the animal is in its swimming position is turned upward. The development of this vertical fin can be traced almost step by step within this division, starting with *Oxygyrus*, and proceeding through *Atlanta* and *Carinaria* to *Pterotrachea*.

In this series, the typical outer appearance of the Prosobranchiate (its shell, visceral dome, mantle and gills which are still retained in *Oxygyrus* and *Atlanta*) gradually disappears owing to development in another direction. *Oxygyrus* still has the characteristics of the Prosobranchiate. The foot consists of (1) a propodium the creeping sole of which has been somewhat hollowed out or deepened; anteriorly it possesses a fin-like outgrowth, which is used as a propelling organ in swimming; and (2) a distinct metapodium directed backwards, like a tail, and bearing an operculum.

The derivation of such a foot from that of certain Prosobranchia, which have distinct propodia and metapodia, such as the salatory *Strombidoe*, is clear. The sole of the foot in *Oxygyrus*, although it can be used for creeping, is looked upon as a sucker. In *Atlanta*, the arrangements of the foot are similar to those in *Oxygyrus*, but the fin-like outgrowth of the propodium has become its most important part, the comparatively reduced sole or sucker appearing merely as an appendage to it. In *Carinaria*, both the foot and the general external appearance of the whole animal are much changed. The metapodium, which here has no operculum, appears as a mere tail-like posterior prolongation of the body. The fin is much broader and longer, and the sucker seems to have shifted backward along its free edge. Finally, the *Pterotrachea*, the sucker (the original sole of the foot) is still further reduced, and only present in the male. The Heteropoda are said to attach themselves occasionally by means of the sucker.

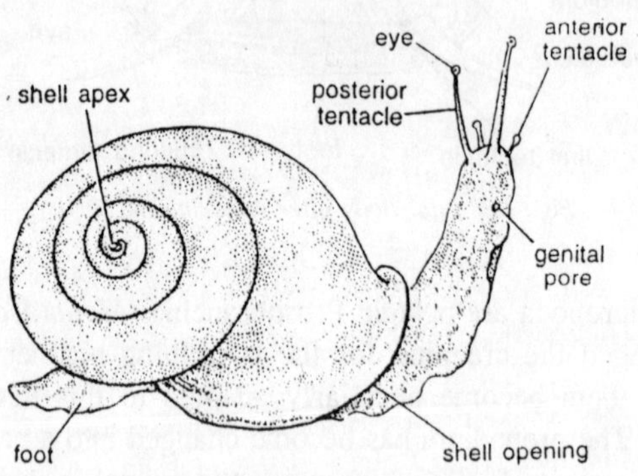

Fig. 7.4. Helix.

Pulmonata

The foot is here almost always undivided, and provided with a large flat sole for creeping. In the few *Auriculidoe* (*Melampas*, *Leuconia*, *Blauneria*, *Pedipes*), it is divided into two portions by a temporary or permanent transverse groove.

Opisthobranchia

In almost all Opisthobranchia the foot has a well- developed sole for creeping. There is no division into parts, and the adult rarely (*Aetcon*) carries an operculum. The epipodium is wanting. The *parapodia*, on the contrary, *i.e.*, lateral lobes or fold- like extensions of the edges of the sole, are highly developed in many Opisthobranchia (*e.g.* the *Elysiadoe*, among the *Ascoglossa*, and very many *Tectibranchia*, such as the *Seaphandridoe*, *Bullidoe*, *Aplustridoe*, *Gastropteridoe*, *Philinidoe*, *Dorididoe*, *Aplysiidoe*, *Oxynocieoe*). The parapodia are often bent back

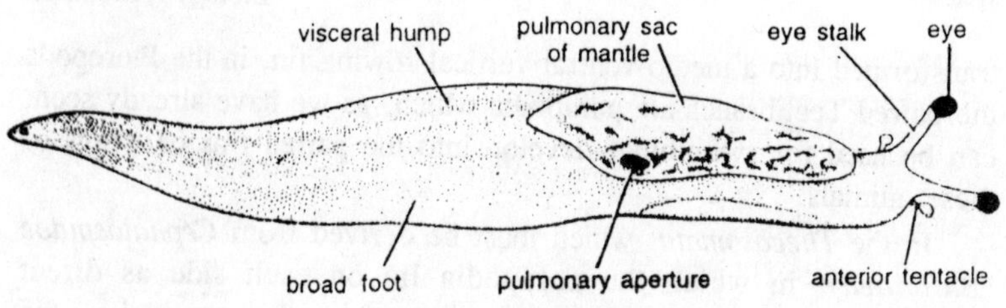

Fig. 7.5. Limax.

over the shell, their edges sometimes touching, so that the shell may be entirely roofed over by them.

In many forms which are provided with parapodia (*Gastropteridoe*, *Philinidoe*, *Doridiidoe*, *Aplysiidoe*) the mantle also bends back over the shell, more or less completely covering it. In these cases the shell is to some extent doubly internal, being covered first by the mantle and then (not in *Philine* and *Doridium*) by the parapodia. The parapodia may fuse posteriorly along their upturned edges (*Aplysiidoe*, *Oxynoe*). In ***Lobiger*** each parapodium is transversely slit, so that two long wing like processes are formed on each side. Many Opisthobranchia (*Aplysiidoe*, *Oxynoe*, *Gastropteridoe*) can propel themselves through the water by means of the waving motion of their parapodia. *Phyllirhoe* is a *Nudibranch* which appears to have become adapted to a pelagic swimming manner of life by the compression of its body into the shape of a long narrow leaf with sharp dorsal and ventral edges; it travels though the water with an undulating motion. The foot has disappeared.

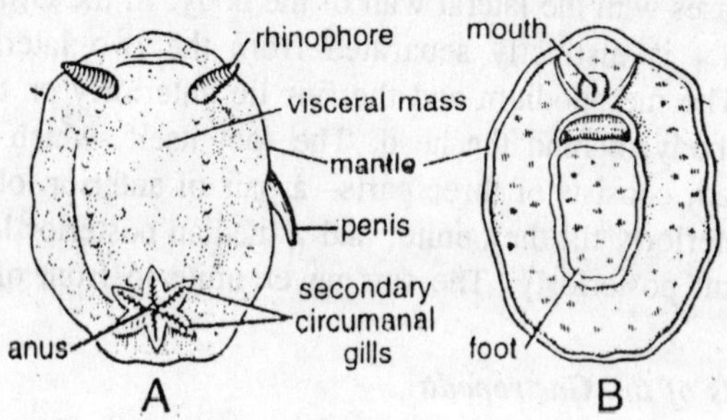

Fig. 7.6. Doris. A—Dorsal view. B—Ventral view.

Pteropoda

The Pteropoda, which are Tectibranchiate Opisthobranchs, have like the *Prosobranchiate Heteropoda*, become pelagie animals adapted for swimming. While in the *Heteropoda* the propodium becomes transformed into a medio-ventral vertical rowing fin, in the Pteropoda the paired Tectibranchiate parapodia which, as we have already seen, can be used for swimming, develop into the paired fins or wings of these animals.

In the *Thecosomata*, which must be derived from *Cephalaspidoe* (*Bulloidea*), in which the parapodia lie on each side as direct prolongations of the reptant surface of the foot, this organ, *i.e.,* the foot, has become confined to the anterior end of the body, and consists of three portions—the median unpaired mesopodium and the two lateral parapodia or fins. The mesopodium is small and the ventral side of it (which corresponds with the sole of the *Cephalaspidoe*, but can no longer be used for creeping) is strongly ciliated. The ciliary movement is from behind forward, *i.e*, towards the oral aperture which lies anteriorly on the foot, and no doubt serves for conveying to it the minute marine animals on which the creature feeds. On the dorsal side of the mesopodium, which projects freely backwards, the *Limacinidoe* carry a delicate transparent operculum, which often becomes detached. The parapodia are large, fin-or wing like, and anteriorly inserted on each side of the median portion of the foot; they unite in front of and above the mouth.

The *Gynnosomata* are to be derived from the *Aplysiidoe*, in which the parapodia are not exactly lateral extensions of the sole of the foot, but arise somewhat above the edge of the sole on each side. This may be explained by supposing that they are fused for a certain distance from their bases with the lateral wall of the body. In the *Gyanosomata*, also, the foot is distinctly separated from the two lateral fins or parapodia. The mesopodium and the fins lie anteriorly on the ventral side of the body, behind the head. The foot itself, which is distinct from the head, consists of three parts—a pair of anterior lobes, which converge anteriorly till they unite, and a median posterior lobe drawn out to a point posteriorly. The fins never unite in front of or above the head.

Pedal glands of the Gastropoda

Many Gastropods, are especially most *Prosobratnchia* and *Pulmonata*, possess, besides the various unicellular glands scattered over the upper and lower sides of the foot, larger multi-cellular localised pedal glands. These belong to two morphologically distinct groups.

1. In the *Prosobranchia* an *anterior pedal gland* opens at the anterior edge of the foot. In those forms in which this anterior edge is divided into an upper and a lower lip, this "labial gland" opens

between the lips. In the *Pulmonata* it opens externally between the head and the foot. It consists of an epithelial tube of varying length, not infrequently as long as the foot itself; this tube runs backward in the median line mostly through the base of the foot; less frequently it lies upon this base, projecting into the body cavity.

This tube serves both as reservoir and duct for the numerous unicellular mucous glands which lie in the surrounding tissue of the foot and open on its walls. It secretes mueus, though it has been incorrectly described as an olfactory organ. It undergoes considerable modifications with regard to its size, the form of its lumen, and the number and arrangement of its glandular cells.

2. Among the *Prosobranchia*, opening on the sole of the foot, there is commonly found an unpaired gland. Its outer slit-like aperture is median, and lies behind the anterior edge of the foot. It leads into a cavity in the foot which serves as a reservoir; the epithelial wall of this cavity projects in the form of folds into its lumen. As in the former case, unicellular glands pure their secretions into it through ducts which pass between the epithelial cells. This sole gland in the *Prosobranchia* has rightly been considered homologous with the byssus gland of the *Lamellibranchia.* It is developed in varying degrees, and not infrequently is altogether wanting. Its slimy secretion forms threads by means of which many *Prosobranchia* attach themselves to objects in the water. Some terrestrial *Pulmonata* also lower themselves from a height (from plants) by means of the tough threads which they secrete.

Besides these two, other pedal glands are occasionally found. Only one need be mentioned, which is found in some *Opisthobranchia* (*Pleurobranchus*, *Pleurobranchora*, *Pleurophyllidia*). It lies at the posterior end of the sole, and consists of glandular eaeca, each of which opens separately.

Scaphopoda

The foot of *Dentalium* is almost cylindrical; it projects downwards into the tubular mantle cavity, and can be protruded through its lower aperture. The free end of the foot is conical; the base of the cone carries on each side a fold or ridge which has been compared, with questionable propriety, to an epipodium. These two lateral folds or ridges encircle the base of a conical end without uniting either anteriorly or posteriorly. A groove runs along the anterior middle line of the foot. In *Siphonodentalium* both this groove and the lateral lobes

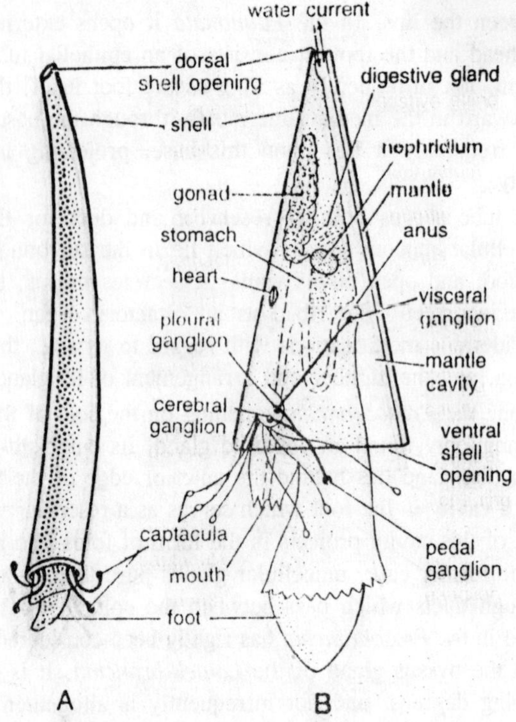

Fig. 7.7. Dentalium. A—External features. B—Internal features.

are wanting, and the anterior end of the foot is broadened into a round disc carrying on its edge small conical papillae.

Lamellibranchia

The foot in this class is, as a rule, laterally compressed, and has a sharp edge directed downwards and forwards, which can be stretched out beyond the shell. It may be called hatchet-shaped (*Pelecypoda*) or linguiform, and is especially suited for forcing its way into mud by means of alternate contraction and expansion. This peculiar shape must be considered as acquired. Originally the foot of the Lamellibranchia also possessed a flat sole for creeping. The *Protobranchia*, in fact, have a foot with a ventral disc and so has *Pectunculus*. The edge of this pedal disc is notched or toothed.

When the foot is retracted, this disc folds down the middle line. The foot in the Lamellibranchia varies much in details, according to the manner of life or a locomotion of the animal, and according to the development of the byssus. One of the special characteristics of the

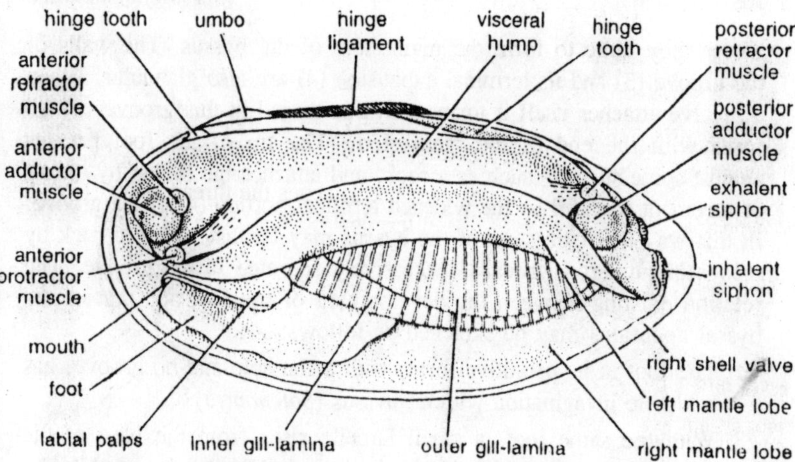

Fig. 7.8. Unio. Internal features after removal of left shell valve.

Lamellibranchiate foot is the gland which secretes the byssus, the latter being a bundle of tough threads varying in thickness, and resembling horn in their physical properties. The Lamellibranch, with these threads, anchors itself to foreign objects.

The byssus can generally be thrown off and replaced by a new one, and many forms can move about on a smooth perpendicular pane of glass by means of alternate attachment and rejeetion of portions of the byssus applied by means of the foot. Stationary bivalves, *i.e.* those attached by one of the shell valves, are in the first instance attached by means of the byssus, for a byssus is, as a rule, present in the young stages of those bivalves which do not possess it as adults. The complete byssus apparatus consists of: (1) a cavity in the foot, into which the byssus gland opens; (2) a duct connecting this cavity with the exterior; (3) a groove which runs from the aperture of the duct along the ventral edge of the foot to its anterior end; and (4) a crescent-shaped or cup-like widening of the groove at its anterior end.

(1) The byssus cavity is divided into narrow shelves by numerous folds, which projects from each side into its lumen. A septum, descending from its roof, further divides it into two lateral parts. The byssus secretion is yielded partly by the cells of the epithelial walls, and partly by glandular cells which lie in the surrounding tissue, their ducts passing between the epithelial cells. The secretion takes the form of the cavity, and is thus held fast as with roots by the numerous lamellae which occupy the shelves. As the amount of the secretion in the cavity increases, these lamellae are pressed into the duct (2),

where they unite to form the main stem of the byssus. The walls of the groove (3) and its terminal expansion (4) are also glandular. When a bivalve attaches itself it forms a byssus thread in this groove, which fuses with the end of the main stem. The tip of the foot presses against some surface, such as a rock, and attaches the thread by means of a cement secreted by the widened terminal portion (4) of the groove. In this way the main stem of the byssus may be fastened to a rock by means of numerous threads successively secreted in the groove. The relation existing between the development of the foot and that of the byssal apparatus may be sketched as follows:—

1. The foot in its primitive form, with a flat sole and no groove, has a simple invagination without byssus (*Solenomya*).
2. With the same foot, a small lamella rises from the base of the simple invagination; the byssus is very slightly developed (*Nucula*, *Leda*).
3. The invagination becomes differentiated into a cavity and a duct, and the byssus and its glands are strongly developed. In consequence of this the foot ceases to be a locomotory organ; its flat sole disappears, and it becomes finger or tongue-shaped often more or less reduced in size, and serves for attaching the byssus. In very many cases the groove is formed from the end of the duct, widening at the tip of the foot as above described. This is especially the case in forms which anchor themselves by the byssus to stones, plants or the shells of other Molluscs. This attachment may be more or less firm, and may be temporary or permanent (*Limidx*, *Spondylidoe*, *Pcetinidoe*, *Mptilidoe*, *Areidoe*, *Corditidoe*, *Erycinidoe*, *Galcommidoe*, *Triducidoe*, *Cyprindoe*, *Veneridoe*, *Glyeymeridoe*, *Myidoe*, etc.).

   When the byssus is very highly developed, some of the pedal muscles become attached to the byssus gland and form the *retractors of the byssus*.
4. Many Lamellibranchs, in the adult state, have neither byssus nor byssus glands, but the cavity the duct and even the retractors (e.g. *Trigonia*) may be retained. The byssal apparatus may be found, in closely-related forms, sometimes with and sometimes without the byssus itself. In the latter ease the foot is generally more strongly developed, and served for locomotion, *i.e.*, for foreing a way forward into sand or mud, which most of these forms inhabit, or for the saltatory motion of *Trigonia*. In these cases it is linguiform, or wedge, or hatchet-shaped (*Acedioe*, *Carditidoe*,

Cyprinidoe, *Tellinidoe*, *Serobiculariidoe*, *Mydoe*, *Cardiidoe*, *Lueinidoe* (foot vermiform, *Donacidoe*, etc.).

5. When the linguiform, or hatched-shaped, and often bent, foot becomes more strongly developed as a fleshy and extensible organ, every trace of the byssus and its apparatus disappears, at least in the adult (*Unionidoe*, many *Veneridoe*, *Cyrenidoe*, *Psammobiidoe*, *Mesodermatidoe*, *Solenidoe*, *Mactridoe*). All these live in mud. The fleshy foot of the *Solenidoe*, which is directed forwards, is so strongly developed that it can often no longer be wholly withdrawn into the shell, which therefore gapes anteriorly. The foot is thick and linguiform in *Solenocurtus*; club-shaped and truncated at the tip in *Pharus*, *Cultellus*, *Siliqua*, and *Ensis*; and cylindrical with an egg-shaped tip, in *Solen*.
6. In forms where one of the valves has become firmly attached to some hard substance, the foot (the byssus being absent) may become rudimentary (*Chamacea*), or may altogether disappear (*Ostrieidoe*). In forms which inhabit mud or excavations made by themselves in stone, etc. and which surround the body with an necessary caleareous tube (*Gastrochoenide*, *Clacagelidoe*), the foot is also reduced to a small, usually finger-shaped rudiment. The series of boring *Pholadidoe* is specially interesting. *Pholas* has a pestle- or sucker shaped foot, which, projecting through the shell cleft, serves to attach the animal while boring. In *Pholadidea* and *Jouannetia* only the young while boring their habitations possess such a foot; as soon as they have finished this work the pedal aperture of the mantle closes, the anterior cleft of the shell is also closed by means of an accessory shell piece called the callum, and the foot completely atrophies, so that the animals are no longer capable of locomotion.

In the attached *Anomia*, also, the foot is small: it is of great importance, however, as bearer of the byssal apparatus. The shelly plug, by means of which the animal is fastened to the ground, and which occupies the deep notch cut by the byssus into the right or under valve, must be regarded as a calcified byssus. Many Lamellibranchs (*Crencella*, *Lima*, *Modiloa*) weave a byssus web which they inhabit like a nest, and which they strengthen by the addition of foreign bodies attached by byssus threads.

Cephalopoda

The question, what part of the body in Cephalopoda corresponds with the foot of other Mollusca, has led to much discussion and careful investigation. It may now be considered as pretty well established that the foot in Cephalopoda forms: (1) the arms, (2) the siphon. The arms are considered as lateral processes of a Molluscan foot which have pushed past the head to the right and left, and have united in front, so that the head is entirely encircled by the foot, and the mouth has come to lie in the middle of the ventral pedal surface, *i.e.* at the centre of the circle of arms or brachial umbrella. That this circle of arms is a derivative of the foot is supported by important anatomical and ontogenetic facts: (1) The arms are innervated from the brachial ganglion, which lies under the oesophagus, and is an anterior division of the pedal ganglion. (2) The arms do not occupy, in the embryo, their definitive position round the mouth, but rise on the ventral side behind the mouth, between it and the anus, in a row on each side.

These two rows shift secondarily forward to form the circle of arms round the mouth. (According to another view, the arms are cephalic appendages, comparable with the cephalic tentacles of the *Pteropoda*). The pedal nature of the siphon or funnel has rarely been doubted. It is innervated from the pedal ganglion. Its two lateral lobes, which in the *Nautilus* remain separate throughout life, but in the

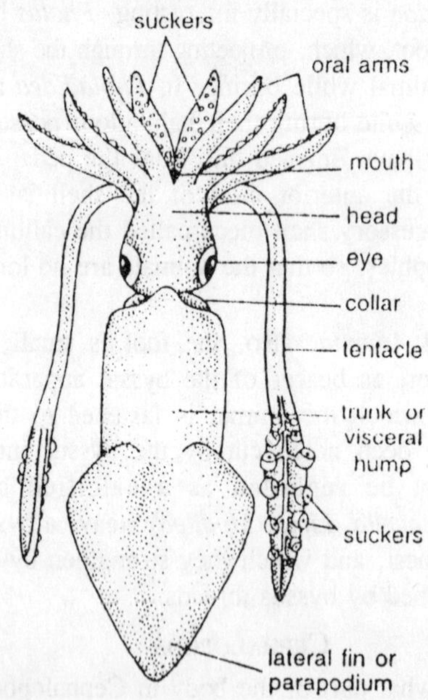

Fig. 7.9. Loligo. Dorsal view.

Dibranchiata overlap, may be considered as epipodia. The accompanying figure of a Cephalopod embryo confirms this opinion; the rudimentary siphon is seen in the typical position of epipodia in the shape of two lateral folds running backward above the foot and under the visceral dome. In *Nautilus* and the *Decapoda* (excluding the *Loligopsidae*) a valve is present within the siphon.

Arms of the Tetrabranchia (Nautilus)

The "head" of the Nautilus carries numerous tentacles placed in a circle round the mouth; these do not rise directly from the integument around the mouth, but stand upon special lobes which are differently developed in the two sexes. These lobes may be compared with the arms of the *Dibranchia*, and the tentacles they carry, perhaps with the suckers on those arms. Each tentacle can be retracted into its own basal portion as into a sheath. If the head be viewed from the ventral side, so that the mouth appears lying in the centre of the extended lobes and tentacles, we see in the female, three inner lobes close to the mouth, two lateral and one posterior.

The posterior inner lobe consists of two fused lateral lobes, the line of fusion being indicated by a lamellated (olfactory?) organ. It carries twenty-eight tentacles, fourteen on each side. Each lateral inner lobe carries twelve tentacles. Besides these three inner lobes, the foot develops a muscular circular fold; this is particularly thick anteriorly, and here forms a lobe, the so-called hood which, when the head is retracted, covers the aperture of the shell like an operculum. The outer circular fold carries nineteen tentacles on each side. Besides these tentacles which belong to the foot, there are two more on each side which probably belong to the head, one lying above and the other below the eye. In the male *Nautilius* (upper figure) the posterior inner lobe is rudimentary. Each of the lateral inner lobes is divided into two portions. In the right lobe, the anterior portion carries eight tentacles and the posterior (antispadix) four, three of which have a common sheath. The anterior portion of the left lobe also carries eight tentacles, and the posterior portions forms the conical spadix, which, instead of tentacles, carries imbricated lamellae. This spadix is looked upon as the hectocotylised limb of the *Nautilus*, and probably takes some part in copulation.

Arms of the Dibranchia

The Dibranchia have either or ten arms, which stand in a circle round the month and carry two longitudinal rows of suckers (acetabula); rows of cirri may accompany the suckers, and the cirri may here and

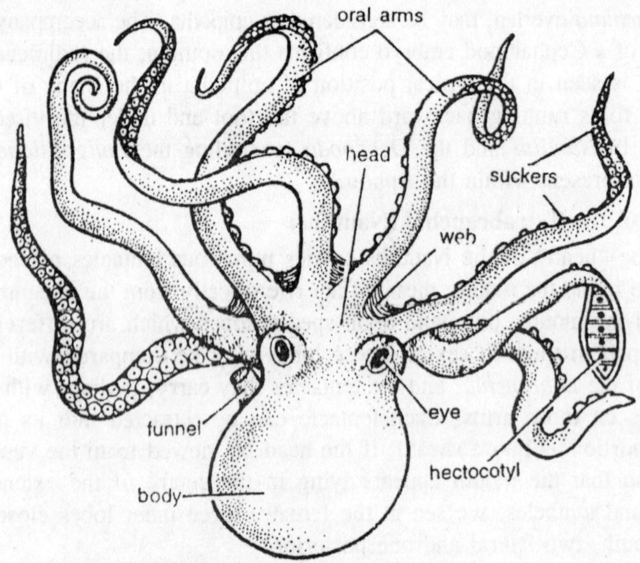

Fig. 7.10. Octopus. Dorsal view.

there become transformed into books or claws (e.g. *Oayehotenthis*). In many *Octopoda*, the long arms are connected by means of membranes near their bases, and occasionally as far as their tips. In the latter case the circle of arms has the appearance of an umbrella, of which the arms are the ribs. The mouth lies in the centre.

The *Octopoda* can creep by means of their circle of arms, the visceral dome standing erect. In this position they must best be compared with snails, the ventral side of the circle of arms functioning like the sole of the snail's foot. The *Decapoda* have ten arms; eight of these correspond with the eight arms of the *Octopoda*, but are shorter and are never connected by membranes. The two others, the prehensile tentacles, are inserted between the third and fourth Octopodan arms on each side and differ from the latter in structure, being long and vermiform, with swollen ends armed with suckers, hooks, etc.

The prehensible tentacles are very contractile, and in many *Decapoda* (e.g. *Sepia*) are concealed in special cavities of the head when the animal is at rest. These cavities probably correspond morphologically with the water pores, which often occur elsewhere at the bases of the arms or on the head. When pursuing prey with *Decapoda* dart these tentacles out of their cavities with great force. One (less frequently two) of the eight or ten arms of the male

Dibranchia is almost always transformed (hectocotylised) to assist in copulation. In some *Octopoda* it even becomes detached from the body and is regenerated.

The hectocotylised arm is, in the *Octopoda*, usually the third arm on the right side, and in the *Decapoda* the fourth on the left. (The arms are counted from before backward). In the female *Argonacut*, each arm of the first pair is widened into a sail-like expansion, which stretches back over the cuter surface of the shell. All Cephalopods, even the more massive Octopoda, are good swimmers. In swimming, the mantle and funnel play the chief parts.

Water is alternately taken into the mantle cavity through the mantle cleft, and expelled from it forcibly through the funnel, the reaction propelling the animal backwards. When the water is being ejected, the mantle cleft is closed by the locking apparatus, so that all the water in the mantle cavity has to pass out through the funnel. Many Decapoda can also swim with the head directed forward, the lower (distal) end of the funnel being bent round, so that the water is expelled in the direction of the visceral dome. In swimming the arms are apposed to one another, so as to diminish the friction as much as possible. Some *Octopoda*, especially those which have interbrachial membranes, assist themselves in swimming by opening and shutting their circle of arms like an umbrella.

Swelling of the Foot (*Turgescence*)

Imbibition of Water

The foot in many *Lamellibranchia* and *Gastropoda* may swell when it has to be protruded from the shell and used for locomotion. Until recently opinions varied very much as to the way in which this swelling or expansion took place. Many believed that water was taken up from without into the blood vascular system or into a special water vascular system, but there was difference of opinion as to the manner in which it was taken in. On the one hand it was said to enter through apertures or pores in the foot, which, however, do not exist, the only pores found being the apertures of the pedal glands (byssus and sole glands).

On the other hand, the water was supposed to enter the foot through intercellular ducts between the epithelial cells, but this theory has also been disproved. Others, again, maintained that the water was conducted by nephridia to the pericardium, and conveyed thence through the blood vascular system; but the pericardium has been shown to the entirely separated from the vascular system. Indeed, many theories on this subject have been put forward and disproved.

It is now the received opinion that, except in the case of one animal, which will be presently described, the foot is swelled by a rush of blood which, flowing into the foot, is prevented from returning to the body by splineter muscles. The exceptional case is that of *Natica Josephina*. In this animal there can be no doubt, that water is taken in to swell the foot. The swelling takes place very quickly—in less than five minutes. When the foot is stimulated it gives out an amount of water which would fill the empty *Natica* shell two or three times. The water is taken in through very small slits, invisible to the naked eye (probably indeed through a single very narrow slit, lying at the edge of the foot), and finds its way to a system of water sinuses, quite distinct from all other cavities of the foot, and also distinct from the blood vascular system (which in *Natica* is closed). There can thus be no question of a direct taking in of water into the circulatory system. The water slits at the edge of the foot can be closed by muscles, with extend from their upper to their lower edges.

8

Locomotion

Molluscs, like flatworms and nemertines, have well-developed mesenchyma, which naturally leads to persistence of at least some muscle cells in a diffused state. At the same time the large size of many molluscs and the considerable amount of musculature in separate parts of their bodies produce a tendency to segregation of muscle cells and formation of muscle tissue; and the development of hard parts, especially a shell, is a stimulus to formation of separate, individualised muscles attached to these hard parts. Therefore most molluscs retain, besides well-separated muscles, diffuse muscle fibres in the mesenchyma, just as we see in them a diffuse neural plexus together with a separate central section of the neural apparatus, and a general network of lacunae together with arterial and venous systems. The most characteristic feature of molluscs is possession of a foot and a shell, with profound effects on the structure of their musculature.

Among lower molluscs the only exception is Solenogastres, which have no shell and either no foot or only a vestige of it evenly developed musculo-cutaneous sac. It is very probable, however, that they lost the foot secondarily, and that the musculo-cutaneous sac is equally secondary in them. The foot is the primary and principal locomotor organ in molluscs. We apparently find a primitive type of foot in Loricata and in most Gastropoda, where it is a flat muscular creeping sole developed along the blastopore side of the body. Traces of the sole are retained in the foot of the primitive bivalves Protobranchia, whereas in most Lamellibranchia it has a cuneiform or digitiform shape and is used less for creeping gastropod foot and its unusual structure in Cephalopoda will be discussed briefly below.

Primitive molluscs are slow-moving, grazing animals which developed heavy protective armour. Most members of the phylum have retained that type of specialisation, but a number of separate groups have taken the path of increased activity, accompanied by lightening or total loss of the shell, development of distant receptors, refinement of the central section of the neural apparatus, and general raising of the level of organisation. Cephalopoda have advanced farthest along that path.

Evolution of the locomotor apparatus has taken different directions in separate groups of molluscs, according to their general trend in development. Loricata have a shell composed of eight movably-articulated plates metamerically arranged along the animal's dorsal side. The shell plates are moved by a number of well-individualised muscles—longitudinal, oblique, and transverse. The fibres of the last-named are, in effect, dorso-ventral, joining the edges of plates that lie one above the other.

The flat foot of Loricata serves as an organ of locomotion and of suction against the substrate. It is permeated by muscle fibres running in all directions—longitudinal, transverse, oblique, and dorso-ventral. The last are attached by their dorsal ends to the plates of the shell

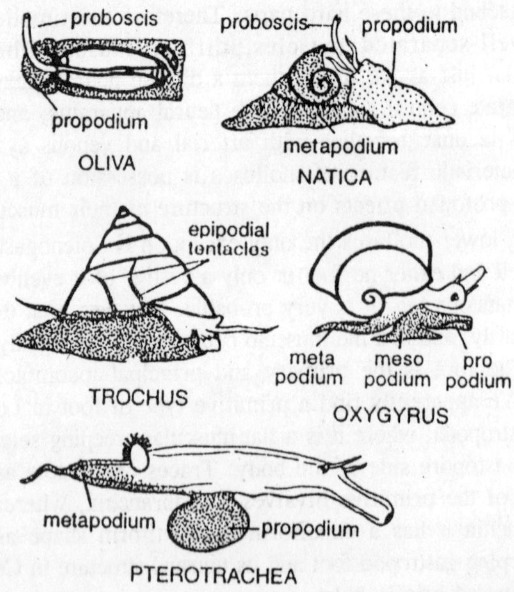

Fig. 8.1. Various types of foot—Oliva, Natica, Trochus, Oxygyrus and Pterotrachea.

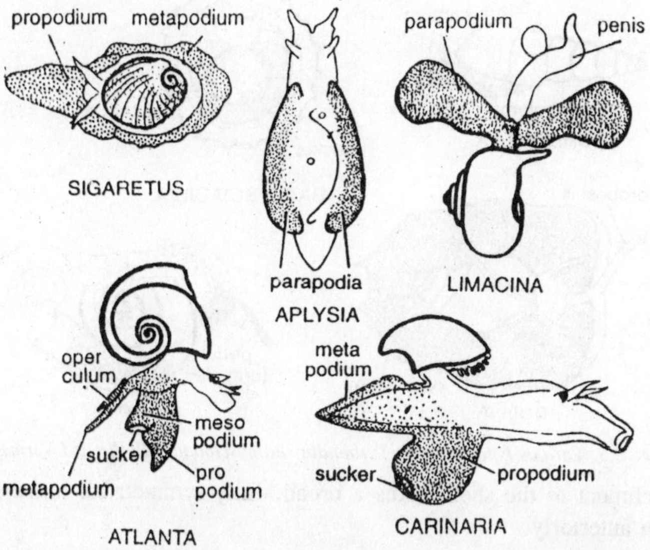

Fig. 8.2. Various types of foot—Sigaretus, Aplysia, Limacina, Atlanta and Carinaria.

and, bending round the medially-located internal organs on both sides, spread fanwise over the foot. The dorsal retractors of the vestigial foot of Neomenioidea appear to be homologues of these muscles. The edges of the mantle, which surround the body of Loricata on all sides, are also permeated in all directions by muscle fibres; consequently when the animal is applying suction to the substrate the edges of the mantle can be closely apposed to the smallest irregularity in the latter. The buccal musculature of Loricata has attained extreme complexity, consisting of 37 paired and six unpaired muscles, as a result of the great complexity of the oral apparatus. Here we encounter for the first time a specialised phytophage, capable of using tough plant food unavailable to lower animals and possessing the adaptations required for that purpose.

Tryblidiida, like Loricata, have paired metameric retractors of the head and foot, connecting these to the shell in two symmetrical rows; Nuculidae still retain four pairs of these muscles, and other Lamellibranchia two pairs. The larvae of lower Gastropoda possess homologues of the paired foot-retractors. The columellar muscle of adult Gastropoda, which is used to retract the foot, is asymmetrical, as it is formed from one of the paired retractors of the larva. In the comparatively-primitive modern Docoglossa, however, its place of

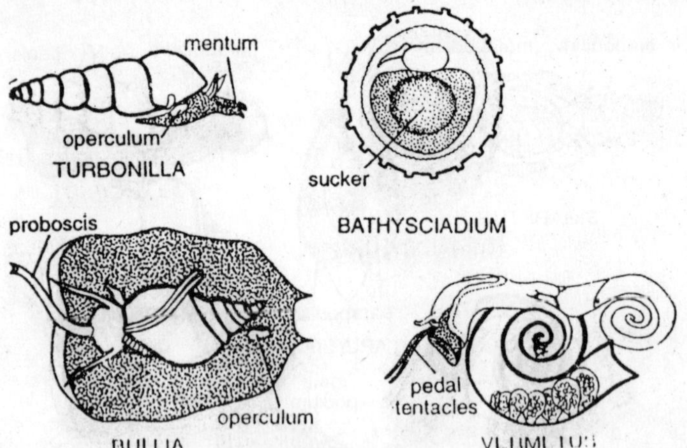

Fig. 8.3. Various types of foot—Turbonilla, Bathysciadium, Bullia and Vermetus.

attachment to the shell forms a broad, fully-symmetrical horse-shoe, open anteriorly.

The Pulmonata with a similar cap-shaped shell (*Siphonaria*, etc.) have a markedly-dissymmetrical columellar muscle. The creeping movement of most Gastropoda convergently resembles that of large Triclada: basically it uses waves of muscle contraction running along the sole, with plentiful lubrication of the substrate with mucus, and in

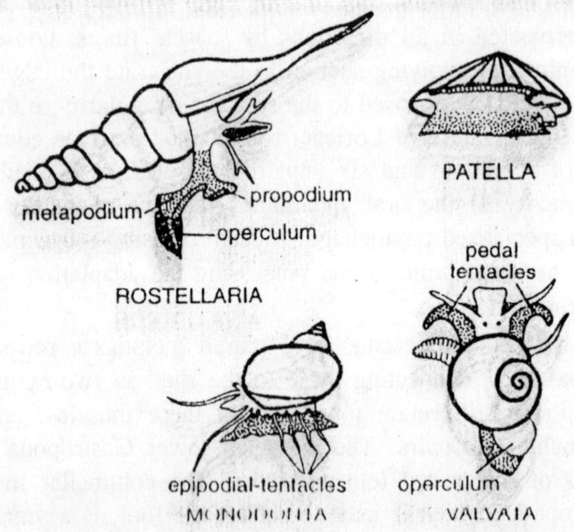

Fig. 8.4. Various types of foot—Rostellaria, Patella, Monodonta and Valvata.

small forms with some participation by the cilia of the ciliated epithelium that always covers the sole. Fissurellidae and Docoglossa practically never emerge at all from the shell, and the range of their possible movements is very limited.

In most Prosobranchia, which have a spiral shell, the columellar muscle is attached by its upper end to the columella, and the lower end partly expands fanwise in the foot and is partly attached to the operculum. Besides helping in creeping it plays the role of a retractor, withdrawing the foot and the head into the shell in time of danger and covering its opening with the operculum. In Pulmonata separate retractors also run from the columellar muscle to the cephalic tentacles and the pharynx. The antagonist of that system of retractors is the cutaneous musculature, especially the annular muscles, whose contraction causes protrusion of the head and trunk.

Consequently the locomotor apparatus of higher Gastropoda enables them to make much more varied and more rapid movements, and, above all, to protrude farther from the shell. When the shell is reduced the columellar muscle may disappear (e.g. in shell-less slugs among Pulmonata and in some Opisthobranchia), or only the ventral parts of its fibres may persist in the form of actual dorso-ventral fibres in the foot.

Among Pectinibranchia extreme development of the foot musculature, and often division of the foot by transverse grooves into two or three parts (propodium, mesopodium, and metapodium) or by a longitudinal groove into two symmetrical halves, leads to the appearance of new forms of creeping, using a walking movement (*Pomatias* among Taenioglossa), or to pulling-up on the protruded propodium, to a peculiar jumping-walking movement, and finally to the swimming of the purely-planktonic Heteropoda (Taenioglossa); in the most specialised Heteropoda (*Carinaria*, *Pterotrachea*) the propodium is converted into a separate fin, flattened in the sagittal plane, and the metapodium forms the posterior part of the fusiform body of the animal; the columellar muscle acts as a swimming muscle. In *Firoloida* the metapodium is reduced, and the whole foot becomes merely a fin.

In the subclass Opisthobranchia the transition to swimming is one of the chief trends in evolution. Many benthic forms among both Tectibranchia and Nudibranchia are able to swim; some purely-planktonic groups have arisen from both of them. Tectibranchia swim by means of paired lamellar expansions of the edges of the sole of the foot, the so-called parapodia. It is strange to see how the large heavy

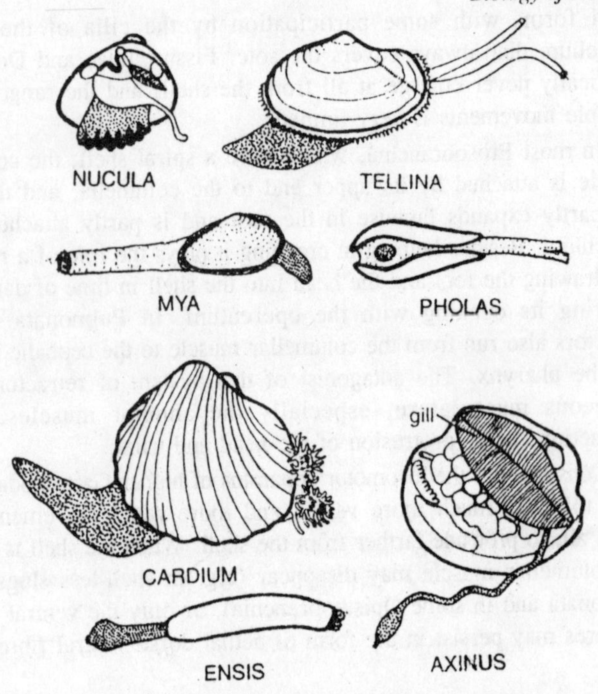

Fig. 8.5. Modification of foot in the class Pelecypoda.

Aplysia, crawling slowly along the floor of an aquarium, suddenly leaves the floor and begins to glide easily through the water by means of undulatory movements of its fins. *Acera bullata*, on transition to swimming, bends its parapodia to the dorsal side, folds them funnelwise, and then uses the umbrella so formed almost as a medusa does, making up to 45 strokes per minute and moving head-foremost.

In *Notarchus* the parapodial lobes grow together above the back, forming a parapodial cavity open to the front; contraction of the parapodial musculature forces the water out and the animal swims with its rear end foremost. In Pteropoda the central (crawling) part of the foot is vestigial and the parapodium forms a pair of wing-like appendages used for swimming. The columellar muscle of Pteropoda Thecosomata is attached to the shell at one end, and at the other end bifurcates and enters the two lobes of the parapodium, forming most of its musculaiture.

Pteropoda are not a natural group: the predatory Gymnosomata are related to Bullomorpha, and the microphagous Thecosomata to Aplysiomorpha. Nudibranchia have no parapodia. If their benthic forms do swim, it is by rapid lateral flexures of the whole body (*Tethys leporina*, *Dendronotus arborescens*). The same method is used by the purely-planktonic *Phyllirhoe*, which has a laterally-flattened fish-like body, with the foot entirely eliminated. Possession of complexly-differentiated musculature of the oral organs is very characteristic of Gastropoda, with also complex musculature of the copulatory apparatus.

The motor apparatus of Bivalvia has two main sections, one associated with the foot and the other with the mantle and the shell. As already mentioned, their foot has four or two pairs of retractors, attached to the shell and homologous with the foot-retractors of Tryblidiida and Loricata. There are also many muscle fibres in the foot, running in different directions. In the digging foot of *Ensis* they form a true musculo-cutaneous sac, resting on the central blood sinus. Before the foot is protruded the heart pumps blood into its sinus, and when the foot is withdrawn the blood is forced out of the sinus into the renal lacunae and the volume of the foot is greatly decreased. Such changes in the volume of organs by means of blood transfer are widespread among molluscs.

In most Gastropoda the edge of the mantle is simple and secretes the external layer of the shell; in *Turritella* a row of tentacles runs along it, preventing mud from entering the mantle cavity. In Bivalvia and in *Neopilina* the edge of the mantle is divided by two grooves running along it into three ridges: the outermost, adjoining the shell, is secretory—its inner surface secretes the periostracum, and its outer surface secretes the prismatic layer of the shell; the central ridge is sensory in Bivalvia, and all the sense organs of the edge of the mantle are formed from it; the innermost ridge (velum) is muscular.

Often the muscles permeating are attached to the shell and are retractors of the velum. The velum is used to close the mantle cavity and to regulate the flow of water by all forms in which the edges of the mantle do not grow together. In a typical case the two halve of the shell are connected by two strong adductor muscles, one anterior and one posterior, which close the shell. The shell opens passively by means of the elasticity of the ligament, which is compressed between the two halves of the hinge when the shell is closed. There is a widespread tendency within the class towards reduction of the anterior adductor and development of unequal-muscled and one-muscled (monomyarian) forms. The posterior adductor, when it alone remains, usually migrates to the centre of the shell, as in *Pecten* or *Ostrea*. The processes of reduction of the anterior adductor, which have developed independently in different groups of Bivalvia, have been brilliantly analysed by C. M. Yonge (1954). In primitive cases the two adductors lie at the ends of the ligament, at the point where the lines of attachment of the pallial muscles of the right and left sides cross each other. The adductors arose there through strengthening of the pallial muscles and fusion of the fibres of the right and left sides, which came into contact through the folding of the mantle along the hinge-line.

In very many Bivalvia the edges of the mantle fuse together over a varying distance, in an extreme case leaving only three orifices leading into the mantle cavity: inhalant and exhalant siphons, and an opening for the foot. In some cases the pallial muscles on the two sides fuse together, as in the formation of adductors, and thus several secondary muscles arise, e.g. the cruciform muscle of Tellinacea, which lies in front of the inhalant siphon. The retractor of the siphons is also formed from the pallial muscles. Other organs of Bivalvia also have muscles, e.g. the gills; the latter are highly developed when the movement of the gills has a hydrokinetic effect, e.g. in Nuculanidae Septibranchia.

In Cephalopoda the foot is derived from several pairs of rudiments; in Dibranchiata four or five pairs move forward and form arms arranged around the mouth, while one pair remains in the rear. The two halves of the latter grow together and form the infundibulum. In *Nautilus*, however, the infundibulum remains in the form of two separate halves. Morphologically the arms of Cephalopoda are homologous with marginal outgrowths of the sole of the foot, but A. Naef (1921-23), following A. Willey, regards the infundibulum as the homologue of the epipodial folds of lower Gastropoda. That interpretation, however, is not the only one. The shell served Tetrabranchiata, and still serves their most recent representative *Nautilus*, as a dwelling.

In Decapoda it is overgrown by the mantle and converted into a skeletal plate lying on the physiologically-dorsal side of the animal. In some Decapoda it has become vestigial, and in Octopoda it has completely disappeared. Besides the shell, or in place of it, an internal skeleton has developed in Cephalopoda, formed of cartilaginous tissue resembling that of vertebrates. The internal skeleton contains the following parts: (*i*) cephalic cartilage, surrounding the brain and the statocysts, and resembling the cartilaginous skull of vertebrates; (*ii*) orbital cartilages, protecting the eyes; (*iii*) occipital cartilage, lying at the morphologically-anterior edge of the mantle; (*iv*) cartilages lying at the base of the arms; (*v*) the cartilages of the fins; (*vi*) the cartilages behind the infundibulum, etc. All these skeletal parts serve as places of attachment for muscles.

Foot in Mollusca

The ventral side of the body, in Mollusca, is characterised by the pronounced development of its musculature, which enables the animal to creep. A fleshy foot, provided with a flat sole, is suited for such a locomotion. The foot is quite distinct from rest of the body, especially from the head. The flat form of the foot with a sole for creeping must be considered the primitive form, such a foot may be found in Chitons among the class-Amphineura, but it is variously modified among various classes of Phylum Mollusca to perform diverse types of functions i.e. swimming, attachment, digging and so on. The foot and its appendages get their nerve-supply from the pedal ganglia or the pedal nerve cord. The foot may be distinguished into various parts as follows:

(i) *Anterior Propodium.* It is the anterior distinctly marked portion from rest of foot.

(ii) *Metapodium.* It is the posterior part of foot and carries operculum.

(iii) *Mesopodium.* It is the central portion between the first two.

Apart from this sometimes there occurs expanded lobe-like margins of the ventral sides of foot called as *parapodium*, which can be folded and united over the head and sometimes there are paired, prominent ridges arising from the side of the base of foot and called *epipodium*, which are provided with sensory tentacles and eye-sports. In caphalopods the foot becomes modified into the tentacles, suckess and siphons. Variations of foot in different groups of molluscs is as follow:

Foot in Monoplacophora

In *Neoplina* the ventral surface bears a prominent foot in the centre. The foot shows a primitive condition. It is broad, circular, with a flat sole on which the animal creeps by muscular waves.

Foot in Amphineura

Class Amphineura is divided in two subclasses. Placophora and Aplacophora.

(i) *Placophora.* It possesses most primitive of foot which is large muscular organ with a flat, creeping surface extending along the whole length of body on its ventral surface. The foot is not differentiated into different parts. The example is *Chiton*. Besides performing locomotary function of creeping, it also serves as an organ of adhesion to the surface of rocks, when at rest.

(ii) *Aplacophora.* The foot is vestigial and is represented by a aciliary ridge running along the base of narrow, median, ventral groove, extending between the mouth and cloacal aperture. In some forms like *Chaetoderma* both, the foot and ventral groove are absent.

Foot in Scaphopoda

Foot is almost cylindrical or conical as in *Dentalium* and projects downward into the mantle cavity through the lower aperture of which it can be protruded out to serve as the digging organ. Engorged with blood, it serves as an anchor. The free end of the foot is conical and trilobed carrying on each lateral side a wing-like fold or *pleat* which is comparable to the *epipodium*. A groove runs along the middle line of foot. In *Siphonodentalium* both, the groove and lateral folds are absent ad the anterior end of foot bears a contractile disc with papillated margins, while in *Pulsellum*, a filiform tenacle springs from the middle of the disc.

Foot in Gastropoda

The foot of gastropods contains scattered unicellular as well as aggregated multicellular glands which secrete mucous. These glands are called *pedal glands*. The pedal glands include: (i) *Labial gland* in a groove at the anterior end, (ii) *Suprapedal gland* between head and anterior end of foot (iii) *Sole gland* opening mid-ventrally in the middle of foot and (iv) *posterior glands* at the hind end of foot.

Foot in subclass Prosobranchia

The foot in different prosobranchiates may becomes variously modified in relation to different conditions of life. In several pectiibranchiates the anterior portion of foot is only well differentiated into *propodium* and sometimes *metapodium* equally developed, e.g., *Natica* and *Xenophorus*. In these forms when the animal is in motion it first advances its propodium and then folds the rest of the foot after it. In some forms propodium also forms a siphon on the left side which helps to conduct the water into mantle cavity. The operculum is always carried by the metapodium.

Epipodium is usually absent among the pectinibranchiates but vestiges of it can be seen in the form of a pair of anterior pedal tentacles in *Vermitus*, in the form of two nuchal lobes in *Ampullaria*. In aspidobranchians a well developed epipodium is always present. A good example is met in the case of *Haliotis* in which epipodium surrounds the base of foot in the form of ridges with digited appendages and long sensory contractile tentacular processes. These may be provided with sensory organs. Broadly speaking, most of he prosobranchiates creep or attach by means of flat sole but in several pelagic pectinibranchiates foot has become especially modified in accordance with swimming mode of life and propodium is changed into a narrow vertical fin and the development of fin can be traced almost step by step starting from *Oxygyus* proceeding through *Atlanta* and *Pterotrachea.*

(a) *Oxygyus*. Foot consist of propodium which anteriorly possesses a fin-like out growth and a distinct metapodium directed backward like a tail and bearing operculum. The sole is mainly used as sucker.

(b) *Atlanta*. The arrangement of foot is similar to that of *Oxygyus*, but the fin-like out growth of propodium has become its most important part. The comparatively reduced sole or sucker appearing only an appendage to it.

(c) *Carinaria*. Both, the foot and general external appearance of entire animal, are much changed. The metapodium which here has no operculum appears as a posterior prolongation of body. The fin is much broader and larger and sucker is present just along 1/3 of its free edge.

(d) *Pterotrachea*. The sucker pesists only in male, while the metapodium loses the operculum and becomes reduced to a short, filament like tail.

Foot in Subclass Opisthobranchiata

While there is no division of foot into various parts and among the epipodia are lacking but parapodia are highly developed. These parapodia are lateral lobes of fold-like extensions of the edges of the sole of foot and they are often bent back over the shell so that their edges sometimes touch each other. Thus the shell is completely covered by them. These parapodia may even fuse along their upturn edges at the posterior end. Such a fusion is remarkable in *Aplysia* and *Notarchus*. Among plegic opisthobranches the parapodia develop into paired fins and foot may become confined the anterior end of the body and two lateral fins are used in locomotion. In nudibranches the foot is no more exist as a differentiated organ but the bcdy of animal is compressed in the shape of a leaf with sharp dorsal and ventral edges to enable the animal to move forward e.g. *Eolis*.

Foot in Subclass Pulmonata

The foot is always undivided and provided with a large, flat sole for creeping. In few forms, however, it is divided into two portions by a temporary or permanant groove.

Foot in Pelecypoda

Foot, like that of majority of mollusc, is a locomotory organ present as a muscular projection from the ventral surface. Usually in this class the foot is painted axe-shaped with sharp edge directed downward and outward. In protobranchiate for examples *Nucula Yoldia* the foot has a flat, creeping surface and is considered the most primitive foot in the class. The foot of lamellibranchiates varies according to the mode of locomotion and according to development of *byssus*. It is secreted by the byssus gland and consists of large number of threads vary in thickness. With the help of these fibres the animal anchors to the foreign bodies. The complete byssus apparatus consists of:

(a) A cavity in the foot into which the byssus gland opens.
(b) A duct connecting the cavity to exterior.
(c) A groove which runs from the opening of the duct along the ventral edge of foot to its anterior end.
(d) A crescent shaped or cup-like widening of the groove at its anterior end where the threads are present.

The relation existing between the development of foot and that of byssus apparatus may be given as follow:

(i) The foot is in its primitive form with a flat sole and no groove and with a simple invagination but without byssus e.g. *Solemya*.
(ii) With the same foot small lamella arises from the base of the simple invagination and byssus is very slightly developed e.g. *Nucula*.
(iii) Invagination becomes differentiated into a cavity and the byssus and its glands are highly developed e.g. *Mytilus*. In *Arca* and *Mytilus* the old byssus may be shed and a new one Formed.
(iv) Many lamellibranches in the adult stage have neither byssus nor byssus gland but the cavity and the duct and even the retractor may be retained. Byssus apparatus may be found into close related forms without the byssus itself and sometimes with it. In the former case the foot is generally more developed and serves for locomotion e.g. *Trigonia*.
(v) The wedge-like or blade-like foot often becomes more strongly developed as a fleshy and extensile organ every trace of byssus and its apparatus are absent e.g. *Unio*, *Cardium*, *Tellina* etc. The anterior end of the foot forms a long slendrical tentacle in *Poromya*, with a swollen free extremity in *Axinus*.

Foot in Cephalopoda

The foot in cephalopoda is represented by arms and siphons. The arms are regarded as the lateral processes of the molluscan foot which have reaches beyond the head and united infront and the head is encircled by the foot and the mouth comes to lie in the centre of this. This circle of arms is derivative of the foot which is proved by its anatomical and ontogenie evidences which are as follows:

(i) The arms or tentacles are innervated by the branchial ganglion which is anterior origin of pedal ganglion and it lies under the oesophagus.

(ii) The arms do not occupied in the embryo their definitive possession from the mouth but arise on the ventral surface behind the mouth and between the mouth and anus in a row found each side. These two rows shifts to form a circle of arms in the adult condition.

Foot in subclass Dibranchiata

In Dibranchiata 8 or 10 such arms are present provided with two rows of suckess. In *sepia* more than two rows on ventral side. These suckers are stalked in Decapoda and sessile in Octopoda. The octopoda has eight arms connected by means of membrane at their bases. In decapods these are ten arms, eight of which correspond to the eight armed octopods but they are not connected by the membrane at their bases. The remaining two arms are longer with swollen ends and their suckers being confined to their free ends. Each of these two arms is situated between 3rd and 4th Octopodus arm on either side. They lie retracted into special cavities of head and thrown out of their cavities with great force, while catching the food. One, rarely both, of arms is modified to assist in copulation and called as hectocotylized arm.

Foot in subclass Tetrabranchiata

The head in *Nautilus* bears several tentacles arranged in circle towards the mouth about 90 in females and 60 in males. These tentacles are present on special lobes which may be taken to correspond to the arms of dibranchiates while tentacles to the suckers. These tentacles are adhesive and prehensile and are retractile within special tentacular sheaths. In the female there are three tentacular lobes in immediate contact with buccal aperture, these are right, left and ventral-anterior lobes. In male the ventral-anterior lobe is absent instead it bears *Owen's organ*, which is supposed to be olfactory in nature. In addition to foot proper the cephalopods possess epipodia which are highly modified to form the funnel or siphon. In *Nautilus* the funnel consists of two symmetrical lateral lobes which are enclosed towards each other and overlap sometimes. But in dibranchiate these lobes fuse with each other to form a complete tube projecting beyond mantle cavity. Usually the anterior lobe of funnel is provided with a velum e.g. *Nautilus* and several decapods but absent in octopods.

9

DIGESTIVE SYSTEM

The alimentary canal is well developed in all Molluscs, and is composed of (1) the buccal cavity; (2) the pharynx or oesophageal bulb; (3) the oesophagus or fore-gut; (4) the mid-gut with the stomach; (5) the rectum or hind-gut with the anal aperture. The mouth originally lies at the anterior, and the anus at the posterior end or side of the body, the latter in the mantle furrow or cavity. The former always retains its original position, but the latter, as central organ in the pallial complex, becomes shifted more or less far forward along the right (less frequently the left) side, in the mantle furrow. When the visceral dome grows out dorsally in such a way that the longitudinal axis becomes shorter than the dorso-ventral axis, as is the case in many *Gastropods* and *Cephalopods* and in *Dentalium*, the mid-gut at least, with its accessory gland, the so-called liver, runs up into this dome, filling the greater part of it. The intestine then forms a dorsal loop, consisting of an ascending portion running up from the fore-gut and a descending portion running down to the anus.

In the *Gastropoda*, where the anus is shifted more or less forward, the descending portion bends forward to the right (rarely to the left) to reach it. Besides this principal visceral loop, which is caused by the development of the visceral dome and modified by the displacement of the pallial complex, the intestine, in nearly all Molluscs, forms secondary loops or coils which add to its length. These loops are found principally in the tubular portion of the mid-gut which follows the stomach. They are as a rule most pronounced in herbivorous animals, which thus have longer alimentary canals than carnivorous forms. The large digestive gland, usually called the liver, enters the

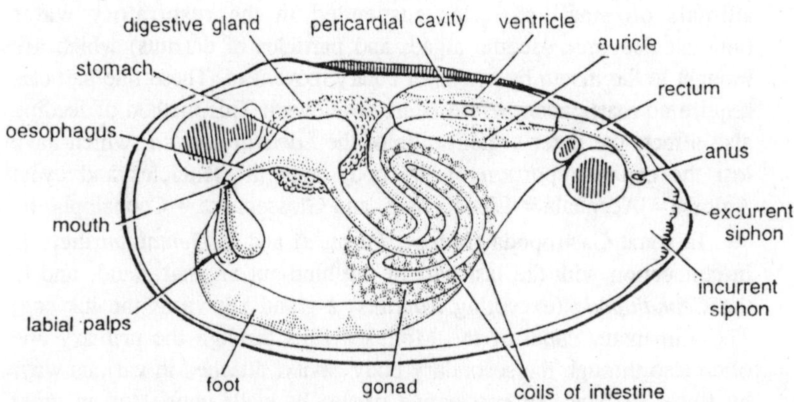

Fig. 9.1. Unio Alimentary canal and digestive gland.

stomachal division of the mid-gut. Functionally, this organ only very slightly corresponds with the vertebrate liver, if indeed it may be said to correspond at all with that organ. It agrees more nearly with the pancreas, and perhaps combines the functions of the different specialised digestive glands of Vertebrates.

There is a radical difference between *Lamellibranchs* and other Molluscs, in the fact that in the latter the anterior portion of the fore-gut which follows the buccal cavity is developed as a muscular pharynx (oesophageal bulb, buccal mass), and carries at its base on a movable lingual cushion a file-like organ, the radula, which is beset with numerous hard teeth composed of conchyolin or chitin. The radula serves chiefly for mastication, but it sometimes used in seizing, holding, and swallowing prey. None of the *Lamellibranchs* have a pharynx provided with a radula, they are therefore called *Aglossa* as opposed to all other *Molluscs*, which are *Glossophora*. Hard jaws, composed of conchyolin, are almost always found in varying number and arrangement in the buccal cavity of the *Glossophora*, but are wanting in all *Lamellibranchs*.

One or two pairs of glands open into the pharynx in the *Glossophora*; these are usually called salivary or buccal glands, although they very slightly if at all correspond physiologically with the glands so named in the Vertebrata. Glands may also open into the buccal cavity. The *Lamellibranchs* have no salivary glands. The absence of a pharynx, tongue, jaws and salivary glands in the *Lamellibranchia* is accounted for by their manner of life. They do not have to seek their food. Some of them are attached and others feed in the same way as attached animals on small particles suspended in the respiratory water (animalculae, mic:oscopic algae, and particles of detritus) which are brought to the mouth by means of ciliary movement. These fine particles require no mastication before being swallowed. This method of feeding also affects the outer organisation of the *Lamellibranchia*, which have lost the cephalic portion of the body with the tentacles and eyes: Aglossa= Acephala= Lipocephala, and Glossophora= Cephalophora.

In some Gastropoda (*Murex*, *Purpura*) and in *Dentalium* there is in connection with the last part of the hind-gut an anal gland, and in the *Cephalopoda* (excepting *Nautilus*) a gland known as the ink-bag. The alimentary canal of the Mollusca runs through the primary and often also through the secondary body cavity, attached in various ways by fibres or bands of connective tissue. Its walls consist of an inner epithelium usually to a great extent ciliated, an outer muscular layer in which longitudinal and circular fibres occur, not always in regular layers, and, where it passes through the primary body cavity, an outer envelope of connective tissue. The pharynx and perhaps sometimes part of the oesophagus, and a part, in all cases very short, of the hind-gut, arise ontogenetically out of the ectodermal stomodaeum and proctodaeum. But the exact limits of the ectodermal and the endodermal portions of the intestine are difficult to determine.

Buccal Cavity, Snout, Proboscis

The alimentary canal has an oral aperture bordered by variously shaped lips, and in many *Glossophora* (in nearly all *Gastropoda*) leads into a vestibule or anterior cavity roofed over by the lips and lined by a continuation of the outer wall of the head. The dermal glands are not unfrequently (many *Opisthobranchia* and a few *Prosobranchia*) more strongly developed on the lips as labial glands. In many *Gastropods*, when the lips open, the mouth is able to seize and hold prey like a sucker. Where the snout is short it is simply contractile (the *Chitonidae*, the *Diotocardia*, most herbivorous *Taenioglossa*, and many *Pulmonata* and *Nudibranchia*). In this case the parts immediately surrounding the mouth are so strongly contractile that when contraction takes place the mouth is drown in somewhat so as to lie at the base of a depression.

An exaggeration of this arrangement, combined with the prolongation of the snout, leads to the formation of the retractile or proboscidal snout. The snout can in such cases be invaginated from its tip, *i.e.* from the oral aperture into the cephalic cavity, the mouth then lying at the base of the invagination (many *Tectibranchia*, *Capulidae*, *Strombidae*, *Chenopidae*, *Calyptraeidae*, *Cyraeidae*,

Lamellariidae, *Naticidae*, *Scalaridae*, *Solariidae*). Finally, in many carnivorous *Prosobranchia* (*Tritoniidae*, *Doliidae*, *Cassididae*, *Rachiglossa*, and a few *Toxaglossa*) a proboscis, often very long and enclosed in a special proboscidal sheath, is developed; this sheath lies in the cavity of the head, which is often prolonged like a snout, and may even stretch back into the body cavity.

The oral aperture lies at the free anterior end of the cylindrical proboscis, and we have to regard the proboscis with its sheath as a very long snout, the base of which, however, is permanently invaginated into itself. In this way the proximal portion of the snout forms the permanent proboscidal sheath, while the distal portions with its terminal oral aperture forms the proboscis. Neither of these portions can be invaginated or evaginated; it is merely a zone lying between them which takes part in the retraction of the proboscis into the body cavity. This zone, when so invaginated, forms a temporary backward prolongation of the proboscidal sheath but when the proboscis in protruded forms the basal portion of the latter. The permanent portion of the proboscidal sheath is connected with the wall of the head by bands which make its evagination impossible, and the inner-wall of the permanent proboscis is connected by muscles or bands with the oesophagus lying within it, so that this portion of the organ cannot be invaginated; the oral aperture can thus never lie at the base of the proboscidal sheath. When the proboscis is retracted, there is therefore an aperture at the anterior end of the snout or the head, which is not the oral aperture, but that of the proboscidal sheath.

When the proboscis is protruded, it projects beyond the aperture of the sheath and carries at its point the oral aperture. The proboscis is retracted by means of muscles attached at the one end to the body wall and at the other to its (invaginable) base. In its protrusion, a flow of blood towards the snout probably plays the chief part, accompanied by contraction of the circular muscles of the head and proboscis. The (carnivorous) *Pteropoda gyanosomata* also have a protrusible proboscis provided with so-called buccal appendages. The same is present in the allied *Aplysiidae*, but is weakly developed. The *Thecosomata* have no proboscis.

The buccal cavity of *Dentalium* is noteworthy. It extends throughout the whole length of the freely projecting egg-shaped snout, which carries leaf-like labial appendages. On each side of the buccal cavity there is a pouch, the so-called cheek pouch, which is lined with glandular epithelium and opens into the cavity anteriorly. An exact comparative investigation of the mechanism of the proboscidal apparatus the contractile snout, etc. of the Prosobranchia is still a desideratum. There are other forms of proboscis, differing greatly from the one just described (*e.g.* that in the *Terebridae*). In the *Heteropoda*, the head forms a long snout which is often described as a proboscis. The name is inappropriate, as this snout is not retractile and the mouth is always found at its anterior end.

PHARYNX AND JAWS, THE TONGUE AND SALIVARY GLANDS

The mouth or buccal cavity is followed in all Molluscs except the *Lamellibranchia* by the pharynx or oesophageal bulb (buccal mass). The pharyngeal cavity opens anteriorly into the buccal cavity, and posteriorly into the oesophagus. The pharynx is characterised by the possession of (1) jaws, which lie anteriorly at the boundary between the buccal and pharyngeal cavities; (2) a lingual apparatus at its base, and (3) salivary glands, which usually open laterally near its posterior boundary.

Jaws

Jaws are almost universal, and are sometimes, especially in carnivorous animals, very highly developed; less frequently they are rudimentary or wanting. They are hard cuticular formations of the epithelium of the anterior pharyngeal region, and no doubt composed of conchyolin or some related substance, in a few cases hardened by calcareous deposits (e.g. *Nautilus*). The jaws serve for seizing prey or particles of food. The great variations in number, form and arrangement of the jaws can best understood by assuming that they originally extended completely round the entrance to the pharynx; and that of this ring

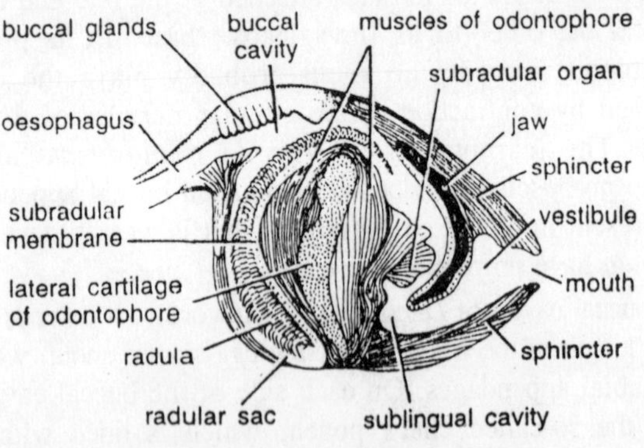

Fig. 9.2. Pila. Buccal mass in V.L.S.

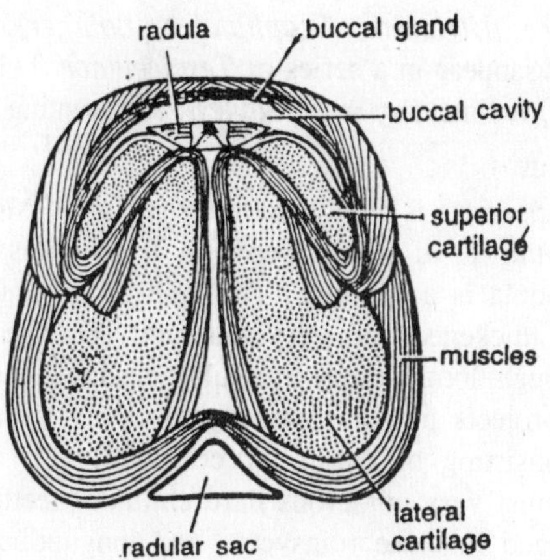

Fig. 9.3. Pila. Buccal mass in T.S.

sometimes only upper and lower or sometimes only lateral portions have been retained. Such a complete circle of jaws is found at the entrance to the pharynx in some forms, such as *Umbrella* and *Tylodina* (*Opisthobranchia*).

The fresh-water *Pulmonates* have an upper and two lateral jaws. Most *Prosobranchia* and *Opisthobranchia* have two lateral jaws. These may approach so near one another as almost to touch (*Haliotis*, *Fissurella*). Terrestrial *Pulmonata* have an upper jaw, and occasionally a weak lower jaw as well. The jaws are particularly strongly developed in the *Cephalopoda*, which have an upper and a lower jaw, the two together resembling in shape the beak of a parrot. In the *Opisthobranchiate* family *Aplysiidae*, *Notarchus*, *Acera*, *Dolabella*, and *Aplysiella* have, besides the lateral jaws, numerous hooks or small teeth on the roof of the pharyngeal cavity.

The hook sacs of the *Pteropoda gymnosomata*, which are wanting only in *Halopsyche*; are perhaps to be derived from these pharyngeal teeth. The hook sacs are paired dorsal outgrowths of the pharyngeal cavity, which vary in length and lie in front of the radula. The walls of the sacs carry hooks projecting inward. When the proboscis of these carnivorous animals is protruded, the sacs are completely evaginated, so that the hooks come to lie outside. Jaws are wanting or rudimentary in the *Amphineura* and the *Scaphopoda*; among the Prosobranchia, in the *Toxoglossa*, *Pyramidellidoe*, *Eulimidoe*, many *Trochidoe*, the *Heteropoda*, and in many *Nudibranchia* (*Tethys*, *Melibe*, *Doridopsis*, *Phyllidia*); in the *Ascoglossa*, and in certain *Tectibranchia* (*Actoeon*,

Doridium, *Philire*, *Utriculus*, *Scaphander*, *Lobiger*). Among the *Pulmonata* they disappear in a series of *Testacellidoe*, being present in *Dandebardia rufa*, rudimentary in *D. Sauleyi*, and wanting in *Testacella*.

Lingual Apparatus

The lingual apparatus is highly characteristic of all Molluscs except the *Lamellibranchia* i.e. of all *Glossophora*. It may be said that every animal with a radula is a Mollusc. The ventral and lateral walls of the pharynx are thickened and very muscular. On the floor of the cavity rises a tough longitudinal muscular cushion, the tongue. Its surface, which projects into the pharyngeal cavity, is covered by a rough cuticle consisting of chitin (or conchyolin ?); on this basal membrane are found very numerous hard chitinous teeth, often many thousands, arranged in close transverse and longitudinal rows. The basal membrane and the teeth together form the radula of the tongue. The anterior end of the tongue projects freely into the pharyngeal

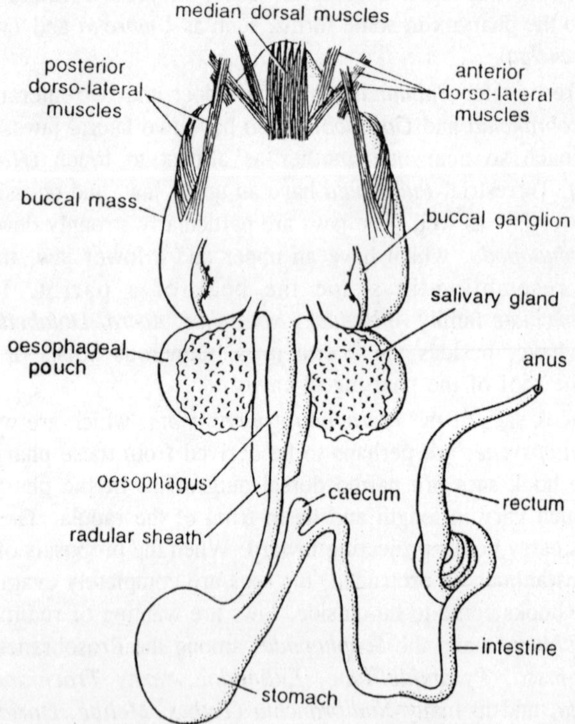

Fig. 9.4. Pila. Alimentary canal (Radular sac shifted backwards below the oesophagus).

cavity, the radula bending down over this end so as to cover for a certain distance its lower surface. Immediately in front of the tongue there is always a depression in the ventral pharyngeal wall, forming a sort of pocket.

The radula, at the posterior extremity of the tongue, sinks into a narrow more or less long tube, the radular sheath, which is an outgrowth of the pharyngeal cavity running downward and backward. The radula, always lying upon the anterior or ventral wall of this sheath, which is anteriorly thickened to form the tongue, extends to the base of the sheath, which is the place of its formation. The tongue with the radula on it is movable, and in most cases its movements can be compared with those of the cat's tongue when licking but are usually slower. This action helps to rasp the food which has been seized, and often also broken up, by the jaws.

The tongue can either move inside the pharyngeal or buccal cavities, or can be extended to the oral aperture or even protruded more or less far beyond it. In or under the fleshy tongue, a lingual cartilage is very commonly found, consisting of two or four or even more pieces. This cartilage forms a support for the radula, and affords firm points of attachment for certain muscles belonging to the lingual apparatus. The musculature of the pharynx, which can be separated into bundles or strands, and is often very complicated, consists first of the muscles which form the wall of the pharynx, and which, being principally developed ventrally and laterally round the radula, determine the special licking movements of the tongue; secondly, of muscles which move the whole pharynx or the whole of the lingual apparatus, evaginating or protruding them.

The second group consists, speaking generally, or protractors and retractors, attached at the one end to the pharynx and at the other to the body wall after running through the cephalie or body cavity. Pressure of blood may also take some part in the protrusion of the pharynx. The tongue and radula further often serve for seizing prey (*e.g.* in the carnivorous *Heteropoda*, in *Testacella*, etc.) The radula is of great importance in classification. Further details concerning it must be sought in special works and in text books of conchology. The points to be specially noticed are (1) the size and form of the whole radula, (2) the number of longitudinal and transverse rows of teeth, and (3) the form of the teeth in each of these rows.

As a rule the transverse rows resemble one another, but exceptional rows differently constituted from those immediately preceding or following them recur at intervals. Three kinds of teeth have been, as a rule, distinguished. First, there is usually a single median longitudinal row of central or rachial teeth. One each side of this row are several rows of more or less similar lateral teeth or pleurae. Finally, at the lateral edges of the radula, there are single or very numerous longitudinal rows of marginal teeth or uncini. Dental formula are used for the radular teeth, in the same way as for the teeth of mammals; in these the number of central, lateral and marginal teeth in a transverse row are given. The reader will find the dental formulae of some of the Molluscs in the Systematic Review. The total number of radular teeth varies very greatly, from 16 in *Eolis Drummondi* to 39,596 in *Helix Ghiesbreghti*. As a rule, the teeth are most numerous and finest in herbivorous animals.

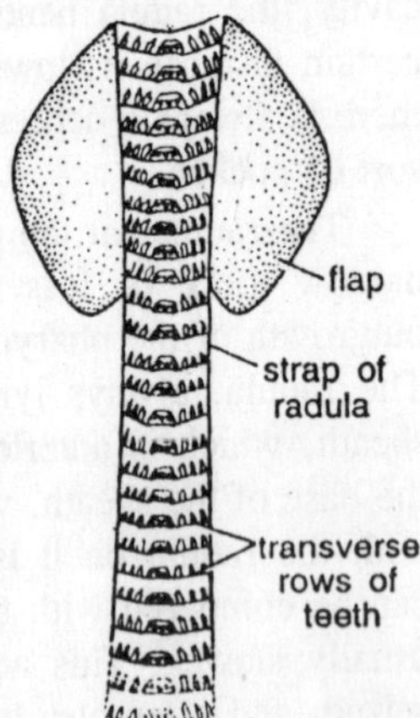

Fig. 9.5. Pila. The radula.

In carnivorous Molluscs we have two extremes: (1) great development of the proboscis, with weak development of the pharynx and radula, and a comparatively small number of teeth (carnivorous *Prosobranchia*); (2) absence of a protrusible proboscis, with great development of the pharyngeal apparatus and the radula, and numerous often large, teeth (*Heteropoda*, carnivorous *Pulmonata* and *Cephalopoda*). The muscular pharynx is most developed in carnivorous *Pulmonates*. In these it may be half (*Daudebardia*) or even more than half as long (*Testacella*) as the whole body, and may occupy a very large part of the body cavity. It is protruded in such a way that the tongue with the radula occupy in anterior end of the evaginated pharynx. In very rare cases (apart from the *Lamellibranchia*) the radula completely atrophies; this is the case in parasitic *Gastropoda* (*Stilifer*, *Eulima*, *Thyea*, *Entoconcha*), in the *Coralliophilidae* (*Coralliophila*, *Leptoconchus*, *Magilus*, *Rhizochilus*) among the *Nudibranchia* in *Tethys* and *Melibe*, among the *Amphineura* in *Neomenia*, and certain species of the genera

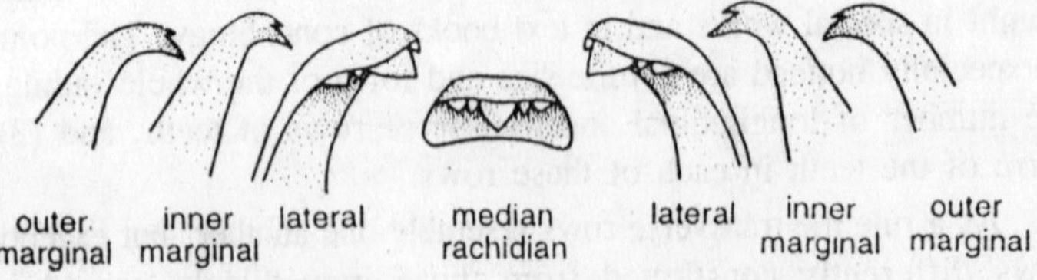

Fig. 9.6. Pila. A single row of radular teeth.

Dondersia and *Proncomenia*. In *Chaetoderma*, a single tooth of the radula is retained. Even in certain carnivorous *Prosobranchia* which are furnished with a proboscis, the above-mentioned reduction of the whole pharyngeal apparatus goes so far that the radula disappears (certain species of *Terebra*).

Formation of the radula

The teeth of the anterior transverse rows of the radula become worn out by use, and we continually being replaced by new teeth which are pushed forward. The formation of new transverse rows of teeth is constantly taking place at the posterior blind end of the radular sheath. In *Pulmonata* and *Opisthobranchia* they appear as cuticular formations secreted by several transverse rows of large epithelial cells—the odontoblasts; the basal membrane which carries the teeth is secreted by the anterior row or rows, the teeth themselves by the posterior rows. Each group of odontoblasts which has formed a tooth is not replaced by another, but continues to produce new teeth behind those already formed, so that for each longitudinal row of teeth there is at the base of the radular sheath a group of odontoblasts which has produced all the teeth belonging to that row.

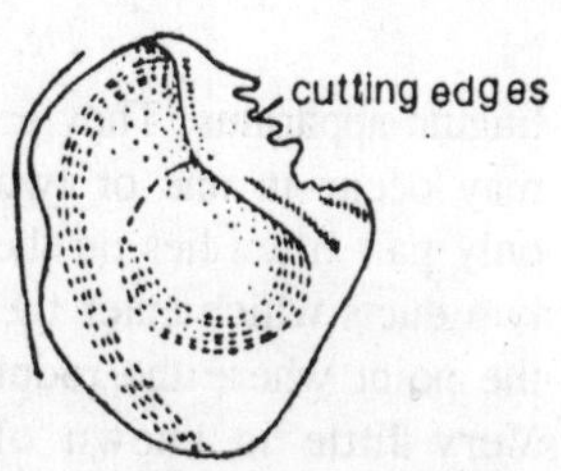

Fig. 9.7. Pila. A jaw.

A layer of "enamel" is deposited on the teeth so formed by the epithelial roof of the radular sheath. In the *Chitonidae*, *Prosobranchia*, and *Cephalopoda*, the odontoblasts are very numerous narrow cells, which form, at the base of the sheath, a cushion divided into as many parts as there are teeth in a transverse row of the radula. The radular sheath in the *Pulmonata*, *Scaphopoda*, *Opisthobranchia*, and *Cephalopoda* is short, and is contained in the ventral and posterior muscular wall of the pharynx, very seldom projecting posteriorly beyond it; but in many *Prosobranchia* it is long and narrow, and reaches back into the cephalic cavity or even right into the body cavity. This latter is especially the case in the *Diotocardia*; in the *Docoglossa* (*Patella*) the sheath, which lies above the foot on the floor of the body cavity, is even longer than the body.

Salivary Glands

Salivary glands (buccal glands, pharyngeal glands) are universally found in *Glossophora*, i.e., in Molluscs which have a pharynx and

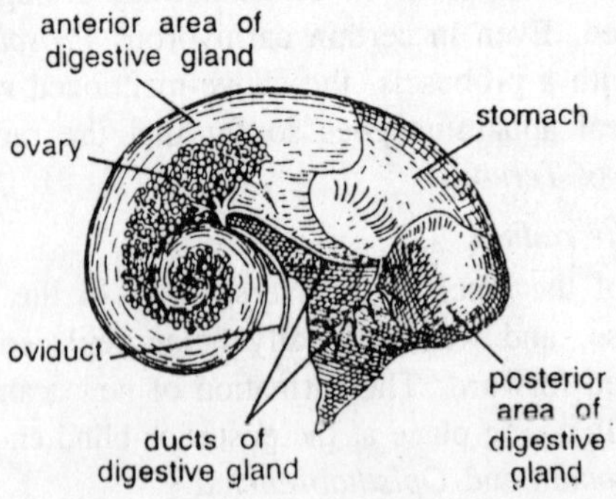

Fig. 9.8. Pila. Digestive gland.

lingual apparatus. They are universally absent in *Lamellibranchs*. They may occur in one or two pairs. The posterior or in other cases the only pair often lies on the wall of the oesophagus, and sends forward two ducts which enter the pharynx laterally, usually somewhat behind the point where the radular sheath opens into the pharyngeal cavity. Very little is known of the function of these glands; an exact morphological comparison of the various pharyngeal glands of the *Gastropoda* is at present hardly possible.

Amphineura

(*a*) *Chiton*. Two small delicate buccal glands lie on the roof of the buccal cavity and open into the mouth. They can therefore hardly be regarded as pharyngeal or salivary glands.

(*b*) *Solenogastres*. Salivary glands are here found in all genera except *Neomenia*, and in *Choetoderma*. They are present in some species but appear to be absent in others. A pair of long glandular tubes with high glandular cells and strong muscular walls lie anteriorly under the intestine and are produced in the form of two narrow ducts, which enter the pharyngeal cavity on the tongue either separately or through a common terminal portion. Besides these there is another pair in some species (*Paramenia impexa*, *Param. palifera*, *Proneomenia vagans*, *Dondersia flavens*); the ducts of these open together through an unpaired terminal portion on the dorsal wall of the pharyngeal cavity, at the point of a papilla which rises from the base of a pit-like depression.

Gastropoda

(*a*) *Prosobranchia*. In most cases there is only one pair of salivary glands. These are usually lobed or branched glandular masses, which lie, in the *Diotocardia*, at the sides of the pharynx, in the *Monotocardia*, at the sides of the oesophagus. In the former case, the ducts are short and do not pass through the oesophageal ring formed by the nerve centres and their connectives and commissures, which in these forms surrounds the anterior end of the pharynx. In the *Monotocardia*, the ducts are long, and generally accompany the oesophagus through the oesophageal ring (which lies behind the pharynx), and open on the posterior lateral wall of the latter. Two pairs of salivary glands are found in certain *Diotocardia* (e.g. *Haliotis*, *Fissurella*), and further in *Patella*, the *Scalariidoe*, *Ianthinidoe*, certain *Purpuridoe*, *Muricidoe*, and in the *Cancellariidoe*. One of the two pairs of glands in *Haliotis* is developed in the form of large lateral glandular sacs covering the pharynx on the right and left.

In the *Ampullariidoe*, the ducts of the salivary glands do not pass through the oesophageal ring, which here, as in the *Diotocardia*, surrounds the anterior end of the pharynx. Whereas the salivary glands are, as a rule, branched tubes or acinose, they are sometimes (*Scalariidoe*, *Ianthinidoe*, *Cancellariidoe*) simply tubular or (*Doliidoe*, *Xenophoridoe*, etc.) sac-like. The passage of the ducts of the salivary glands through the oesophageal ring in the *Monotocardia* may have come about by the shifting back of the ring along the pharynx from its former position in the *Diotocardia*, where it encircles the anterior end of the pharynx in front of the apertures of these ducts. The salivary would thus necessarily become surrounded by the ring.

The ducts in the *Monotocordia* become the longer the further the nerve ring shifts back from the mouth and pharynx. They are very long in animals provided with a protrusible proboscis, where the ring lies far back on the oesophagus, behind the non-evaginable portion of the proboscis. The ducts here run along the whole length of the latter. But in those cases in which the oesophageal ring has shifted back more quickly than the ducts have lengthened, the glands lie in front of the ring. In the event of the subsequent lengthening of their ducts, the glands might stretch back outside the ring. The arrangement of the glands in the *Toxoglossa* and *Rachiglossa* would thus be explained; here the greater part of the glands lies behind the ring, although the ducts are said not to pass through it. The acid secretion of the salivary glands of certain *Prosobranchia* (species of *Dolium*, *Cassis*, *Cassidaria*, *Tritonium*, *Murex*) and *Opisthobranchia* (*Pleurobranchus*, *Pleurobranchidium*) contains 2.18-4.25 per cent of free sulphuric acid. These carnivorous animals are able, by means of their proboscides, to bore into other Molluscs and *Echinoderms* which are protected by calcareous skeletons. The sulphuric acid in their glands probably serves for transmuting the carbonate of lime into sulphate of lime, which can then easily be worked through by the radula.

(*b*) *Pulmonata*. Two salivary glands are always found, their ducts entering the pharynx to the right and left of the boundary between it and the oesophagus. The glands lie on the oesophagus and the anterior part of the stomach in the shape of long, lobate, jagged leaves. In some cases they are acinose or round and compact.

(*c*) *Opisthobranchia*. The salivary glands, of which only one pair is almost always found, here vary in size and shape still more than in the *Pulmonata*. These glands, which enter the pharynx, must not be confounded with other glands which in many *Opisthobranchia* enter the buccal cavity, are sometimes more strongly developed than the salivary glands.

Dentalium

Dentalium has no salivary glands opening into the pharynx, for the glandular "cheek pouches" enter the buccal cavity, and two diverticula which lie further back belong to the oesophagus. The *Cephalopoda* have a posterior and an anterior pair of salivary glands. Were the fore-gut, which here rises vertically in the visceral dome, to occupy the horizontal position it has in the *Gastropoda*, the anterior pair would lie dorsally and the posterior ventrally with regard to it. The two posterior glands are always present (except in *Cirrhoteuthis* and *Loligopsis*, in which they are said to be wanting), and lie on the oesophagus. Each gland has a duct, which soon unites with that from the other gland, forming a terminal portion which accompanies the oesophagus through the cephalic cartilage, and opens above the radula into the pharyngeal cavity. The posterior glands occasionally (*e.g.* in *Oegopsidoe*) fuse behind the gullet, in which case the duct is single throughout its whole length.

The anterior salivary glands are specially well developed in the *Octopoda*, and lie on the pharynx, into which they empty their secretions by a duct, which seems always to be unpaired. In the *Decapoda* the anterior glands are much smaller or rudimentary; they are generally represented by a single gland hidden within the muscular wall of the pharynx. *Nautilus* has no posterior salivary glands, but there are glandular outgrowths of the pharyngeal cavity on each side of the tongue, which perhaps correspond with the anterior salivary glands of other Cephalopods. The Cephalopoda (without exception) have an additional acinose lingual gland, opening into that part of the pharyngeal cavity which lies between the tongue and jaws. The *Lamellibranchia*, as already mentioned, have neither pharynx, jaws, tongue, nor salivary glands. In the *Nuculidoe*, however, which are rightly considered to be primitive forms, the mouth leads into a widening of the intestine, on each side of which a glandular pouch opens. These pouches perhaps correspond with the oesophageal pouches of the *Chitonidoe* and *Rhipidoglossa*, which will be described later.

Natica, which bores through the shells of living *Lamellibranchs* and feeds on their bodies, has a sucker-like organ on its proboscis. The epithelium of the concave side of this organ, which is applied to the shell attacked, forms a gland for secreting acid—probably sulphuric acid—which serves for dissolving the carbonate of lime of the bivalve shell, which is then at once thrown out in the form of powdered sulphate of lime.

Oesophagus

That portion of intestine which lies between the pharynx (or the mouth in *Lamellibranchs*) and the stomach is called the oesophagus, the stomach being here used as the name of that widening of the intestine into which the gland of the mid-gut opens. It is always easy to detect the anterior boundary of the oesophagus. In *Lamellibranchs* it lies at the mouth, but in the *Glossophora* at the posterior and upper end of the pharynx. The posterior boundary, however, can often only arbitrarily be defined, as the oesophagus, which is usually narrow and tubular, often widens vary gradually into the stomach, the structure of its walls at the same time gradually changing.

In other cases, widenings of the alimentary canal occur before the stomach, and it is difficult to decide whether these are anterior divisions of the stomach or posterior widenings of the oesophagus. In *Lamellibranchia*, terrestrial *Pulmonata*, most *Opisthobranchia*, and the *Cephalopoda Decapoda* the oesophagus is a simple ciliated tube running to be stomach, being often provided with longitudinal folds and therefore extensible; in other divisions, however, complications occur, which are caused by glandular outgrowths or muscular enlargements. In a few *Solenogastres* (e.g. *Proneomenia*), on the boundary between the short oesophagus and the mid-gut, a more or less long blind diverticulum occur; this is single, and runs forward dorsally to the pharynx, and may extend over the cerebral ganglion to the end of the head.

The *Chiton* there are two lateral glandular sacs (sugar glands) connected with the short oesophagus; their inner glandular walls project into the lumen in the form of villi, and their secretion changes boiled starch into sugar. Similar glands, which communicate with the anterior part of the oesophagus, are found in the *Rhipiodglossa* (e.g. *Haliotis*, *Fissurella*, *Turbo*). The glandular epithelium in these also projects in the form of villi or folds into the lumen. The so-called crop of the *Docoglossa* (*Patella*) no doubt corresponds with the two lateral oesophageal sacs in the *Chitonidoe* and *Rhipidoglossa*. This is a saccular widening of the oesophagus, which, on account of the constitution of its walls, has been compared with the psalterium of a Ruminant. A similar widening of the oesophagus is found in *Cyproeidoe* and *Naticidoe*, which must be countcd among the most primitive of the *Monotocardia*.

In those *Monotocardia* which are provided with a proboscis, the length of the thin oesophagus is in proportion to that of the proboscis. The mouth lies at the tip of the proboscis, then follows a short and often rudimentary pharynx, and then the long oesophagus, which runs through the whole length of the non-protrusible portion of the proboscis, passes through the oesophageal ring, and may be even further prolonged posteriorly. When the proboscis is retracted, the posterior portion of the oesophagus becomes coiled; when the proboscis is extended, it lies in the protruded or evaginated basal portion. Not infrequently in carnivorous *Monotocardia* there is a glandular widening in that section of the oesophagus which follows the long proboscidal portion. The oesophagus is most complicated in the *Rachiglossa* and many *Toxoglossa*, where this widening, in the form of a large compact accessory gland, can become separated from the intestine (Leiblein's gland, poison gland) and where other glands are widenings may occur.

It seems probable that in certain *Prosobranchia* digestion and resorption takes place even in the fore-gut. In the *Pulmonata* and *Opisthobranchia*, there is sometimes a widening (crop, fore-stomach) anteriorly to the stomach, and in the same way the short oesophagus of the *Scaphopoda* has a glandular widening, or two lateral glandular diverticula. Among the *Cephalopoda*, the *Decapoda* have a simple thin tubular oesophagus; the oesophagus of the *Octopoda*, however, is provided with a lateral pouch, the crop, whose walls are not glandular. This may serve as a reservoir of food when the stomach is already full. In *Nautilus*, the crop is a very large saccular widening of the oesophagus, larger than the stomach itself.

THE MID-GUT WITH THE STOMACH AND DIGESTIVE GLAND (LIVER)

The oesophagus leads into a wider portion of the alimentary canal, the stomach. Into this the ducts of a gland open; this gland is strongly developed in nearly all Molluscs, and is usually called the liver, but may be more appropriately named the digestive gland, since it in no way fulfils the functions of the vertebrate liver. As far as it present known, it functions rather as a pancreas, or it combines the functions of the various digestive glands of the vertebrate intestine, no such through division of labour as is found in the Vertebrates having taken place. The digestive gland is, in most cases, a richly-branched tubular or acinose gland, which to the naked eye appears a compact lobate body of a brown, brownish-yellow, or reddish colour. Its glandular epithelium consists of three sorts of cells—hepatic, ferment, and calcareous cells. In many *Nudibranchia* the gland breaks up into branching intestinal diverticula, which spread through the body almost like the gastro-canals or intestinal branches in the *Turbellaria*, and run up into the dorsal appendages of the body (cladohepatic *Nudibranchia*).

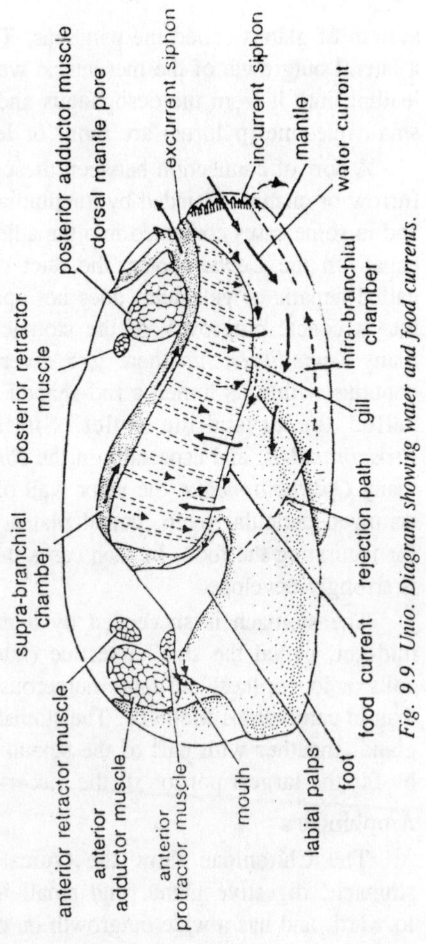

Fig. 9.9. Unio. Diagram showing water and food currents.

Choetoderma, among the *Solenogastres*, has a simple midgut diverticulum, which may correspond morphologically with the digestive gland of other Molluscs; but in *Proneomenia*, *Neomenia* etc., the straight mid-gut is provided throughout its whole length with narrow lateral glandular sacs arranged closely one behind the other at right angles to it. A part of the mid-gut gland (the part nearest to the point where the duct leaves it) and the glandular epithelium of the duct may be specially differentiated in *Cephalopoda*, and may, finally, form a distinct system of glands called the pancreas. The stomach is not infrequently a lateral outgrowth of the mesenteric wall, so that the aperture (cardia) leading into it from the oesophagus and that leading out of it into the small intestine (pylorus) are more or less near one another.

A sort of connection between these apertures may arise, a ciliated furrow or channel bounded by longitudinal folds running between them, and in some cases continued into the adjoining sections of the alimentary canal. In the *Cephalopoda*, the duct of the digestive gland (the so-called hepatic or gall duct) does not open direct into the stomach, but into a coecal outgrowth of the stomach, the spiral coeeum. In very many *Lamellibranchia* there is a diverticulum of the stomach which contains within its lumen a rod-shaped gelatinous cuticular formation, called the crystalline stylet. Similar structures occur in the *Prosobranchia*, and especially in the *Rhipidoglossa* and *Toxoglossa*. In many *Opisthobranchia*, the inner wall of the stomach carries variously-arranged cuticular teeth, dental plates, jaw plates, etc., which serve for triturating the food. In such cases the muscular wall of the stomach is strongly developed.

The stomach is succeeded by a narrower tubular section of the mid-gut, called the small intestine (intestinum), which usually forms coils or loops; these are more numerous in herbivorous or detritivorous than in carnivorous Molluscs. The stomach, small intestine, and digestive gland, together with part of the sexual organs, compose the whole or by far the largest portion of the visceral dome, where this is present.

Amphineura

The Chitonidae show the typical division of the mid-gut into stomach, digestive gland, and small intestine. The stomach lies far forward, and has a wide outgrowth on one side, which is, functionally, a reservoir of secreted matter. The cardia and the pylorus lie near one another. The digestive gland is paired; the larger liver to the right has four apertures, while the smaller one to the left has only one principal aperture into the stomach. The small intestine is more than four times as long as the body, and it forms many loops which are constant in their arrangement. *Chiton* feeds on small or even microscopic algae. Unlike the *Chitonidoe*, the *Solenogastres* show no separation of the mid-gut into stomach and small intestine.

The mid-gut runs straight through the body, the greater part of which its fills. The glandular lateral coeca found in *Neomenia*, *Proneomenia* etc. and called hepatic diverticula, and caused by the projection into the lumen from each side of narrow septa arranged at right angles to the gut, or transversely; in these septa, muscle fibres run down to the rudimentary foot, and blood lacunae abound. In *Proneomenia Sluiteri*, septa of the first, second, third, or fourth order can be distinguished, as seen in the figure. The septa on the right alternate with those on the left side of the body. In the dorsal middle line the mid-gut forms a narrow ciliated longitudinal groove which cuts deep into the gonad, cilia are also found on its medio-ventral surface.

Gastropoda

The digestive gland of the Gastropoda falls into two or more lobes, between which the stomach and the coils of the small intestine lie embedded. One, two, or more ducts or the gland may open into the stomach. The walls of the digestive gland show the same division into layers as the wall of the alimentary canal. For details as to the ferment, hepatic, and calcareous cells forming the epithelium of the gland, and their physiological constitution, the reader must be referred to special histological physiological treatises. In the *Nudibranchia*, as already mentioned, the digestive gland breaks up into a system of glandular diverticula (the so-called "diffuse liver").

The *Aeolidiadoe* (e.g. *Tergipes*) afford an instructive instance of this. Three diverticula rise from the stomach, two anterior and lateral, and one posterior and unpaired. These ramify in the body cavity, and finally send up their last ramifications or lobes into the dorsal appendages. The contents of the intestine can penetrate into these last ramifications of the "diffuse liver". Further, within the *Nudibranchia* the breaking up of the compact digestive gland to form a "diffuse liver," *i.e.* the loosening from one another, and the spreading out of the glandular tubes which are in close contact in the compact gland, can be followed almost step by step. In the *Tritonidoe* the gland is a great compact mass.

In other families, such as the *Tethymelibidae*, *Lomanotidae*, *Dendronotidae*, *Bornellidae*, *Scylloeidae*, it divides into two anterior accessory livers and a posterior principal liver, from which diverticula run up into the dorsal appendages. Finally, the accessory and principal livers break up into separate "hepatic branches" (*Aeolidoe*), which in some cases anastomose. The posterior principal branch of the "diffuse liver" gives off specially numerous lateral branches; it often widens out to a pouch, and may then be compared to an extended gall bag, or a posterior diverticulum of the stomach. In *Phyllirhoe*, a pelagic form, without dorsal appendages, the "diffuse liver" is simplified, consisting of four unbranched blind tubes, the two anterior opening into the stomach separately, the two posterior entering it together. The stomach of many *Opisthobranchia* consists of two divisions separated by a constriction. In some forms, such as the *Bullidoe* among the *Teetibranchia*, the *Pteropoda thecosomata*, and the *Techymelibidoe*, *Bornelidoe*, *Scylloeidoe*, among the *Nudibranchia*, it is armed with hard chitinous plates, spines, teeth, etc., occurring in varying number and arrangement on its inner wall.

Scaphopoda

The mid-gut of *Dentalium* consists of a looped stomachal tube bent back on itself, and of a small intestine lying in a tangled coil behind the oesophagus. Two digestive glands, lying in the upper part of the body, open through wide apertures into the stomach.

Lamellibranchia

In the Lamellibranchia the oesophagus, which lies under the anterior adductor, widens at the anterior base of the foot to form the stomach. This descends somewhat into the foot. At the posterior base of the stomach lie two apertures; one of these is the pylorus, and leads into the small intestine which runs more or less coiled within the base of the foot; the other leads into a tubular diverticulum, the sheath of the crystalline stylet. The large richly-branched acinose digestive gland (liver) opens through several apertures into the stomach, with which it lies in the anterior part of the pedal cavity. In *Pholas*, *Jouannetia*, and *Teredo*, the stomach has another coecum besides the sheath of the crystalline stylet. In all bivalves there is, on the inner wall of the stomach, a gelatinous cuticular structure (dreizackiger körper, flêehe tricuspide), which varies in thickness, and is continued into the gelatinous crystalline stylet. This latter is secreted in concentric layers as a cuticular structure by the epithelium of the sac in which it lies.

A plausible suggestion has recently been made as to the use of these gelatinous structures, viz. that they serve for surrounding with a slimy envelope foreign particles, such as sharply-pointed grains of sand, which enter the alimentary canal with the food; injury to the delicate walls of the intestine is thus avoided, and the travelling of such particles along the digestive tract is facilitated. The point of the crystalline stylet projects freely into the lumen of the intestine. In some forms it does not lie in a separate sac, but in a groove (*Najada*, *Cardium*, *Mytilus*, *Pecten*, etc.).

The tricuspid body and the crystalline stylet appear temporarily, and are renewed periodically. Similar structures have been observed in the stomachs of various *Gastropods*, *Haliotis* has a stomachal coecum which can be compared with the sheath of the crystalline stylet. In the lower *Lamellibranchs*, the *Nuculidoe* and *Solenomyidoe*, the stylet is either very slightly developed or wanting. In the *Arcidoe* also, it is only slightly developed. The *Septibranchia* (*Poromya*, *Cuspidaria*) are distinguished from all other *Lamellibranchia* by the absence of coils, and the consequent shortening of the small intestine (*cf.* on the intestine of the *Lamellibranchia*).

Cephalopoda

The stomach in the Cephalopoda always lies in the dorsal portion of the visceral dome in the shape of a sac with a strong muscular wall. It always has a coecal appendage (stomachal or spiral coecum), which varies in shape and size; into this the digestive gland (liver) opens. This coecum is a reservoir for the secretion of the digestive gland. Food never enters it, there are even valves at the point of entrance into the stomach, which allow the secretion collected in the coecum to pass into the stomach, but prevent the entrance of the contents of the latter into the coecum. In *Nautilus*, the coecum does not open into the stomach, but into the commencement of the small intestine, and is in the form of a small round vesicle with lamellae projecting into its lumen.

In *Sepia* and *Sepiola* also, the coecum is more or less round; in *Rossia*, it is slightly developed; in *Loligo* and *Sepioteuthis*, very long

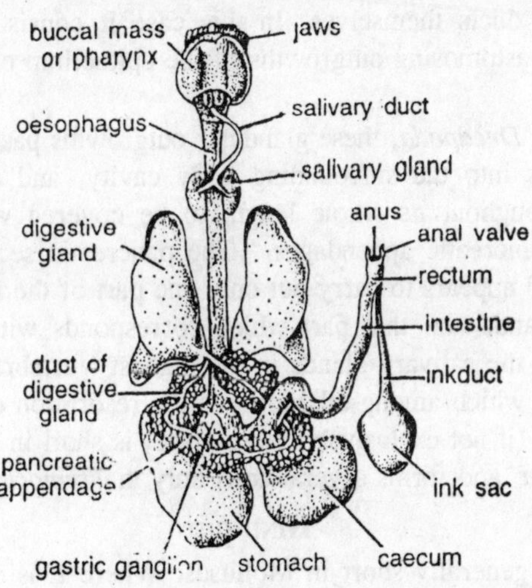

Fig. 9.10. Sepia. Alimentary canal.

and ending in a point; in all *Oegopsidoe* and *Octopoda*, more or less spirally coiled at the blind end. The well devcloped digestive gland seems to arise as a paired organ, even when unpaired in the adult. The whole of the much branched gland is surrounded by a common integument, and it thus outwardly appears to be compact. The digestive gland of *Nautilus* consists of five lobes (four paired and one unpaired), which lie around the crop. They have two ducts, which enter the coecum through a short common terminal portion. In the *Dibranchia* also, the digestive gland always lies on the ventral side of the stomach, close to the ascending oesophagus. It is undivided, and round or oviform in the *Octopoda*, *Oegopsidoe*, and *Sepiola*.

In *Loligo* and *Sepioteuthis*, it is traversed by the oesophagus and the aorta; in *Enoploteuthis*, its dorsal half is cut into two points by these organs; and the same is the case in *Rossia*. In *Sepia* and *Spirula*, the gland forms two lateral lobes which are distinct in *Sepia*, but connected along the middle line in *Spirula*. There are always two ducts (gall ducts) which rise near the median plane from the upper part of the gland, and open into the stomachal coecum separately or through a common terminal portion. The following facts have been ascertained as to the function of the so-called pancreas of the Cephalopoda.. It is originally a specially differentiated portion of the digestive gland, and is easily distinguishable in the *Octopoda* by its different colour; it lies near that part of the gland from which the ducts spring. In *Loligo*, the pancreas is found in the much thickness wall of the ducts themselves. In this case it consists of numerous glandular anastomosing outgrowths of the epithelium of the ducts into their wall.

In other *Decapoda*, these glandular outgrowths pass from the wall of the ducts into the surrounding body cavity, and then each duct appears throughout its whole length to be covered with acinose or ramified "pancreatic appendages." The pancreatic secretion contains diastase, and appears to carry out only one part of the functions of the digestive gland, viz, that part which corresponds with the digestive functions of the salivary glands in the highest Vertebrates. The small intestine, in which among all Molluscs the resorption of the digested, food chiefly (if not exclusively) takes place, is short in the carnivorous *Cephalopoda*, and forms several coils only in *Tremoctopus violaceus*.

Hind-gut

This is generally short in Molluscs. Where it is sharply marked off from the small intestine, it usually differs from the latter in being thicker and more muscular. In the majority of *Lamellibranchs*, and in nearly all *Diotocardia*, the rectum traverses the ventricle; this fact, with many others, supports the relationship of these two groups. In certain Molluscs, viz. the *Scaphopoda*, a few *Prosobranchia* (*Muricidoe*, *Purpuridoe*), and the *Cephalopoda*, the hind-gut has an accessory (anal) gland, which is well known in the *Cephalopoda* as the ink-bag. The rectal gland in *Dentalium* is a branched acinose gland opening into the hindgut, according to one account through six separate ducts, and according to another through one single duct. Eggs and spermatozoa have been met within the lumen of this gland, and it has been supposed that they have been accidentally drawn out of the mantle cavity by the swallowing like action of the hind-gut, which has been observed in *Dentalium*.

The anal gland found in some *Rachiglossa* (*Monoceros*, *Purpura*, *Murex*) is always dark in colour (brown or violet), and is either tubular with many bulgings of its wall, or acinose with an axial duct. It always enters the hind-gut near the anus. A gland has been found near the rectum in the *Pteropoda thecosomata* (*Clio*, *Cavolinia*) and the *Bullioda*, and has been described as an anal gland, but this requires further investigation. The ink-bag of the *Cephalopoda*, which is wanting only in *Nautilus*, is a much developed anal gland. It enters the hind-gut near the anus. The ink or sepia pigment secreted by it consists of extremely minute particles which are ejected with vehemence from the bag and discharged through the funnel. The pigment quickly mixes with the water and envelops the animal in a pigment cloud, thus screening it from its enemies.

Form and Position of the Ink-bag

The typical position of the ink-bag is in front of the rectum, *i.e.* in the loop formed by the intestine in ascending from the mouth and then descending to the anus. In *Spirula*, *Enoploteuthis*, and *Sepioteuthis*, the ink-bag is very small; it progressively increases in size in series both of *Decapoda* and of *Octopoda*, its division into a saccular portion and a duct opening into the hind-gut in front of the anus becoming more and more distinct. In the *Octopoda*, it lies embedded in the upper part of the liver within the muscular hepatic capsule. It is still found in this position (between the liver and the rectum) in *Sepiola*. In other *Decapoda*, however, the ink-bag is found shifting higher and higher in the visceral dome, its duct at the same time intrescing in length. Finally, in *Sepia* (and the fossil *Dibranchia*), it is found at the top of the visceral dome, behind thc gonad. Its duct runs along the right side of the hind-gut, bending round somewhat before reaching the anal section of the rectum so as to enter the latter anteriorly. Ontogenetically, however, even in *Sepia*, the ink-bag arises as an *anterior* outgrowth of the rectum.

Structure of the Ink-bag in Sepia

The ink-bag in this instance consists of three parts: (1) the pigment gland which secretes the "ink"; (2) the pigment reservoir and the duct, which forms (3) an ampulla with a glandular wall near its aperture. The pigment gland is a sac at the base of the ink-bag on its anterior wall (that turned towards the gonad). It projects into the cavity of the ink-bag, which serves as reservoir and duct for the pigment. The latter, after being formed in the gland, passes through an aperture in its wall into this reservoir.

The cavity of the gland is traversed by numerous perforated and richly vascularised lamellae of connective tissue, which are inter-connected in such a way as to form a kind of sponge like structure. New lamellae are continually being put forth by the formative zone of the gland, which is a narrowed portion bent back downwards, while the oldest lamellae, which lie nearest the aperture of the gland, become

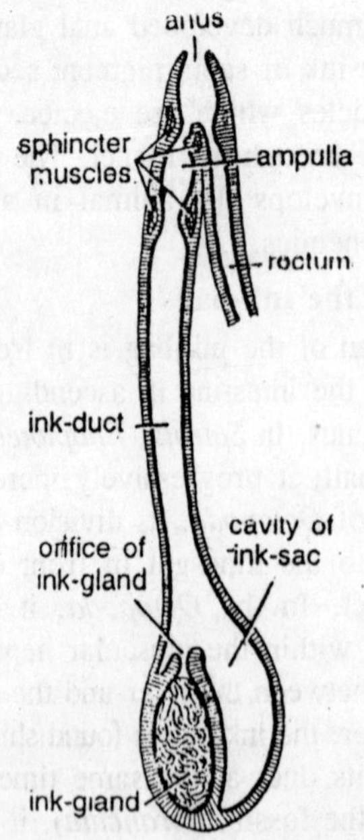

Fig. 9.11. Sepia. Ink-sac in M.L.S.

detached and degenerate. All the lamellae are covered by a glandular epithelium and the formation of the pigment can be traced in all its stages from its appearance in the epithelial cells of the formative zone to its condition in those of the oldest lamellae. In the formative zone, the young glandular cells are at first colourless.

In the succeeding lamellae, however, pigment granules increase in number and from the older lamellae are emptied into the cavity of the gland, the epithelial cells then becoming detached and breaking up. Both the gland and the reservoir are surrounded by a vascularised integument of connective tissue; the same integument forms the framework of connective tissue running through the lamellae or trabeculae within the gland. The ink-bag is further enveloped as a whole in tough integument consisting of three layers: (1) an inner glittering silvery layer (argentea), similar to the corresponding layer in the outer integument; (2) a central muscle layer (inner longitudinal and outer circular muscles; and (3) an external layer of connective tissue.

The terminal ampulla has, at its two narrow ends, folds projecting inward and functioning as valves; it can be closed at these parts by sphincter muscles. The ampulla itself also forms longitudinal folds on its inner surface between which glandular tubes open. The anus, in the Cephalopoda, always carries two lateral projecting appendages, which are often lancet-shaped. The short and narrow hind-gut of the *Solenogastres* opens into the dorsal, portion of a cavity, the cloaca, which lies at the posterior end of the body; this, again, communicates with the exterior by means of a ventral and very extensible longitudinal slit. Into this cloaca the ducts of the genital organs, which are morphologically to be regarded as nephridia, also open. In the *Lamellibranchia*, after the hind-gut has traversed the heart, it runs straight backward over the posterior adductor, to open through the anus into the posterior and upper portion of the mantle cavity (anal chamber).

10

Respiratory System

The requirement of oxygen presents little difficulty to small organisms in which diffusion is adequate to carry requirement of oxygen to the cells. It is among the molluscan animals with their relatively large size and covered body that we first meet distinct specialized respiratory structures. Primitively these structures were probably simple projections of the epithelium of mantle but in most present molluscs these projections called as *ctenidia*. The ctenidia are greatly modified in the direction of increase in surface area and greater vascularity. The second important development among respiratory organs found in connection with the ability of certain gastropods to live permanently on ground. In pulmonates ctenidia has disappeared and the walls of mantle cavity have folded to form a sac-like structure which is highly vascular. So in molluscs there are three types of respiratory organs:

1. Ctenidia
2. Pulmonary sac
3. Skin.

Ctenidia

Ctenidia are the organs of aquatic respiration and is mostly present in aquatic molluscs. The pulmonary chamber or sac is the organ of aerial respiration and is present in terrestrial molluscs. But in certain cases like *Pila*, *Planorbis* etc. both the methods of respiration are employed. Some animals may be regarded as in process of change from an aqueous to tessestorial life and as we see in *Siphonaria*, the ctenidia are reduced to a great extent since the animal comes out of water on the average 22 hours out of 24 hours.

Structure of Ctenidium

The ctenidium has a *central axis* one or two rows of *lamellae* or *filaments*. The entire gill is surrounded by cilia. The ctenidium receives venous blood through an affarent branchial vein and after oxygenation in filaments it is send back through an efferent branchial vein.

Number of Gills

The number of gills vary from 1-80 in different classes of Mollusca.

Types of Gills

In molluscs following types of ctenidia are seen:

1. *Monopectinate*. The filaments or lamellae are arranged in a single row on one side of the axis.
2. *Bipectinate*. The gill has two rows of gill filaments on the sides of central axis.
3. *Plicate gills*. These are transverse folds of integument.
4. *Holobranchs*. The ctenidia extend all over the body.
5. *Merobranchs*. When the ctenidia are restricted to a particular area of the body.
6. *Secondary gills*. The true gills are absent. But certain other structures funchion as gills. They are the following. (i) *Anal gills*—Gills present around the anus. (ii) *Cerata*—These are slender vascular bodies located on the dorsal surface. (iii) *Pallial gills*—Gills arranged in single row around the foot.

In Monoplacophora

In class Monoplacophora, 5 pairs of gills are present. These are unipectinate, composed of finger-like lamellae. The gills are present in pallial groove. Vestigial gills are supposed to be present on the opposite side of the ctenitial axis. The gills are probably not true gills, more likely they represent prectenidial structure e.g. *Neoplina*.

In Amphineura

In order Placophora (Chitons) the ctenidia are arranged in two rows. The number vary from 6-80. According to species when their number is few, the ctenidia are mostly confined to the posterior end. It is not easy to decide whether the multiple form modification found in Chiton is a simple or evolved one. In the animals of order Aplacophora the foot is absent and the mantle Cavity is greatly reduced in the form of a cloacal Cavity at the posterior. In *Chaetoderma* there are two small, feather-like gills symmetrically placed on either side of the anus. In *Neomenia* the gills are further reduced and consist of single bunch of filaments lying below the anus. Such gills restricted to a limited space at the posterior end of the body are called *Merobranchial type*.

In Gastropoda

The gills show a good number of variations started from bipectinate paired structure to monopectinate single structure due to the asymmetry and in some cases certain secondary structures or adaptive gills are developed which subserve the function of respiration. The gills are shifted in front of the body within the mantle cavity as a result of torsion. In Prosobranch symmetrically paired gills occur only in *Haliotis* and *Fissurella*. In all others the asymmetry of the body have caused one of the branchi to become reduced and consequently it is only one branchi usually left in vast majority of prosobranchs. Even in *Haliotis* the right branchi is rather smaller than the left. In Aspidobranchs (*Trochus*, *Turbinus*, *Acmaea* etc.) the gill is bipectinate but in Pectinibranchs like *Triton*, *Pila* etc. the gill is monopectinate.

In opisthobranchs a true ctenidium only occurs in the *Aplysia*. It lies on the right and usually more or less external being partly covered sometimes quite shallow and sometimes by a fold of membrane. The Nudibranchs (*Doris*, *Aeolis*) have altogether lost the true ctenidium with mantle cavity and respire by various secondary structures such as anal gills (*Doris*) and Cerata (*Aeolis*). In *Patella* (Prosobranch) bears a series of adaptive gills occurs in a row on each lateral side in the pallial groove. Such gills are called pallial gills. They may be found together with the ctenidium in *Pneumoderma* (Ospisthobranch).

In Pelecypoda

The ctenidia are two in number right and left each consisting of a horizontal axis bearing two rows of filaments. They lie one on each side of foot. They lined a chamber known as branchial chamber. Leading into this chamber and behind it are two siphons, one of which conducts water to the branchial chamber called *inhalant siphon* and other carries the water outside is called *exhalant siphon*. In many cases like *Mya* both the siphons are exceedingly long, sometimes longer than the whole shell. In some cases the two siphons are free throughout their entire length while in others they become fused.

In other genera which do not burrow like *Pecten*, *Mytilus* etc. the siphons are either reduced or altogether absent. In Protobranchiata (*Nucula*) the gills are relatively smaller, situated at the back of the mantle cavity, and their filaments relatively few, form small, compressed and horizontal disposed, triangular leaflets, free from one another. The leaflets are arranged in two opposite rows in a gill. The arrangement divides the mantle cavity into a large inhalant chamber below each gill and a small exhalant chamber above.

In Filibranchia, the gill-filaments become elongated and thread-like. In *Arca* each gill filaments is bent upon itself to from an elongated 'V'. Thus each gill forms a 'W' in section and consists of two 'V' shaped demibranchs. Each demibranch is two lemellae thick made up of a descending limb and an ascending limb. In *Mytilus* the gills are strengthened by the development of delicate non-vascular bar or inter-lamellar junctions between the two limbs of each filament. Thus the single ctenidium of *Nucula* is replaced by two plate-like laminae in *Mytilus*. Each lamina is made up of an outer and an inner lamina.

In Psendolamellibranchia (*Pecten*, *Ostrea*), the gills have a greater cohesion than in filibranchs and the reflected distal ends of filaments become fused laterally with the mantle. In Eulaniellibranchia (*Unio*, *Anodonta*) the ciliary interfilamentar junctions are replaced by vascular cross connections with narrow opening between them called *ostia*. The interlamellar junctions also become large and vascular, so that the interlamellar space is partitioned into vertical water tubes. The incoming water enters the water tubes through these ostia. In Septobranchia (*Poromya*, etc.) the gills are represented by a muscular septum which divides the mantle cavity into the dorsal and ventral parts. This septum is perforated at certain places, which allows the communication between dorsal and ventral chambers.

In Cephalopoda

The ctenidia are situated in the pallial cavity on either side of the visceral mass. In Dibranchia there are only one pair of gills. A gill is composed of lamellae whose number varies in different forms. Each lamellae is thrown into transverse folds which are inturn folded so that the respiratory surface is largely increased. The gills are attached dorsally to the mantle (*Sepia*, *Loligo* and *Octopus*). In Tetrabranchia there are two pairs of gills which are free throughout their existence (*Nautilus*).

Histology of gill-filament

The gill-filaments are hollow structures and each filament has two sides. One side facing towards the mantle cavity and other towards the inter-lamellar cavity. The gill filaments are provided with Chitinouss rods to give a sort of rigidity to the gill filaments. They are covered with ciliated epithelium. The epithelial cells are larger on the side facing mantle cavity and secrete mucous. The presence and arrangement of cilia is remarkable, on the sides are *lateral cilia*, on the outer ridge-like faces of filaments are *frontal cilia* and those lying in between are *latero-frontal cilia*.

Pulmonary Sac or Cavity

It is an organ of aerial respiration and is found in land molluscs, mostly in pulmonates and some amphibious prosobranchia. When we use term 'lung' it must be remembered that this organ in mollusc does not correspond morphologically with spongy, cellular lungs of vertebrates. It simply performs the same function. The lung in mollusc is a cavity or pouch livcd by blood vessels which are disposed over its surface in various ways forming network. In adaptation to terrestrial mode of life the pulmonary respiration resulted. Therefore it is found in vary different groups of Gastropoda, but different stages of evolution are best studied in Streptoneura. In this group certain aquatic forms, though they possessed Ctenidia, required the habit of living for short or longer time out of water. Consequently certain modifications have taken place in the internal structure of mantle. So that in certain forms such as *Ampullaria* and *Pila*, both Ctenidia and pulmonary sac are present and the mantle cavity is divided into right pulmonary chamber and left branchial chamber by an incomplete septum. In others the ctenidia has totally absent and the pulmonary chamber is fully developed (*Gadinia*).

Outer Skin

Respiration by moist-skin is the simplest method of respiration in molluscs. In certain aquatic (Scaphopoda and terrestrial Opisthobranchiata), terrestrial (*Vaginula*) and parasitic (*Ontodax* and *Entococha*) neither ctenidia nor localized respiratory organs are present. Respiration occurs through entire body surface or soft-skin of the body. In *Oncidium* respiration is, however, carried by skin as well as by dorsal papillae present on skin. Skin function as respiratory organ even when ctenidia, as well as pulmonary cavity is present.

Respiratory Organs

The most important of the pallial organs in the Mollusca is the gill, for it is in order to protect it that the mantle, and with it the pallial cavity, develop. The gill found in the mantle cavity is throughout all the divisions of the Mollusca a homologous organ, to the derived from the gill of a common racial form. But since this gill is wanting in certain Mollusca (*e.g.* many *Opisthobranchia*), and is functionally replaced by new organs which are morphologically altogether unconnected with it, it has been found useful to distinguish the primitive Molluscan gill by the name of etenidium. The word, therefore, has a special morphological significance.

The ctenidia of the Mollusca are originally paired and symmetrically arranged ciliated processes of the body wall, carrying two rows of branchial leaflets, and projecting into the mantle cavity. Venous blood flows into the gills through afferent vessels (branchial arteries) and after becoming arterial by means of the respiration, flows through efferent vessels (branchial veins) back to the body, passing first through the heart. At or near the base of each etenidium there always lies a sensory organ, which is considered as olfactory, the so-called osphradium or Spengel's organ.

Such primitive ctenidia are met with first in that group of the Mollusca which has undoubtedly retained more primitive characteristics than any other, viz. the *Chitonidoe* among the *Amphineura*. They are further, found in all other Mollusca which have retained the original bilateral symmetry of the body, such as the *Lamellibranchia*, the *Cephalopoda*, and–a point of great importance–also in the primitive *Gastropoda*, the *Zeugobranchia*. In the latter, however, the left etenidium was originally the right and *vice versa*, but this will be dealt with more in detail later. With regard to the number of gills originally present on each side of the body, opinions are divided. Those who hold that there were several seem justified by the arrangement in

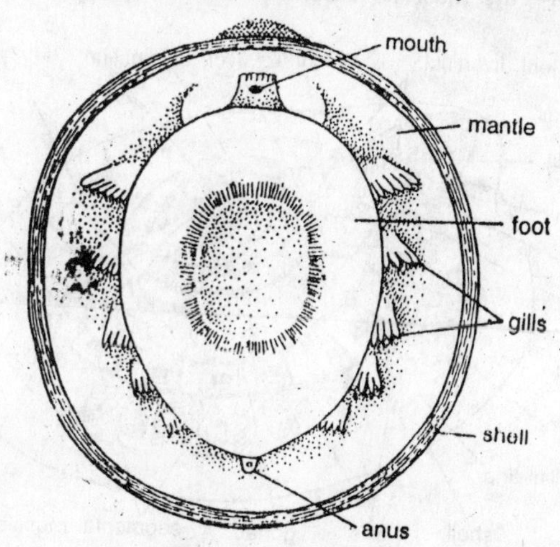

Fig. 10.1. Neopilina galatheae. Ventral view.

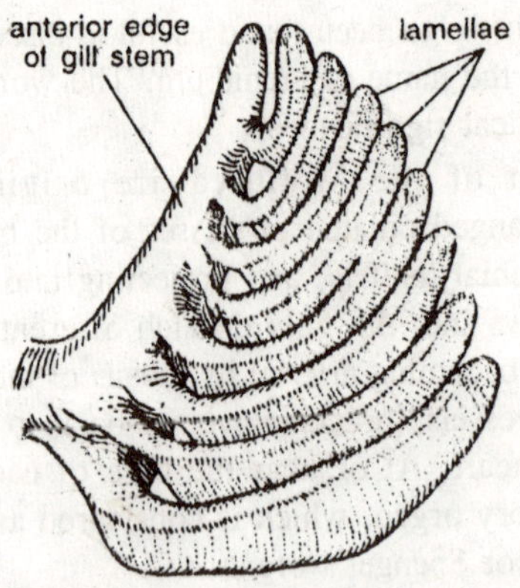

Fig. 10.2. N. galatheae. A gill in ventral view.

Chiton, where numerous consecutive ctenidia lie in a longitudinal row in the branchial furrow (mantle cavity) on each side, and also by that in the *Nautilus*, which is rightly considered the most primitive of extant *Cephalopods*, where four gills are found (*Tetrabranchia*). We shall, however, see later that the other view, viz. that the *Mollusca* originally possessed only one pair of ctenidia has, to say the least, equal claim to be accepted. In all other Mollusca with paired ctenidia, including the *Lamellibranchia*, there is only one pair at the posterior part of the body. Further, in the racial form of the *Prosobranchia*, a single pair of gills must be assumed to have occupied a posterior

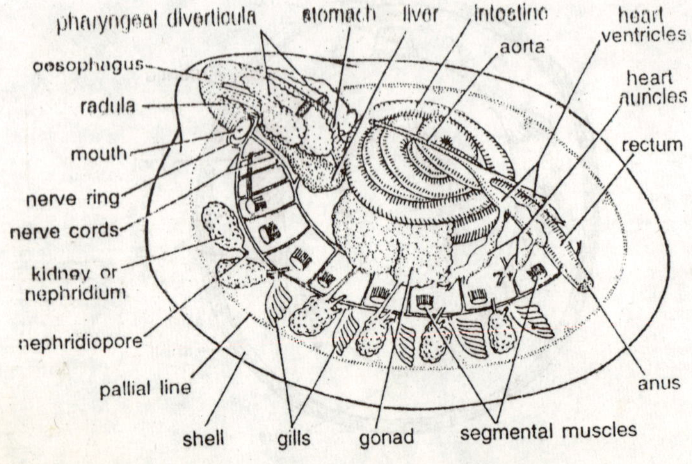

Fig. 10.3. Neopilina. Internal anatomy in side view.

position in a mantle cavity which, with them, shifted forward later to the anterior position. The *Zeugobranchia* still retain this single pair of gills.

In most *Prosobranchia*, the asymmetry of the body is also seen in the gills, only the left gill of the two in the *Fissurellidoe* and *Haliotidoe* being retained, the right completely disappearing. In the forms which most resemble the *Fissurellidoe* and *Haliotidoe*, the single-gilled *Diotocardia* (*Turbinidoe*, *Trochidoe*, etc.) the gill is still feathered on both sides, but in all *Monotocardia* it has only a single row of leaflets.

In one division of the *Opisthobranchia*, the *Tectibranchia*, one ctenidium is still retained, that on the right side. Other *Opisthobranchia* have lost the true ctenidium together with mantle cavity; it may be replaced by analogous (but not homologous) respiratory organs, such as adaptive gills.

The *Pulmonata* in consequence of their adaptation to aerial respiration, have lost the ctenidia. The blood, which has become arterial in the ctenidia, reaches the heart through the auricle, and passes into the body through the arteries. It is therefore evident that a close relation must exist between the gills and auricles. This relation is briefly as follows: where the gills are paired, the auricles are paired, and unpaired gills are accompanied by a single auricle on that side of the body on which the gill is retained. Where gills are paired, there is almost always only one pair, and then there is one right and one left auricle. The *Nautilus* has four gills, and, to correspond, two right and two left auricles. The *Chitonidoe*, on the other hand, in spite of their numerous pairs of gills, have only one right and one left auricle. The *Seaphopoda* possess neither true *ctenidia* nor any other localised gills. Respiration may take place at the various soft-skinned surfaces which come in contact with the water, such as the inner surface of the mantle, the tentacles, etc.

Amphineura

Chitonidae—A single ctenidium of a *Chiton* may serve as a type of the Molluscan gill with its two rows of leaflets. The plumose etenidium rises freely from the base of the branchial groove (mantle cavity). The axis here takes the shape of a thin septum. At each side, on the broader surface of the septum, extending from base to tip, there is one row of smooth, delicate branchial leaflets. In outline they are more or less semicircular, and stand crowded together in great numbers almost like the leaves of a book. The entire surface of the branchial epithelium is eiliated; on the axial epithelium, the cilia are

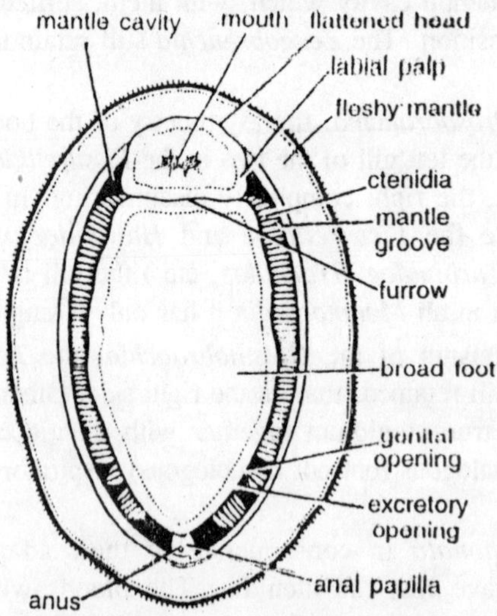

Fig. 10.4. Chiton—ventral view.

remarkably long. One that side of the axis which is turned towards the foot, a blood-vessel runs from base to tip, conducting venous blood to the gill (afferent branchial vessel). On the opposite side, which faces the mantle, another vessel, the branchial vein, runs from the tip to the base of the gill and carries the blood, which has become arterial by respiration, to the general branchial vein, and through it to the auricle. These vessels have to special endothelial walls, but are surrounded by circular muscle fibres. The branchial vein is accompanied by a powerful longitudinal muscle. At the base of each branchial leaflet, the blood flows out of the branchial artery through an aperture into the narrow cavity of the leaflet and passes through a similar aperture on the opposite side of the axis to enter the branchial vein. Nerves are supplied to the ctenidium from the pleuro-visceral nerve which runs close to its base. The number of ctenidia in each row varies very much in the different species of *Chitonidoe*; it ranges from 14 to 75. The row extends along the whole length of the branchial furrow, or else (in *Chiton loevis*, *Pallasii*, and *Chitonellus*) is confined to its posterior half.

Solenogastres—(Proneomenia, Neomenia, Choetoderma)

The mantle cavity, in these forms, is much reduced, consisting only of the groove on each side of the rudimentary foot; it opens into the cloacal cavity, or rather widens to form that cavity. The cloaca is thus the posterior portion of the mantle cavity. In *Choetoderma*, the foot has disappeared, and the mantle cavity is reduced to the cloaca, in which one typical gill lies on each side of the anus. These gills are regarded as the last ctenidia of the rows found in the *Chitonidoe*, which in *Chitonellus* and some species of *Chiton* are already confined to the posterior half of the body. In *Neomenia*, there is no longer a pair of etenidia, but a mere tuft of filaments rising from the wall of cloacal cavity, and in *Proneomenia*, there are only irregular folds of the cloacal wall. On the relation of the gills in the *Chitonidoe* to certain patches of epithelium, which may perhaps be considered as osphradia, see the section on Olfactory Organs.

Gastropoda

The *Fissurellidoe* among the *Prosobranchia* stand nearest to the racial form of the Gastropoda. The mantle cavity is anteriorly placed; into it from behind and above project two long gills feathered on each side; these lie symmetrically to the middle line, and to the right and left of the anus. The posterior portion of their axes is connected by a band with the floor of respiratory cavity, while the anterior pointed portion projects freely. The fact that in the *Fissurellidae* (and related forms) the gills are paired and symmetrical is very significant. It points to the primitive character of these forms, and enables us to compare their gills with those of the lower *Lamellibranchia*, i.e. the *Protobranchia* and of the *Cephalopoda*. We must however, again emphasise the generally-assumed fact that the left gill of *Fissurella* answers to the right gill of the *Lamellibranchia* and *Cephalopoda*, and the right gill of the former to the left of the latter, these latter having retained their primitive symmetry in this respect. This assumption becomes the more plausible when we consider that the mantle cavity with its organs originally lay posteriorly on the body, and shifted forward secondarily along its right side.

The *Haliotidae* are closely connected with the *Fissurellidae*. Their spacious mantle cavity is, however, forced to the left side by the great development of the columellar muscle. There are two gills, feathered on both sides, of which, the right is the smaller. The axis of each gill has united, for nearly its whole length, with the inner wall of the mantle, and only its anterior end is free; its tip even projects a short distance beyond the respiratory cavity. Although the *Fissurellidae* and *Haliotidae* still possess two gills, other *Diotecardia*, have retained only the left and larger gill of *Haliotis*. This gill, is however, still feathered on both sides, although this characteristic is obscured in a

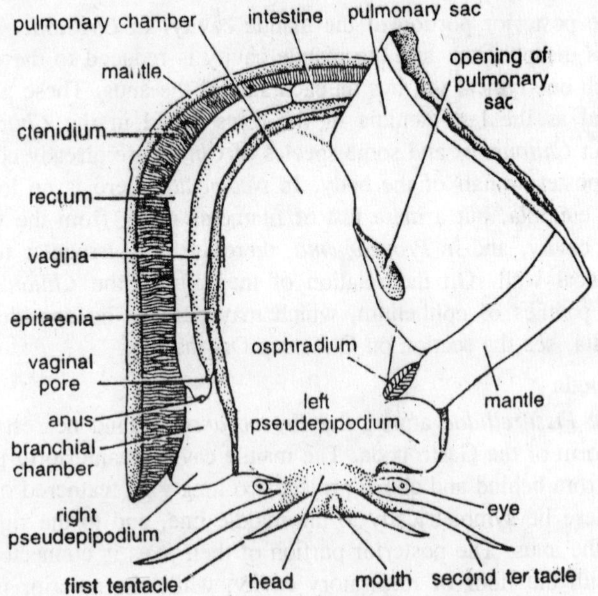

Fig. 10.5. Pila. Mantle cavity and pallial complex.

peculiar manner. The septum or axis of the gill, to the broader surfaces of which the branchial leaflets are attached, and one edge of which had, in *Haliotis*, already fused with the inner wall of the mantle, becomes attached to the mantle by its other edge also (viz. that along which the branchial artery runs), somewhat to the right of the first line of concrescence.

In this manner, which is illustrated by the accompanying diagrammatic sections, the mantle cavity is divided by the branchial septum into two unequal parts, which open into one another anteriorly. Into the much smaller upper division the one row of smaller branchial leaflets projects, while the opposite row of larger leaflets hangs down into the lower and larger chamber. The anterior end of the gill, however, is still free its point projecting anteriorly (*Teochidae*, *Turbinidae*, *Neritidae*).

In the *Docoglossa* (*Patellidae*) the arrangement of the gills is very varied. While the *Leptidoe* have no gills whatever, we find in *Patella* a single row of numerous small branchial leaflets right round the body, on the inner or under side of the short encircling mantle fold, between it and the foot. This row is broken only in one place anteriorly on the left. It is, however, evident that these gills, which somewhat resemble those of the *Chitonidae*, are no true etenidia, from the fact that there are *Docoglossa* (*e.g.* some forms of *Tectura* and *Seurria*), which possess, in addition to this marginal row of leaflets, a typical ctenidium corresponding in every way with that of the *Turbinidae*, *Trochidae*, etc. Other forms, such as *Acmaea*, have only the true ctenidium and no marginal branchial leaflets.

In the large second division of the *Prosobranchia*—the *Monotocardia*—the arrangement of the gills is, on the whole, remarkably uniform. There is only a single gill feathered on one side, united to the mantle along almost its whole length; this gill corresponds with the left gill in *Fissarella* and *Haliotis*, and the single gill in *Turbo* and *Trochus*. It generally lies quite to the left in the mantle cavity. The rise of this gill can best be explained by recalling the arrangements already described in *Turbo* and *Trochus*. We have only to assume that the row of small leaflets turned towards the mantle in *Turbo* disappears,

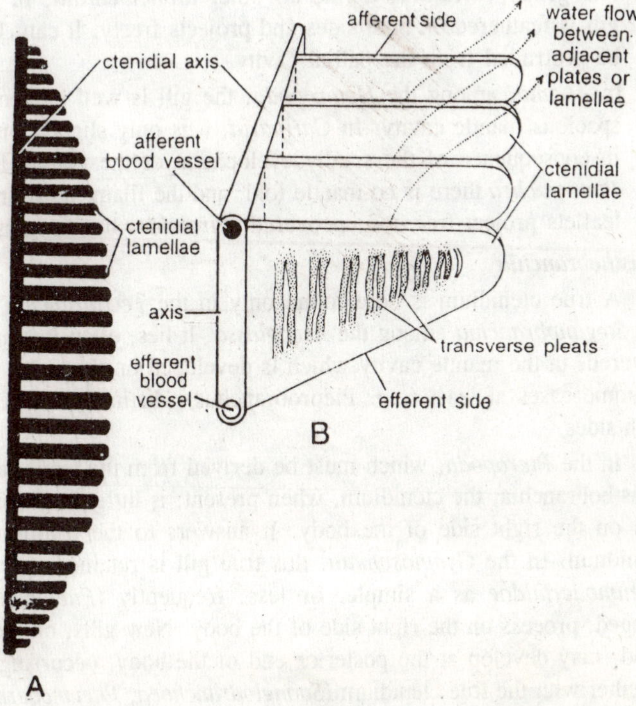

Fig. 10.6. Pila. Respiratory organs. A—A monopectinate ctenidium. B—Stereogram to show flow of blood within it.

and that the branchial septum unites with the mantle across its whole width. A few anomalous forms alone require special mention.

1. In a series of terrestrial *Monotocardia*, aerial respiration has taken the place of aquatic respiration, and the ctenidium has disappeared (*Acicula*, *Cyclostoma*, *Cyclophorus*, etc.).
2. The *Ampullaria* are amphibian *Prosobranchia*. A doubling of the mantle gives rise to a very spacious pulmonary sac, on the inner surface of which the respiratory vascular network spreads out. The lower wall of this pulmonary sac, which forms at the same time the roof of mantle cavity, is perforated by an aperture for the inhalation and exhalation of air. The ctenidium is placed to the extreme right of the mantle cavity, a position which is in some way connected with the great development of the pulmonary sac. It nevertheless answers to the left gill in other *Monotocardia*, as can be seen from its innervation.
3. The genus *Valvata* is unlike all other *Monotocardia*, in that its gill is feathered on both sides and projects freely. It can, further, be protruded from the pallial cavity.
4. In *Atlanta*, among the *Heteropoda*, the gill is well hidden in the spacious mantle cavity. In *Carinaria*, it is only slightly protected in consequence of the small development of the mantle fold. In *Pterotrachea* there is no mantle fold, and the filamentous branchial leaflets project free and uncovered. *Firoloides* has no gills.

Opisthobranchia

A true ctenidium is here found only in the *Tectibranchia* and in the *Steganobranchia* among the *Ascoglossa*. It lies, often incompletely covered, in the mantle cavity which is developed on the right, and is, in some cases at least (*e.g.* Pleurobranchus), distinctly feathered on both sides.

In the *Pteropoda*, which must be derived from the tectibranchiate Opisthobranchia, the ctenidium, when present, is little developed, and lies on the right side of the body. It answers to the tectibranchiate ctenidium. In the *Gymnosomata*, this true gill is retained only in the *Pneumodermidoe* as a simple, or less, frequently (*Pneumoderma*), fringed, process on the right side of the body. New gills, on the other hand, may develop at the posterior end of the body, occurring either together with the true ctenidium (*Spongiodranchoea*, *Pneumoderma*), or alone (*Clionopsis*, *Notobranchoea*) until they in their turn disappear (*Clione*, *Halopsyche*). Among the *Thecosomata*, the *Cavolinidae* alone possess a gill which rises in the form of a series of fold-like elevations of the body wall in the pallial cavity, and which, running in a wavy line, forms a semicircle, open anteriorly, the greater portion of it, however lying on the right side.

Lamellibranchia

The Lamellibranchia also possess typically two symmetrically placed gills, each provided with two rows of branchial leaflets. The opinion which until lately was common, that the Lamellibranchia possessed two gills on each side of the mantle cavity, has been shown to be incorrect—these two gills in reality answering to the two rows of branchial leaflets of one typical gill. It is worth while to follow, step by step, the interesting series of modifications undergone by the original gill in the Lamellibranchia.

Protobranchia

The primitive arrangement is found in the *Protobranchia*. Taking *Nucula* as an example, we find a gill like that of *Fissurella*, consisting of an axis along which the branchial artery and the branchial vein run, and which is attached by a short membraneous band to the posterior and upper portion of the body or visceral dome, and to the posterior adductor muscle. On this axis are attached two rows of short flat branchial leaflets. These two plumose gills coverage posteriorly, and project with their free tips into the mantle cavity. The leaflets of both rows are directed somewhat downwards, so that they are at right angles to one another.

In *Malletia* and *Solenomya*, on the contrary, they lie in the same plane, the two rows standing out on opposite sides of the axis. In *Malletia*, this plane is horizontal, but in *Solenomya* it trends downwards and inwards. The number of leaflets on the very slender gill of *Malletia* is much smaller than on that of *Nucula*; they are consequently neither so crowded nor so flattened. Each leaflet contains a blood sinus, which is a continuation of the branchial artery. Two rods of connective tissue run along the lower edge of each leaflet from the axis to its tip, and serve from its support. Similar, supports are found in almost all Lamellibranchia and in many *Gastropods*. The epithelium of the branchial leaflets is beset with long cilia—(1) at the ventral edge; (2) on both (anterior and posterior) surfaces, near the ventral edge. The first-named cilia form, with regard to the whole gill, a longitudinal row along the free ventral edge of each row of leaflets, and bring about a current in the water along this edge from behind forward. The other cilia mentioned above, mingling together like the bristles of two

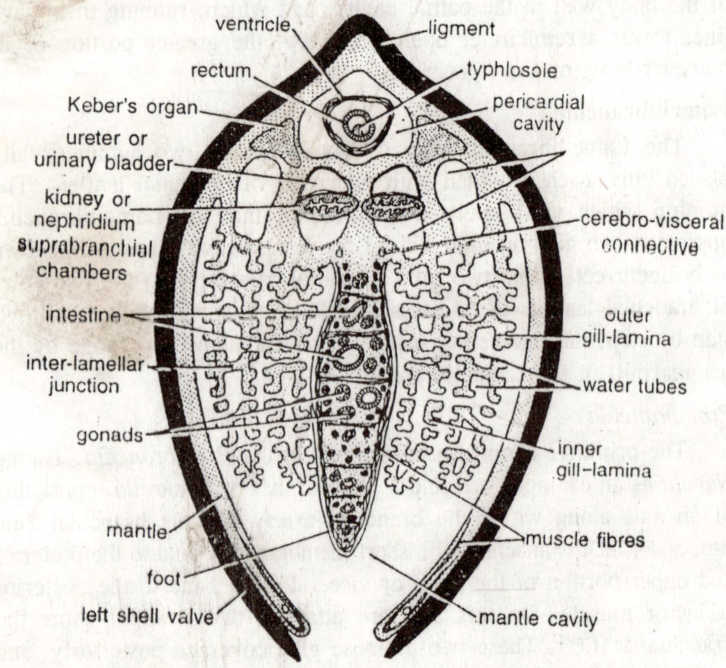

Fig. 10.7. Unio. T.S. Body through middle region of gills.

brushes which are pressed together, form a loose connection between the successive leaflets of the row.

Filibranchia

In the *Filibranchia* the leaflets in each of the two rows are very long and filamentous, and hang down far into the mantle cavity. The branchial filaments of the two rows are recurved and bent back upon themselves, so that in each filament a descending and an ascending portion can be distinguished. The prolongation of the filaments corresponds with a necessary increase of the respiratory surface. By this bending back of the filaments, the gills make the most of the limited space afforded by the mantle cavity. Each filament of the outer row is bent outwards, and of the inner row inwards. The filaments of each row may be so crowded together that the whole row looks life a leaf or fringe. This branchial leaf consists of two closely contiguous lamellae, one the descending and the other the ascending the two passing into one another at the lower edge of the leaf. The descending lamella is formed by the descending portions of the filaments, and the ascending by the ascending portions.

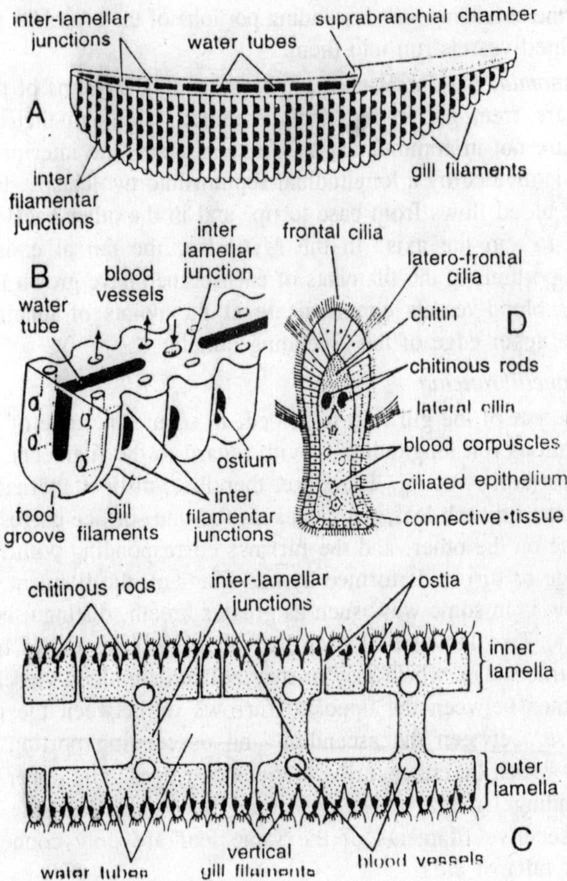

Fig. 10.8. Unio. Structure of gill. A—A demibranch seen from outside. B—A portion of gill (enlarged). C—A portion of gill lamina in T.S. D—Cross section of one gill filament.

On the outer leaf, the ascending lamella is the outer one, on the inner leaf the inner. In the Filibranchia, the separate branchial filaments retain their independence—they are free, *i.e.* the separate filaments of a series are unconnected with one another, and the descending and ascending portions of one and the same filament are in no way united. There are, however, on both the anterior and posterior sides of the filaments places covered with long cilia closely crowded together. These ciliated tufts on adjoining filaments mingle, and so give rise to a sort of connection between the filaments of each leaf. In the *Mytilidae*, so-called interfoliar junctions or trabeculae occur at certain points between the ascending and descending portions of the branchial filaments but no blood-vessels run into them.

In *Anomia*, thc dorsal ends of the ascending portions of the outer lamella are free, but in the *Areidoe* united, although their internal cavities are not in communication. In such cases, the interior of each filament is divided by a longitudinal septum into two canals. In one of these the blood flows from base to tip, and in the other back from tip to base, *i.e.* to the axis. In the *Mytilidae*, the dorsal ends of the recurved portions of the filaments of each branch have grown together, and their blood-vessels communicate at the points of junction, *i.e.* along the upper edge of the ascending lamella.

Pseudolamellibranchia

Each leaf of the gill is here folded, to secure increase of surface. The plications run longitudinally with regard to the filaments, and are thus almost dorso-ventral. There are, therefore, distinct alternate ridges and furrows on each leaf, the ridges on the one surface corresponding with those on the other, and the furrows corresponding with furrows. Each ridge or furrow is formed by one filament; the filament forming the furrow is in some way, such as greater breath, distinguished from the others. Two lamellae of each leaf of the gill are united here and there by trabeculae, which may or may not contain blood-vessels. They occur either between the opposite furrows or between the opposite ridges, *i.e.* between the ascending and descending portions of the filaments which lie either in the furrows or ridges. The upper edge of the ascending lamella of the outer leaf may unite with the mantle. The consecutive filaments of the same leaf are only connected by means of tufts of cilia.

Eulamellibranchia

The branchial leaves are either smooth or folded, but there is always organic connection, by means of numerous vascularised junctions, not only between the ascending and descending lamellae, but between the successive filaments. The junctions are therefore both interfoliar and interfilamentar. This leads to the entire disappearance of the original filamentous structure of each leaf, which now becomes an actual leaf or lamella with perforations or slits, the remains of the spaces between the original filaments, leading into an internal system of sinuses or canals, which in their turn are in remains of the spaces between the ascending and descending lamellae. This peculiar arrangement was formerly considered typical of the Lamellibranchia, and was the origin of their name. It was supposed that the animals of this class had two leaf-like gills on each side of the mantle cavity, *i.e.* four altogether, but we now know how the two branchial leaves on each side arose, that they are in fact the two, modified, rows of leaflets of the original plumose gill of the *Protobranchia*.

The Lamellibranchia in reality possess only one gill on each side in the mantle cavity. The blood now no longer flows through the primitive filaments of the lamellae of the gills and back again, but the afferent and efferent channels lie in the trabecular network between the two lamellae of a branchial leaf. Instead of the two leaves of a gill hanging down into the mantle cavity parallel to one another, the outer leaf may stand up dorsally in the cavity, so that the two come to lie in the same plane (*Tellinidoe* and *Anatinacea*). The ascending lamella of the outer leaf may be wanting (*Anatinacea*, *Lasoea*), and in fact the entire outer leaf may be absent (*Lucina*, *Corbis*, *Montacuta*, *Cryplodoa*).

In all Lamellibranchia, with the exception of the *Protobranchia*, and further, of the *Areidae*, *Trigonidoe*, and *Pectinidoe*, the gill and mantle unite, the dorsal edge of the ascending (outer) lamella or, where this is wanting, the free edge of the single lamella of the outer leaf becoming fused with the mantle. In the same way, the dorsal edge of the ascending (inner) lamella of the inner leaf may become fused with the upper part of the foot. If the two gills, which have fused with the foot, fuse with each other behind the foot in the middle line of the mantle cavity, they form a septum which, uniting with the septum formed by the mantle between the inhalent and exhalent siphons, divides the cavity into an upper and a lower chamber. The water flows through the lower (inhalent) siphon into the large lower chamber, bathes the gills, and, streaming forward, conveys the particles of food it contains to the mouth. It then flows back along each side of the foot in the upper chamber of the mantle cavity (which is itself divided into two canals by the line of insertion of the gill) into the single posterior and upper chamber behind the foot, and escapes through the upper (exhalent) siphon.

Septibranchia

These Mussels were formerly erroneously considered to be gill-less. As a matter of fact, the branchial septum just described has in them been much modified in structure, and has become a muscular septum, running across the mantle cavity in a horizontal direction and joining the siphonal septum posteriorly, while anteriorly it passes round the foot. This septum is broken through by various perforations and slits, while allow of communication between the upper and lower chambers of the mantle cavity, and very in the different genera.

Cephalopoda

The gills of the Cephalopoda are always feathered on both sides. Those of the *Dibranchia* have been the most thoroughly investigated. In *Sepia*, each gill has the shape of a slender cone, its whole length being applied to the visceral dome in the mantle cavity, in such a way that the base is directed dorsally towards the apex of the visceral dome, and the point ventrally towards the free edge of the mantle fold or the mantle cleft. The points of the two gills diverge. The two rows of flat triangular branchial leaflets are carried by the two branchial vessels, each leaflet being attached by one end of its base to the branchial artery and by the other to the branchial vein. In the axis of the gill between the two vessels, and also between the bases of the two rows of leaflets, a channel is formed which communicates by a slit between each successive pair of leaflets with the mantle cavity; through this canal the respiratory water freely flows. The slits in this axial channel are arranged alternately on each side, like the leaflets between whose bases they lie. The branchial vein forms the posterior support of the gill turned towards the mantle, and the branchial artery the anterior support turned towards the visceral dome.

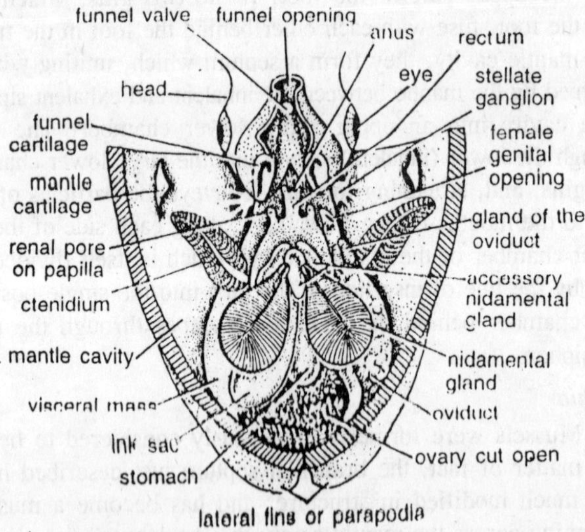

Fig. 10.9. Sepia. Mantle cavity of a female specimen.

The artery is united along its entire length with the integument of the visceral dome by a membrane of connective tissue. The anterior edge of each leaflet (that facing the visceral dome) is connected with this membrane, which may be called the gill-suspensor, by means of another triangular membrane. A special vein runs along the posterior free edge of each leaflet, and enters the general branchial vein at its base; and a special artery runs along the anterior edge, *i.e.* along that edge of the leaflet which is fastened to the suspensor. Each leaflet is wrinkled in such a way that the folds on the two surfaces alternate, each fold being creased in its turn. These two systems of folds cross each other at right angles, and serve to increase the respiratory surface.

At the point where the suspensor of the gill passes into the integument of the visceral dome, its contains a cellular body, which is traversed by a system of inter-cellular blood-channels. This may perhaps be a blood-making gland. It receives venous blood from branches of the principle branchial artery and of the special arteries of the leaflets, and returns the same along two veins which run back to the base of the gill, there, with others, to open into the venous sinus of the renal organ; from this organ the blood passes for the second time along the branchial artery into the gill. We thus find that not all the venous blood which is conducted by the branchial artery towards the gills enters the leaflets for purposes of respiration; part of it streams through the "blood-making" gland, and returns to the venous branchial heart still unpurified.

There are, further, certain find branchings of the branchial artery which serve for nourishing the gill and its suspending membranes. The blood in these returns to the venous sinus through a special vessel which runs parallel to the branchial artery on its anterior side. A powerful nerve enters the gill at its base and ramifies through it. A muscle spreads over the surface of the "blood-making" gland, and a special musculature brings about the contractions of the principal branchial vein.

The gills of the *Octopoda* differ considerably, though not essentially, in structure from those of the *Decapoda*. The branchial channel is much larger, and the leaflets are not only folded, but have on each side alternating lamellae, which in their turn may carry similar lamellae of the second order, and so on till in some cases the seventh order of subsidiary lamellae is reached. The leaflet is thus an extremely complicated, folded, or feathered structured with its surface increased to an extraordinary degree.

Adaptive Gills

The *Scaphopoda* and many *Gastropoda* possess no true ctenidia. In the *Pulmonata* and the few air-breathing *Prosobranchia*, the ctenidia, as organs adapted for aquatic respiration, have disappeared. It is, however, at present difficult to determine the cause of their disappearance in *Opisthobranchia* which inhabit water, and in the gill-less forms of the *Pteropoda*, all the more so, as in most *Opisthobranchia* they are replaced by adaptive gills, which are new structures in no way comparable morphologically with ctenidia. These adaptive gills may even appear (*Pneumoderma*) before the true ctenidia have disappeared. The *Scaphopoda* and many *Opisthobranchia* have no gills whatever, and in these respiration evidently takes place at various suitable parts of the surface of the body. In many cases, also, where epipodial or parapodial processes are developed as well as gills, or the mantle possesses extensions, these may help the gills in the functions of respiration.

Adaptive gills are found in most *Ascoglossa* and in the *Nudibranchia*; also, as mentioned above, in the *gymnosomatous Pteropoda*. In the latter, they consist of small fringed or plain ridges at the posterior end of the body; these may be of various shapes; a description of them would be of no special interest to the comparative anatomist. The principal forms of adaptive gills of the Nudibranchia are: (1) the anal gills of the *Orididoe*; (2) the longitudinal rows of branchial leaflets to the right and left under the mantle fold of the so-called *Phyllidiidoe*; (3) the dorsal appendages or cerata of the *Nudibranchia* and most *Ascoglossa*.

Anal Gills

These take the form of delicate leaflets, generally feathered on both sides, which, in the *Dorididae*, form a rosette round the anus, which has a median dorsal position towards the posterior half of the body. Cerata may occur with the anal gills (*Polyceridae*). The view that these gills are ctenidia has as yet no sufficient foundation.

Longitudinal Rows of Branchial Leaflets

These organs, which lie to the right and left of the body in the *Phyllidiidoe* and *Pleurophyllidiidoe*, bear the same relation to the (lost) true ctenidium as do the respiratory structures of the *Patellidoe* above described to the same organ, which in them is sometimes present, sometimes wanting. The longitudinal rows consist of numerous small lamellae which project from the lower side of the enveloping mantle fold into the shallow pallial cavity. There is either one long row of these lamellae running along the whole length of the mantle fold and only interrupted anteriorly (*Phyllidia*), or a row interrupted posteriorly as well (*Pleurophyllidia*); or again, the rows of lamellae are confined to the posterior end of the mantle fold (*Hypobranchioea*). The genus *Dermatobranchus* has no gills.

Dorsal Appendages (Cerata)

These processes vary very much in form, being sometimes simple and sometimes branched; they differ also greatly in number and arrangement. At their tips there are often enidophore sacs; these are invaginations of the ectoderm in which stinging cells with stinging capsules are developed. Diverticula of the intestine (digestive gland) enter the cerata, and may open outward at their tips. The cerata are generally striking and beautiful both in colour and markings. In some cases they may serve for protection and concealment, in others, where the brilliant colouring is combined with stinging properties, they may serve as a warning. They often break off easily at the base (as a protective arrangement), and are always quickly regenerated. They no doubt assist, like the rest of the body surface, in respiration, especially where they are much branched and richly supplied with blood-vessels. Certain *Opisthobranchia* and altogether gill-less, *e.g.* the *Elysiidae*, *Limapontidae*, and *Phyllirrhoidae*. Among the *Pulmonata*, the shell-less genus *Onchidium* has developed adaptive gills. The species of this genus are amphibious, living on the sea-coast, within reach of the tide. Their pulmonary cavity is very small; respiration therefore takes place by means of the richly vascularised dorsal integument, and especially of the simple or branched dorsal papillae, in which there is a rich vascular network, which receives the blood from an afferent vessel and gives it off to an efferent vessel.

Lungs

The total disappearance of the typical molluscan ctenidium is characteristic of the *Pulmonata*, and is connected with their terrestrial life and aerial respiration. Instead of water, air enters and escapes from the mantle cavity which lies either anteriorly or laterally on the visceral dome, and thus the mantle cavity becomes a pulmonary cavity. The free edge of the mantle fold, which forms the roof of this cavity, unites with the nuchal integument beneath it, except at one point on the right, where the respiratory aperture, which can be closed at will, allows of the entrance and egress of air. Along the line of its concrescence with the integument, the edge of the mantle is much thickened, forming the mantle, border and is very rich in lime secreting glands. The inner delicate surface of the mantle, which forms the roof of the cavity, is overspread by a close respiratory vascular network. A circular vein runs along the mantle collar. From it spring numerous fine anastomosing vessels which ramify on the mantle. These vessels are again collected into larger trunks, which enter the large pulmonary vein. This vein runs upwards are backwards, along the right side of the pulmonary cavity, to the left of and almost parallel with the rectum, and enters the auricle.

The circular vein contains venous blood, but the pulmonary vein conducts blood which has become arterial through respiration in the vascular network, to the heart. Since, in most *Pulmonata*, as in the *Prosobranchia*, the respiratory organ and the pallial cavity in which it is found lie in front of the heart, this order is prosopneumonic. An account of the opisthopneumonic condition of certain *Pulmonata*, which results from the displacement of the visceral dome and mantle to the posterior end of the body. Certain *Pulmonata* (*Limnuridae*) have become readapted to aquatic life, but their respiration is the same as that of the terrestrial forms, they rise periodically to the surface of the water to take in air. The respiratory cavity is, however, filled with water when the animal is young, and it is then a water breather.

In *Limnoea abyssicola*, a deep-water form found in the lake of Geneva, this form of aquatic respiration continues throughout life, and the pulmonary chamber, in no way modified, is constantly filled with water. In certain terrestrial *Prosobranchia* (*Cyclostoma*, *Cyclophorus*, etc.) the respiratory cavity becomes transformed, as in the *Pulmonata*, into a pulmonary chamber, and its roof is covered with a respiratory vascular network. But there is here no concrescence of the edge of the mantle with the nuchal integument. *Cyclostoma* still retains a rudiment of a prosobranchiate gill, but this is lost in *Cyclophorus*. The amphibian *Ampullaria* possess both a gill and a pulmonary sac, and can breathe either water or air.

11

CIRCULATORY SYSTEM

Molluscs differ abruptly from annelids in that their supporting and connective tissue is mostly cellular in nature. That mesenchymatous tissue contains a large number of figures and lacunae of schizocoelic origin, filled with blood. Phylogenetically the vascular system of molluscs has developed from these cavities. Its development consisted in gradual *vascularisation* of the lacunae; the originally-irregular cavities gradually took the form of sinuses or canals, along which the blood could easily move in a definite direction: later the sinuses decreased in diameter and their walls, which were formed of intermediate substance of connective tissue, became organised and acquired regular contours—the sinuses had become converted into blood vessels. Endothelium has also been described in some molluscan blood vessels. The stimulus to the above development of the lacunar system was the commencement of fluid circulation in it. We find the simplest form of vascular apparatus among molluscs in Solenogastres. It contains no vessels.

There are two longitudinal sinuses, ventral and dorsal. The posterior part of the dorsal sinus is invaginated into the dorsal side of the coelom and possesses musculature derived from the coelom wall. It is therefore contractile, and forces blood forward along the dorsal sinus. This contractile part of the dorsal sinus is called the ventricle of the heart. Solenogastres have no auricle of the heart; blood is pumped into the ventricle from the efferent branchial sinuses, which still have no proper walls. The heart, as in all molluscs, is covered externally with peritoneal epithelium. On the side turned towards its cavity the wall of the heart is formed directly by its musculature. The same heart-wall structure is found also in other molluscs. The structure of the rest of the vascular apparatus is very imperfect, and it is able to produce only very slight blood circulation. Among Loricata, *Nuttalochiton hyadesi* also has no definite blood vessels except a heart.

In all other Loricata the dorsal sinus has become a true blood vessel with well-formed walls, and is called the dorsal aorta. The aorta runs straight forward and discharges into the cephalic sinus, a system of schizocoelic cavities surrounding the brain. On the way the aorta gives off a number of branches, to the genital gland, etc. The cephalic sinus is separated from the rest of the body by a diaphragm, a tubular invagination of which forms the visceral artery, which supplies blood to the gut and the liver. The rest of the blood circulation is purely lacunar. The blood collects from small into large lacunae, mainly into the three longitudinal sinuses of the foot; thence it flows into the afferent branchial sinuses, which run along the inner margin of the mantle cavity. Passing through the gills, the oxygenated blood collects in the efferent branchial sinuses and through them reaches the auricle. The muscular auricles pump the blood into the ventricle.

The combination of thick-walled ventricle with thin-walled auricle is widespread in the animal kingdom. The physiological significance of this adaptation is: to create blood flow in the arterial system, a ventricle with fairly thick muscular walls is required; considerable pressure, which cannot exist in the venous system, is required to stretch it during diastole. The auricle has musculature strong enough to stretch the relaxed walls of the ventricle and thin enough to be capable of stretching under the action of venous pressure. Thus we see in Loricata the beginning of canalisation of the systems of sanguiferous lacunae, which takes two directions: first, organisation of an arterial system carrying blood from the heart; second, organisation of a venous system carrying blood first to the gills and then to the heart. The parts of both systems nearest to the heart are organised first, and those farther away later.

Among Gastropoda we still find forms with relatively-short and poorly-branched arterial and venous trunks; the cephalic sinus is most fully retained in the most primitive Aspidobranchia (*Fissurella*, *Haliotis*). At the same time higher Gastropoda have much-branched arterial and venous systems. In most Gastropoda venous blood collects in a system of perivisceral sinuses surrounding the gut, the liver, and the genital gland. Some of the blood flows thence directly into the afferent branchial sinus or vessel; some goes there after passing through the kidney; and some goes directly into the efferent branchial vessel and through it into the auricle. Thus in most Gastropoda the blood in the heart is predominantly, but not entirely, arterial.

In some Opisthobranchia, however, (e.g. *Gastropteron*), all the venous blood passes through both the kidney and the gills before entering the heart. Therein they resemble Cephalopoda. The location of the afferent and efferent branchial vessels depends entirely on the structure of the branchial apparatus, and varies greatly in the different groups of Gastropoda. In Pulmonata and other higher Gastropoda the blood movement produced by withdrawal of the body into the shell is much more forcible than any that could be produced by contraction of their hearts, but on the other hand the transfer of blood from one part of the body of these animals to another as a result of heart action is an important component element in their locomotor activity.

The blood of Gastropoda sometimes contains dissolved respiratory pigments: haemoglobin in *Planorbis*, haemocyanin in *Helix*. The blood of *Loligo* (Cephalopoda) also contains haemocyanin. Among Cephalopoda capillaries appear in *Nautilus*, but only in the skin, and only there does the blood pass from arteries into veins, bypassing the lacunae. Dibranchiata, especially Decapoda, have a capillary system not only in the skin but also in the musculature. In the head region, however, they have a large sinus that collects venous blood from the head and the foot. That sinus gives rise to a well-formed cephalic vein that ascends into the visceral hump and there divides into two venae cavae that enter the branchial hearts, contractile sacs lying at the base of the gills. The venae cavae receive a number of veins from the viscera. Passing near the kidneys, the venae cavae and other venous trunks enter them by botryose expansions; they are used to cleanse the blood of excreta. The branchial hearts force the blood through the vessels of the gills, after which it enters the vessels through the branchial vein and flows from there into the auricles and then into the ventricle. All blood entering the ventricle passes first through the kidneys and the gills.

Octopoda differ from Decapoda only in details. Thus Cephalopoda have very complete and perfect blood circulation. Like fish, they have only one blood circuit, but whereas in fish the heart contains venous blood, in Cephalopoda it contains arterial blood. The difficulties caused by the need to force blood through two systems of capillaries, body capillaries and gill capillaries, have been overcome in Cephalopoda by their possession of supplementary branchial hearts. The perfected blood-circulation system and, in particular, the fine ramification of blood vessels in Cephalopoda, reaching partially the level of closed blood circulation, is linked with the general high level of organisation of the group and is one of the features that permit some of them to reach gigantic size. Very large animals can exist only if they possess a capillary system, since only then is the nutrition and respiration of massive organs possible. In general, the maximum body size attainable by any group depends directly on the degree of perfection of their circulatory apparatus.

Vertebrates, which possess a highly-developed circulatory apparatus including capillary blood circulation, produce the largest animals on earth. Cephalopoda, and in particular Dibranchiata, possessing partial capillary circulation and a highly-developed blood-circulation system, take second place with such giants as *Architeuthis*. As we shall see later, this rule also obtains in various groups of arthropods. The centre of organisation of the circulatory system of molluscs is the activity of the heart.

All molluscs have a heart, except a few of the most reduced forms. The heart lies in the pericardium (coelom) in all molluscs except Octopoda and *Anomia*. The pericardium provides freedom of heart movement and at the same time plays a hydraulic role: during systole of the ventricle the volume of fluid within the pericardium decreases, so that negative pressure is created; this produces a flow of blood from the veins into the auricle, which also lies within the pericardium. The heart usually lies dorsally to the hind-gut, and rarely ventrally to it; in Rhipidoglossa and most Lamellibranchia the ventricle of the heart is penetrated by the hind-gut, i.e. it has a peri-intestinal location. The latter type of structure is probably primary. We may assume that the ventricle of molluscs is homologous with the intestinal sinus of annelids, as described in *Dinophilus* where, in the absence of post-larval segments, it is located in the larval body, like the mollusc heart. Having decreased in size and assumed in the role of a powerful pumping organ, the intestinal sinus has become the heart ventricle. The latter has become free of the gut in most molluscs, usually moving dorsally, rarely ventrally. It is characteristic that forms retaining the peri-intestinal position of the ventricle include Rhipidoglossa (and *Neopilina*), which are so primitive in many other respects. As a rule two auricles open symmetrically into the ventricle.

In *Neopilina* and *Nautilus* there are four, and in a number of Gastropoda only one. The auricles are muscular sections of the efferent branchial vessels. In most molluscs a single anterior aorta (aorta cephalica) arises from the heart: these include Amphineura, Gastropoda, *Neopilina*, *Nautilus*, and a few relatively-primitive Lamellibranchia. In Gastropoda the aorta gives off a large branch, the arteria visceralis. In Cephalopoda Dibranchiata and most Lamellibranchia two aortae, anterior and posterior, arise from the heart. In some Cephalopoda several other arteries, which usually arise from the aorta, arise directly from the ventricle.

Circulation in Different Groups

Amphineura

Chitonidae (Polyplacophora)

The heart is symmetrical, with two lateral auricles. The ventricle and the two auricles are long tubes. The auricles are in open communication with the ventricle about the middle of their length. Besides this, the two auricles pass into one another posteriorly, the posterior end of the ventricle also opening into them at this point. The ventricle lies against the dorsal wall of the pericardium, to which it is attached by a median band of endothelium. The ventricle passes into an aorta which allows the blood to flow into the coelom through apertures in its wall.

With the exception of the pedal arteries, the rest of the circulatory system is lacunar; there are no vessels with walls of their own. The venous blood is collected from the lacunar system of the body (primary coelom) into longitudinal channels which run on each side under the pleurovisceral cords. From these channels it flows into the gills, where it becomes arterial, and returns through other longitudinal channels which run above the pleurovisceral cords. Two transverse channels in the region of the heart convey the arterial blood into the auricles. The two pedal arteries lie laterally and ventrally with regard to the pedal cords; they probably draw their blood from the aorta and pass it on to the lacunar system of the foot.

Solenogastres

The heart lies above the hind-gut on the dorsal side of the pericardium. It does not lie freely in the latter, nor is it suspended by an endothelial band, but simply projects into the pericardium from above, so that only its under surface is covered by the pericardial endothelium. The presence of two auricles has not been proved. The rest of the circulatory system is purely lacunar. Specially large blood channels lie in the depths of the principal septa which project into the

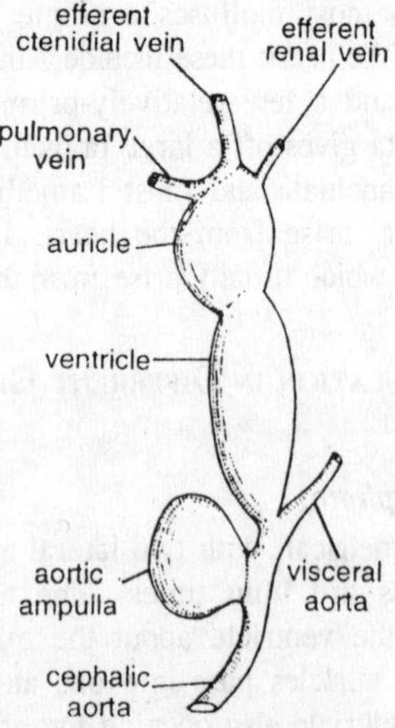

Fig. 11.1. Pila. Heart.

mid-gut, and bulge these out. Large blood sacs are also occasionally found in folds which project into the pharyngeal cavity from its wall, and there are more or less large sinuses in the folds, which, in *Neomenia* and *Choetoderma*, project into the cloaca and may be regarded as gills. In all these parts the intestinal epithelium separating the sinus from the intestine is ciliated, and respiration no doubt takes place.

Gastropoda

The lowest Gastropods, *i.e.* the *Diotocardia* among the *Prosobranchia*, have a heart with two auricles. This is not only the case in the *Zeugobranchia* (*Fissurella*, *Haliotis*, etc.) which have two gills, but also in the *Azygobranchia* (*Turbinidoe*, *Trochidoe*, *Neritidoe*), in which only the left (originally the right) gill has been retained. No branchial vein then enters the smaller (rudimentary) auricle on the right, the veins having atrophied with the gill. In the *Zeugobranchia*, the long ventricle lies in a line with the hind-gut, which runs lengthwise through it. In the *Azygobranchia*, the ventricle lies transversely with respect to the hind-gut which runs through it, the left auricle lying in front of the ventricle, and the right auricle behind it. The left branchial vein enters the anterior (left) auricle. If we suppose the posterior (right) auricle to have disappeared altogether, as is the case in all other Gastropoda, the heart consists of a ventricle and one auricle lying in front of it, which receives the branchial or pulmonary

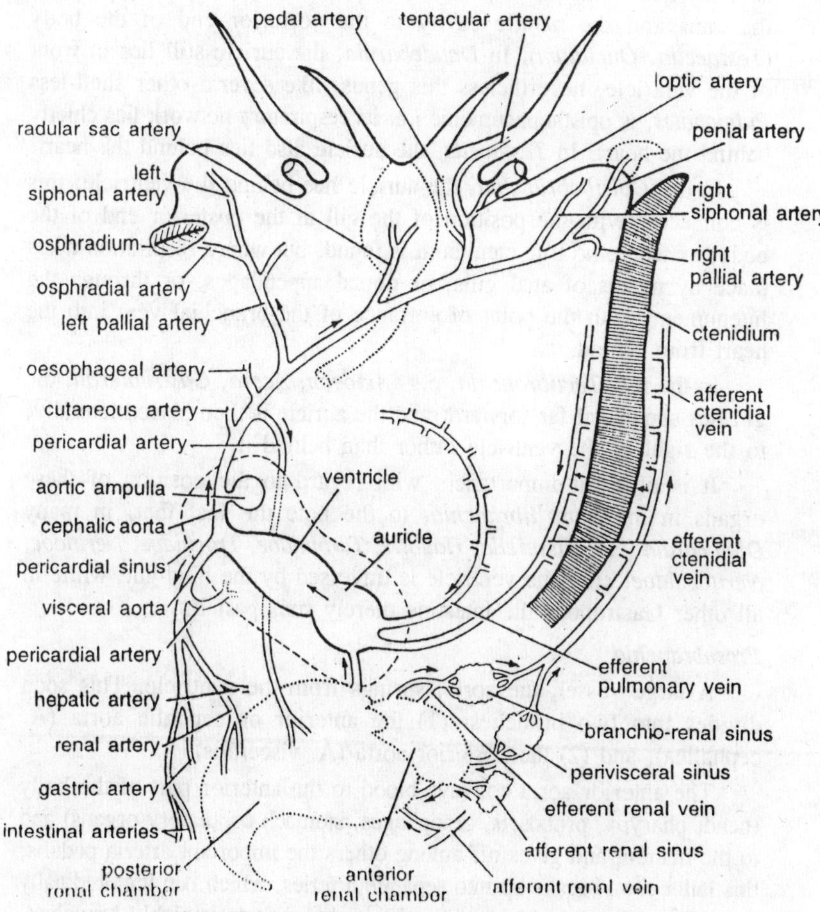

Fig. 11.2. Pila. Heart and blood-vascular system.

vein from the gill or lung in front of it. This serial order of the ventricle, auricle, branchial or pulmonary vein and respiratory organs is characteristic of the *Azygobranchia*, *Monotocardia*, and most *Pulmonata*.

The *Docoglossa* (*Patella* and allied forms) have only one auricle; the ventricle in *Patella* (not in *Aemoea*), however, is divided into two parts. Among the *Monotocardia*, only *Cyproea* (as far as is at present known) has a rudimentary right auricle, closed on all sides except at its aperture into the ventricle. Among the *Pulmonata* there are forms in which the auricle lies behind the ventricle. This must be regarded as a secondarily acquired position, determined by the shifting back of the anus and the mantle cavity to the posterior end of the body (*Testacella*, *Oncidium*). In *Daudebardia*, the auricle still lies in front of the ventricle, nevertheless this genus, like several other shell-less *Pulmonates*, is opisthopneumonic *i.e.* its respiratory network lies chiefly behind the heart. In *Testacells*, the auricle also lies behind the heart.

In the *Opisthobranchia*, the auricle lies behind the ventricle; this is connected with the position of the gill at the posterior end of the body, or where no true ctenidium is found, but where respiration takes place by means of anal gills, or dorsal appendages, or through the integument, with the point of entrance of the branchial vein into the heart from behind.

In the few *Tectibranchia*, e.g. *Actoeon*, *Acera*, *Gastropteron*, the gill lies somewhat far forward, and the auricle is then placed laterally, to the right of the ventricle rather than behind it.

It is of great importance, with regard to the position of these organs in the *Lamellibranchia*, to the note the fact that, in many *Diotocardia* (e.g. *Fissurella*, *Haliotes*, *Turbinidoe*, *Trochidoe*, *Neritidoe*, *Neritopsidoe*, etc.) the ventricle is traversed by the hind-gut, while in all other Gastropods the intestine merely runs past it.

Prosobranchia

A large vessel, the aorta, springs from the ventricle. This soon divides into two branches: (1) the anterior or cephalic aorta (A. cephalica), and (2) the posterior aorta (A. visceralis).

The anterior aorta conveys blood to the anterior part of the body (head, pharynx, proboscis, oesophagus, stomach copulatory organs) and to the mantle, and gives off among others the important arteria pedalis; this latter soon breaks up into separate arteries, which run longitudinally through the foot. In some cases the cephalic aorta is richly branched, breaking up into numerous fine vessels which spread out in and on the above-mentioned organs; in others, the arteries, without branching, open into arterial sinuses. Among these, the large cephalic sinus into which the anterior aorta opens (*e.g.* in *Haliotis*) deserves special mention. Where the cephalic aorta runs beyond the oesophageal ring formed by the central ganglia and their commissures, it passes through this ring.

The aorta visceralis supplies the organs which lie in the visceral dome, especially the digestive gland, the genital glands, and the mid-gut. The venous blood collects in the lacunar spaces of all parts of the body, and flows into a large venous sinus, *i.e.* into the space in which the stomach, salivary glands, intestine, digestive gland, and genital organs lie. This space or primary body cavity is somewhat spacious round the stomach, but very limited in the visceral dome, where the lobes of the digestive gland, the walls of the intestine, and the genital glands with their accessory parts are so crowded together as to leave very narrow spaces between them. The blood passes out of the large venous sinus back into the heart by three channels.

1. A large part of it flows through lacunae or vessels into the paired or unpaired branchial artery (afferent branchial vessel). In the course of branchial respiration the blood becomes arterial, and collects in an efferent branchial vessel (*cf.* section on the respiratory organs), which, as branchial vein, conducts it to the auricle of the heart. Where there are two gills, there are naturally two branchial arteries and two branchial veins, the latter conducting the arterial blood to the two auricles.
2. Another part of the venous blood flows through the kidney, then again collects in lacunae or vessels which lead to the gills, and finally reaches the heart through the bronchial veins. Less frequently, the venous blood, after passing through the kidney, enters the auricle more or less directly, *i.e.* without passing through the gills, and there mixes with the arterial blood coming from the gills.
3. A certain part of the venous blood, passing by both the kidney and the gill, flows direct into the branchial veins leading to the auricle. The arterial blood in the heart is thus mixed with venous blood.

Pulmonata

The blood vascular system is like that of the *Monotocardia*. The only important deviation is caused by the occurrence of pulmonary respiration. Various veins collect the venous blood out of the large body sinus and the lacunar system and unite to form one large vein, which accompanies the hind-gut, and as vena circularis, runs along the thickened edge of the mantle which concresces with the nuchal integument. From this vein spring numerous venous vessels which spread out on the under surface of the mantle, *i.e.* on the roof of the mantle cavity, and there form a delicate respiratory network. In this network the blood becomes arterial, and is next conducted through many vessels into the large pulmonary vein (vena pulmonaris), which runs back almost parallel to the rectum along the roof of the mantle cavity, to enter the auricle.

The vessels of the respiratory network form projecting ribs on the surface of the mantle. The pallial epithelium in the mantle cavity is ciliated. The efferent pulmonary vessels, which, near the kidney, run along the right side of the pulmonary vein, first enter the kidney and break into a fine vascular network before passing into that vein. The cephalic aorta does not pass through the oesophageal ring, but runs between the pedal and visceral ganglia; this is said to be the case in most *Opisthobranchia*.

In *Opisthopneumonic Pulmonata* (e.g. *Daudebardia*, *Testacella*), in which the small or rudimentary visceral dome has shifted to the posterior end of the body, and the organs elsewhere found in the dome (liver and genital organs) now lie in the body cavity above the foot, and thus in front of the posteriorly placed heart, the posterior aorta (A. visceralis) is much reduced, but the anterior aorta (A. cephalica) is strongly developed. The posterior aorta supplies only the posterior lobes of the liver and the hermaphrodite gland, and the anterior aorta (cephalic aorta, A. ascendens) has thus to supply the anterior lobes and even part of the genital organs, which usually receive their blood from the posterior aorta. In *Oncidium*, there is an arteria visceralis corresponding with the posterior aorta, which branches off soon after the aorta leaves the heart, but it here runs anteriorly.

Opisthobranchia

Here also the arrangement is essentially the same as in the *Prosobranchia*, though modified by the different position of the gills, as has been already briefly noted. *Gastropteron* affords a good illustration of the circulatory system of the *Tectibranchia*. The heart, which is enclosed in a spacious pericardium, lies to the right, in front of the above the base of the gill. It lies transversely, the larger and more muscular ventricle to the left, the auricle to the right. Out of the ventricle springs the aorta, which at once divides into a posterior and an anterior aorta. The anterior aorta enters the cephalic cavity, giving off as its principal arteries: (1) the artery of the copulatory organ. (2) The two large pedal arteries, each of which again soon divides into two branches, viz. (*a*) an anterior artery, which branches richly in the parapodia; (*b*) a posterior artery, which runs back on each side parallel to the median line of the foot. (3) The arteries of the cephalic disc. (4) The arteries of the oesophageal bulb and of the oesophagus. (5) The anterior end of the aorta itself branches in the tissues surrounding the mouth. The following are the chief branches of the posterior aorta: (1) The gastric artery. (2) The hepatic arteries. (3) The genital arteries. The venous blood flows back from all parts of the body through richly branched channels into two large venous sinuses, one of which represents the cephalic and the other the body cavity.

Wide but short vessels convey the venous blood out of these sinuses into the kidney, which contains a rich venous lacunar system. From the kidney it flows direct into the afferent branchial vessel, becomes arterial in the gills, and collects in the efferent branchial vessel, which, as the branchial vein, soon enters the auricle. All the venous blood in *Gastropteron*, therefore, on its way back to the heart, passes first through the kidney and then through the gill, so that only arterial blood flows through the heart. This is, however, not by any means the case in the *Tectibranchia*. For example, in *Pleurobranchus*, a large part of the venous blood passes from a dorsal circular sinus through a very short but wide passage direct into the branchial vein close to its point of entrance into the auricle, passing by both the kidney and the gill.

Dorididae

Without going into details as to the circulatory system of this group, it may be mentioned that part of the venous blood passes directly through two lateral vessels into the auricle. Another part flows into an inner venous circumanal sinus, which lies at the base of the circle of gills. From this the blood rises into the gills, becomes arterial, flows back into an outer circumanal vessel, and thence back through the branchial vein into the auricle.

Nudibranchia

The heart, enclosed in the pericardium, almost always lies in front of the centre of the body, in the median line. The aorta, which springs from the ventricle, divides into an anterior and a posterior aorta, each of which breaks up into an arterial system, the arteries having walls of their own. The finer branches of these arteries open into the lacunar system of the body, which occasionally forms canals resembling vessels, and is connected with the large cephalic and visceral sinuses. Veins, apparently with walls of their own, run from the lacunar system of the dorsal appendages or the integument, and carry the arterial blood back to the auricle. The blood usually finally enters the heart through three "branchial" veins,—two lateral and one median posterior,—which open into the posteriorly-placed auricle.

Scaphopoda

The circulatory system of *Dentalium*, but for the recently-discovered rudimentary heart, is entirely lacunar, consisting of systems of canals, sinuses, and spaces, the special arrangement of which cannot here be described. The pericardium with the heart lies on the posterior side of the body, dorsally to the anus. If we imagine the intestine of *Dentalium* straight and horizontal, the heart would occupy the typical position on the dorsal side of the hind-gut. It has no auricles, and is merely a sea-like bulging into the pericardial cavity of its anterior wall. It is connected by fine slits with the surrounding sinuses of the body.

Lamellibranchia

Heart

In nearly all bivalves, the heart, which is traversed by the hind-gut, possesses two lateral auricles, and lies in a pericardium. There are, however, isolated exceptions to this rule. In *Nucula*, *Arca*, and *Anomia*, the ventricle lies over (dorsally to) the hind-gut. This dorsal position must be regarded as the primitive position of the Lamellibranchiate heart, since the above genera are among the most primitive bivalves, and further, since the heart of the *Amphineura*, the *Scaphopoda*, and the *Cephalopoda* also lies over or behind the hind-gut. The perforation of the heart by the hind-gut must have arisen by the bending of the ventricle down round the latter.

The heart in the above-mentioned genera is further distinguished by the fact that the ventricle is more or less elongated in the transverse direction, its lateral ends being swollen, while the central part, which lies above the intestine, becomes narrower and thinner. This modification goes furthest in *Arca Nooe*, where there seem to be two

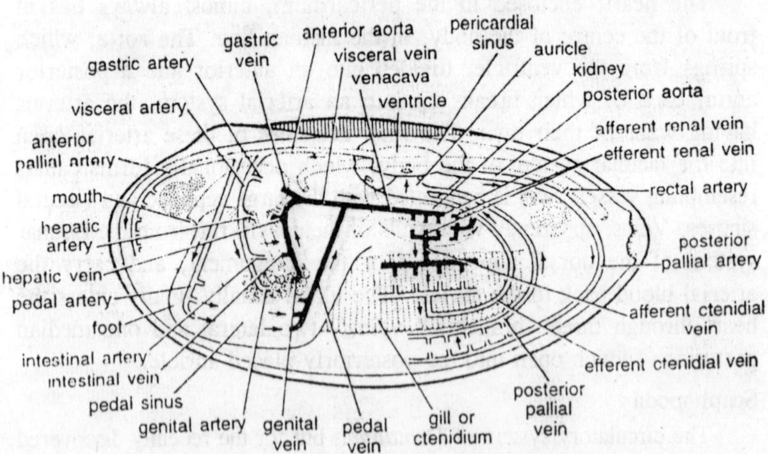

Fig. 11.3. Unio. Blood vascular system.

lateral ventricles unconnected by a central portion. This separation of the ventricle into two lateral parts has here brought about a separation of the two aorta. The two anterior as well as the two posterior branches, however, after a comparatively short separate course, unite to form an unpaired anterior and an unpaired posterior aorta. Although these genera have, as a rule, a heart lying above the hind-gut, in some specialised forms the heart is placed under the hind-gut, e.g. *Meleagrina*, *Ostrea*, *Teredo*. The cause of this modification must lie in the increasing distance between the base of the gills and the original region of the heart, the auricles and the ventricle having shifted with the latter.

The auricles, however, no longer lie laterally to the ventricle, but are drawn down to its lower side, where they grow together, communicating through a more or less large aperture. *Pinna*, *Avicula*, and *Perna* exhibit the consecutive stages in the displacement of the heart to the lower side of the hind-gut. The shifting of the gills from the original region of the heart just mentioned is caused by the shifting forward of the posterior adductor, which grows more and more massive and finally reaches a median position on the shell valve. It has already been mentioned that this posterior adductor, by the continuous reduction and final disappearance of the anterior adductor, becomes the one adductor of the *Monomyaria*. In *Teredo* also, the heart lies on the under side of the hind-gut. This is connected with the approximation of the hind-gut with the anus to the mouth dorsally, while the gills, remaining in their original position, retain the heart on the lower side of the hind-gut.

Circulation

The arteries have walls of their own, and branch into fine vessels, which discharge the blood into the lacunar system of the body. The venous system seems to have no distinct vessels with walls of their own, although it forms more or less wide channels resembling true vessels. An anterior and a posterior aorta spring, as a rule, from the ventricle. The anterior aorta runs forward above the intestine and breaks up into various arteries. The arteria visceralis supplies the intestine, the digestive gland, and the genital gland; the pedal artery supplies the foot; the anterior pallial artery spreads out over the anterior part of the mantle and the oral lobes (labial palps). The posterior aorta leaves the ventricle posteriorly and runs along the lower side of the hind-gut. It soon divides into two large lateral arteries,—the posterior pallial arteries.

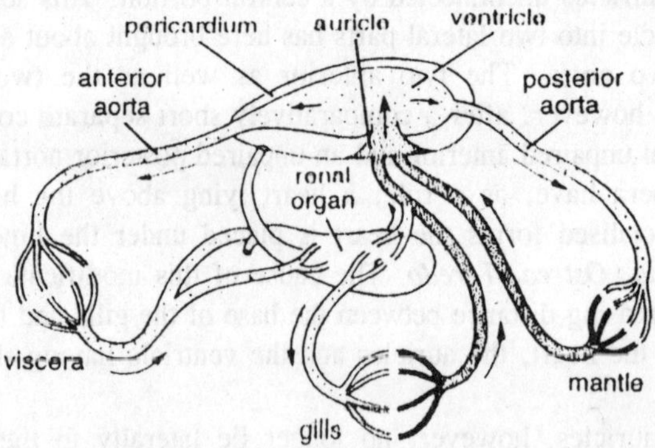

Fig. 11.4. Unio. Course of blood circulation.

The principal branches of the anterior and posterior pallial arteries run along the free edge of the mantle on each side and then unite, forming together the arteries of the pallial edge. From the roots of the posterior pallial artery smaller arteries spring, which supply with blood the hind-gut, the pericardium, the posterior adductor, the retractors of the siphons etc. The venous blood is collected out of the lacunar system of the body through converging channels into one longitudinal venous sinus; this lies under the pericardium. From this sinus, the greater part of the blood flows through the complicated system of venous canals in the kidneys, after which it is collected on each side into a branchial artery which runs along the base of the gills, and thence enters the two branchial lamellae. It becomes arterial through respiration in the gills, flows as arterial blood into a branchial vein parallel with the branchial artery, and thence into the auricle. Part of the venous blood, however, passes by direct channels out of the venous sinus into the branchial artery (passing by the kidneys), and part even flows direct into the pericardium.

In this way some venous blood comes to be mixed with the arterial blood flowing through the heart from the gills. Not all Lamellibranchia have an anterior and a posterior aorta springing out of the heart. In the lower groups of the *Protobranchia* and *Filibranchia* there are numerous forms (*Nucula*, *Solenomya*, *Anomia*, *Mytilidoe*) in which only one anterior aorta leaves the ventricle; this soon, however, gives off the arteria visceralis, which supplies blood to those parts which, in other *Lamellibranchia*, are fed by the aorta posterior. In their possession of a single aorta rising from the ventricle, the above lower Lamellibranchiates agree with *Chiton* and the *Gastropoda*. The rise of this aorta from the posterior end of the ventricle in the *Prosobranchia* and in most *Pulmonata* is a secondarily acquired arrangement, caused by the shifting forward of the pallial complex.

It must further be noted that in a very specialised bivalve, *Teredo*, the posterior aorta fuses with the anterior, and thus the two leave the heart as one vessel. In those Lamellibranchiates which have siphons, a muscular and contractile widening occurs in the posterior aorta near the point where it leaves and ventricle; this is called the bulbus arteriosus. Its special function is perhaps that of bringing about pressure of blood, to assist in the extension of the siphons. The backward flow of the blood into the ventricle in the contraction of the bulbus arteriosus (systole) is prevented by a linguiform valve which projects from its anterior wall.

Cephalopoda

Heart

We must here again point out the important fact that *Nautilus* has a heart with four auricles, while the *Decapoda* and *Octopoda* a heart with only two auricles. This difference is connected with the difference in the number of the ctenidia: four in *Nautilus* (*Tetrabranchia*), two in the *Decapoda* and *Octopoda* (*Dibranchia*). In *Nautilus*, the heart is an almost square sac drawn out to two points on each side; the four auricles which open into the four points of the ventricle are long tubes, more like widened branchial veins than auricles. The strongly muscular ventricle of the *Dibranchia* is almost always elongated into a tube.

In the *Octopoda* it lies transversely, the two auricles being in the same plane with the ventricle. In the *Ocgopsidoe*, the ventricle lies along the longitudinal axis of the body, *i.e.* it is elongated dorso-ventrally, and the auricles are at right angles to it. The heart of the *Myopsidoe* occupies a position halfway between those just mentioned. The heart here described is the arterial heart, which corresponds with the heart of the other Mollusca. It is called arterial to distinguish it from the venous hearts, which will be described below.

Circulation

It is important to note that the circulatory system is at least partially closed. There is not only a richly-branched arterial, but a richly-branched venous system, the vessels of which have walls of their own. These two systems pass into one another in certain parts of the body, *e.g.* the integument and certain muscle layers, through a system of capillary vessels. In other parts, however, the arterial

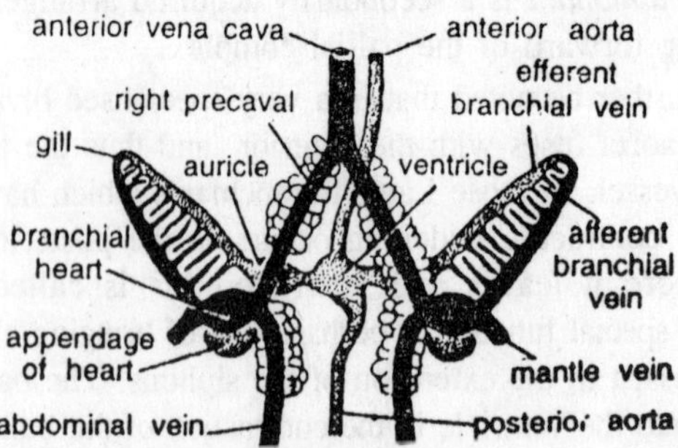

Fig. 11.5. Sepia. Circulatory system.

branches conduct the blood into a lacunar system; when it has become venous, the blood collects out of this into sinuses (especially into a peripharyngeal cephalic sinus), and flows to the gills through veins with walls of their own. Two aorta rise-from the ventricle: (1) the aorta cephalica, which runs downward (upwards in the figure) to the head, and (2) the aorta abdominalis, which runs up towards the apex of the visceral dome.

The former is much stronger than the latter. The aorta cephalica first gives off branches to the mantle and to the anterior wall of the body, and then provides the stomach, the pancreas, the digestive gland, the oesophagus, the salivary glands, and the funnel with arteries. After accompanying the oesophagus, it divides in the head into two branches, which run to the bases of the arms, and there break up into as many arteriae branchiales as there are arms. The aorta abdominalis supplies with arteries the hind-gut, the ink-bag, the genital organs, the dorsal part of the body wall, and the fins, when these latter are present. Only in the *Oegopsidoe* are the aorta limited to the two, above described, springing from the heart.

In the *Octopoda* and the *Myopsidoe*, there are other arteries rising out of the ventricle, and running to the same part of the body as the aorta, abdominalis in the *Oegopsidoe*; among these are the arteria genitalis, which runs to the genital glands, and, in the *Myopsidoe*, a fine vessel called the arteria anterior. At certain places, the arteries may swell out to form small muscular and contractile widenings, called peripheral arterial hearts. In the venous system of *Sepia*, the venous blood in each arm collects (partly through capillaries and partly through lacunae) into a vein running down the inner side of the arm.

All the brachial veins convey their blood to a circular cephalic sinus surrounding the buccal mass, which is the reservoir for collecting the venous blood from the whole head region. Out of this sinus springs the large vena cephalica, which runs up along the posterior side of the oesophagus and the liver into the visceral dome, collecting on the way venous blood from the liver, the funnel, etc. A little below the stomach it forks, forming the two venae cavae, which open into the two contractile venous hearts at the bases of the gills. From the upper part of the visceral dome the blood collects into several abdominal veins, the most important of which are an unpaired vena abdominalis, opening into the vena cephalica exactly at the point where it divides into the venae cavae, and two lateral abdominal veins, which open into the latter near their point of entrance into the branchial hearts. In the region of the heart, all these veins carry acinose or lobate appendages (venous appendages), which are hollow, and communicate at many points with the veins, so that they are richly supplied with blood.

The cavity into which these appendages project is that of the renal sacs, and the epithelium which covers them belongs to the epithelial wall of the kidneys. We thus see that here the blood flowing back from the body has abundant opportunity of giving off its excretory constituents to the kidneys. Appendages are found on both the branchial hearts; these are the pericardial glands, which will be further described later. The two branchial hearts, by their contraction, drive the venous blood into the afferent branchial vessel.

The blood, which has become arterial in the gills, flows through, the efferent branchial vessel (the so-called branchial veins) into the auricles of the heart, and thence into the ventricle (on the branchial circulation). In the Cephalopoda, unlike the other Mollusca, the whole of the blood, in returning from the body, flows through the gills, so that the heart contains only arterial blood. By far the greater part of the blood, before entering the gills, comes into contact with the kidneys in the venous appendages. In the *Octopoda*, the venous system shows some not unimportant modifications. In *Octopus*, two veins, connected with one another by anastomoses, run along the outer side of each arm and collect the venous blood. At the bases of the arm these veins become connected in pairs and unite later in such a way as to form on each side a lateral cephalic vein.

These two veins unite to form the large vena cephalica which runs up in front of the funnel and behind the oesophages. The brachial veins do not here, as in *Sepia*, convey their blood first to the venous cephalic circular sinus, but are directly connected with the cephalic vein. A cephalic sinus nevertheless exists in *Octopus*; it is not, however, connected with the vena cephalica, but with a large sinus which fills the whole visceral dome, and is, in fact, the primary body cavity, in which the viscera lie bathed by the venous blood. The latter flows out of this large venous sinus through two wide veins, the so-called peritoneal tubes, into the upper part of the vena cephalica, near the point where this divides into the two venae cavae. *Nautilus* is chiefly distinguished by the absence of the branchial hearts. Further, each of the two venae cavae divides into two branches, which run, as afferent vessels, to the gills.

12

EXCRETORY SYSTEM

The organs which serve for excretion are homologous in all Mallusca. They consist typically of two symmetrical sacs, which, on the one hand, open into the mantle cavity, through the two outer renal apertures, and on the other are connected by two inner apertures (renal funnels, ciliated funnels) with the pericardium or coelom. The nephridia always lie near the pericardium. Their walls are richly vascularised, indeed a large part of the venous blood, in returning from the body, flows through the renal walls and gives off excretory matter before it enters the respiratory organs. The renal walls are traversed exclusively by venous blood. The nephridia are paired in all symmetrical Molluscs, and also in those *Gastropoda* which have paired gills and two auricles (*Diotocardia*). In all other *Gastropoda*, along with the original right ctenidium (which, in the *Prosobranchia*, lies to the left), and the corresponding auricle, only one kidney (the corresponding one) is retained.

Nautilus, which has four gills and four auricles, has also four kidneys; only two of these, however, communicate with the visceropericardial cavity. A relation between the nephridial and genital systems similar to that found in the Annelida exists in the *Solenogastridae*, the nephridia functioning as ducts for the genital products, the latter passing from the hermaphrodite gland (genital chamber of the coelom) into the pericardium. In a few *Lamellibranchia*, *Diotocardia*, and in the *Scaphopoda*, there is a relation between the genital glands and the nephridia, the former opening into the latter; so that a certain part of the nephridium functions not only as renal or urinary duct, but also as efferent genital duct. In all *Diotocardia*, it is the right nephridium which functions as genital duct. In the *Monotocardia*, in which the right nephridium of the *Diotocardia* has atrophied as such, its duct persists as genital duct. In all other Molluscs the genital ducts are entirely distinct from the urinary passages.

AMPHINEURA

The Kidneys of the *Solenogastridae* and the *Chitonidae* differ greatly from one another in structure.

1. In the *Solenogastridae*, two canals spring from the pericardium, embrace the hind-gut, and open into the cloaca beneath it through a common terminal portion. These canals function as ducts for the genital products. It is also certain that they correspond morphologically with kidneys of other Molluscs, even though their excretory activity has not been proved. They are covered with an extraordinarily deep epithelium of long filiform glandular cells.

   In some *Solenogastridae*, an accessory gland opens into each nephridial canal.
2. In the *Chitonidae*, the strongly-developed paired nephridia function exclusively as excretory organs.

Each nephridium consists of a wide canal shaped like a long Y, the diverging portions being directed backward, and the undivided portion forward. These Y-shaped kidneys run longitudinally along each side of the body through its whole length. One of the paired limbs of the Y opens outward into the posterior part of the mantle cavity, the other into the pericardium, which also lies in the posterior part of the body. In this way the pericardial and outer apertures of the kidney lie near one another. The third limb of the Y ends blindly anteriorly. Secondary lobules or lobed canals open into all the three parts of the kidney, and are specially abundant in its anterior portion. Except in the terminal portion of the efferent branch, the epithelium of the limbs as well as that of the lobes is cubical and ciliated.

GASTROPODA

Prosobranchia

Diotocardia

Among all the Gastropoda, *Fissurella* alone possesses a symmetrical excretory apparatus, in the sense of having two nephridia opening into the mantle cavity to the right and left of the anus. The left nephridium is, however, much reduced, while the right, which is strongly developed, sends its lobes everywhere into the spaces between the lobes of the

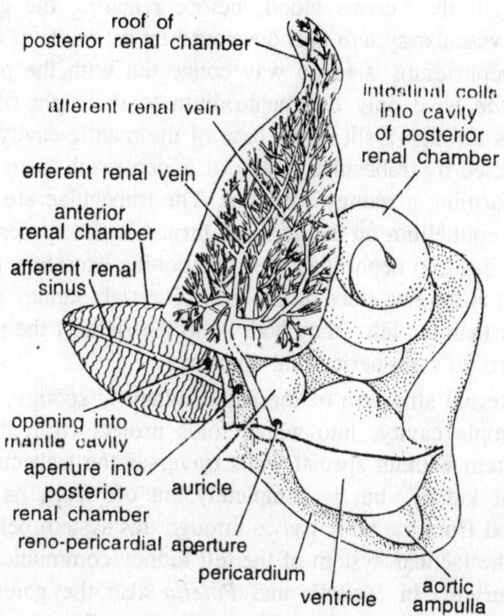

Fig. 12.1. Pila. Excretory organs.

liver, the intestine, and the genital organs. There are no reno-pericardial openings. The genital gland does not open direct into the mantle cavity, but through the right kidney. In *Haliotis*, *Turbo*, and *Trochus*, both nephridia are present. The left nephridium has, however, almost entirely lost its excretory function, but is still connected both with the pericardium and the mantle cavity. It is called the papillar sac, its walls projecting into its cavity in the form of numerous large papillae. The blood lacunae which penetrate into the papillae communicate direct with the auricles, and are thus supplied with arterial blood. In these lacunae of the papillae a crystalloid substance (albumen ?) is deposited.

It has been thought that these papillar sacs serve as reservoirs of nutritive material (in the form of the crystalloids just mentioned), and when needed yield it up to the blood. The right nephridium is exclusively excretory in functions. It is divided into two lobes, one behind the other, which communicate by means of a wide aperture; the anterior lobe lies under the floor of the mantle cavity, bulging it upward. A spongy network, covered with excretory epithelium, rises from part of its wall into the cavity of the nephridial sac. The meshes of the network are penetrated by a system of vessels with walls of their own.

Nearly all the venous blood, before reaching the gills, passes through the vascular system thus developed on the walls of the kidneys. The right nephridium is in no way connected with the pericardium. The *Neritidae* have only one nephridium to the right of the heart, which opens through a slit in the base of the mantle cavity. The renal sac is traversed by trabeculae, many of which reach from one wall to the other, forming a spongy structure, The trabeculae are covered by a glandular epithelium on the surfaces turned to the spaces of the sac. *Patella* still has two nephridia, both functioning excretory organs. The apertures lie at the two sides of the anus. The right kidney is, however, much larger than the left. They both lie to the right of the pericardium, but there are no reno-pericardial apertures.

The internal structure of the right kidney is spongy, but the left forms a simple cavity, into which folds project from the walls. A lacunar system without special walls traverses the trabecular network of the right kidney, but is completely cut off from its cavity; the venous blood from the body passes through this system before entering the gills. The lacunar system of the left kidney communicates directly with the auricle. In *Haliotis* and *Patella* also the genital products pass, as in *Fissurella* and the *Diotocardia* generally, out of the genital gland into the right kidney, and are ejected through the right renal aperture.

Monotocardia

The Monotocardia have only one nephridium functioning as an excretory organ, viz, the left of the *Diotocardia*. This takes the form of a sac lying immediately below the mantle cavity on the right side of the pericardium, directly under the integument. It is generally found to the left of the hind-gut; less frequently (*Cassidaria*, *Tritoniidae*) the kidney is traversed by the rectum, or the latter runs forward below it. The slit-like pallial aperture of the kidney, however, is always found to the *left* of the hind-gut, quite at the base of the mantle cavity. This position of the kidney, and especially of its outer apertures, had already led to the assumption that the Monotocardian nephridium corresponds with the left kidney of the *Diotocardia*, before this fact was established. The assumption was all the more plausible because of the occurrence of a gland called the anal kidney in a few *Monotocardia* (e.g. *Dolium*); this gland opens to the right near the anus, and might represent the right kidney of the *Diotocardia*.

The kidney is always connected by means of a canal (the reno-pericardial canal) with the pericardium Lamellae or trabeculae, covered with the glandular epithelium of the kidney, project inward from the lateral walls of the renal sac. These are especially strongly developed in fresh-water Prosobranchia (excepting *Valvata*) traverse the whole kidney, and impart to it a spongy structure. The venous blood always flows through the whole of the glandular part of the kidney, either in special vessels or in lacunae, before passing on to the gills; but an open communication with the renal cavity is never found. In the *Toenioglossa Proboscidifera* the kidney forms two lobes similar in structure. In *Natica* and *Cypoea* the lobes begin to differ, and among the *Stenoglossa* this difference becomes more and more marked in a way which need not here be described.

In *Paludina* and *Valvata* the kidney no longer opens into the posterior base of the mantle cavity, but is continued as a urinary duct (ureter), which runs forward in the mantle and opens at its edge. The above-mentioned theory that the single kidney of the Monotocardia corresponds with the left kidney of the *Diotocardia* has recently been ably opposed, another theory being put forward in its place. Attention is specially drawn to the fact that in the *Diotocardia* the left kidney is always the smaller, that in *Patella* it is shifted to the right side of the pericardium, and that in *Haliotis*, *Turbo*, and *Trochus* (as papillar sac) it is not excretory in function.

In *Haliotis*, *Turbo*, *Trochus*, and *Patella* the lacunar system developed in the wall of the left kidney is in direct communication with the auricles. In most Monotocardia there is a differentiated part of the kidney, viz. that which is called the *nephridial gland*. This consists of two principal parts : (1) canals, covered with ciliated epithelial cells and opening into the kidney. These are merely protrusions of the renal wall, which project into the organ; their epithelium is a continuation of the renal epithelium. (2) Between these canals, the organ is filled with cells of connective tissue and muscles, and contains blood lacunae, one of these being specially large and communicating with the auricle. This latter portion of the organ perhaps plays the part of a blood-forming gland. This nephridial gland may perhaps be the persistent excretory portion of the lost nephridium, *i.e.* the right of the *Diotocardia*. The duct of this lost nephridium is now known to persist as genital duct. As we saw above, all *Diotocardia* discharge the genital products through the right nephridium.

Pulmonata

The Pulmonata have only one kidney, which lies in the mantle at the base of the pallial cavity, between the rectum and the pericardium.

The renal sac is of the so-called parenchymatous type the excretory epithelium of its wall projecting into the cavity in the form of numerous folds and lamellae in such a way as to leave hardly any central free space. The kidney always communicates by means of a ciliated canal (renal funnel or renal syringe, "Nieren-Spritze") with he pericardium. The position of the kidney and the morphology of the urinary duct have already been explained.

Opisthobranchia—Tectibranchia

Only one kidney is found in the usual position on the right side of the body, with the pericardium in front of it and the hind-gut behind it. It is of the parenchymatous type, and is connected by a ciliated canal with the pericardium. It opens at the base of the gill in front of the anus. In the *Pteropoda* the delicate-walled kidney is not parenchymatous, but is a simple hollow cavity lined with epithelium, and always communicates with the pericardium, against which it lies.

Nudibranchia

The kidneys of the Nudibranchia are strikingly different in form from those of the *Tectibranchia*. The unpaired kidney is here somewhat similar to the paired kidney of the *Chilonidoe*. It is a somewhat wide tube (renal chamber) traversing the cavity of the body, to a greater or less extent; branches entering it from all sides. This tube is connected at one end with the pericardium by a duct (renal syringe, pyriform vessel), which varies in length, and at the other opens outward through a ureter at the base of or near the anal papilla. It is said that *Pleurobranchoea*, a *Tectibranchiate*, from which the Nudibranchia may perhaps be derived, possesses a Nudibranchiate kidney. In *Phyllirhoe*, the urinary chamber has no branchings, it runs back from the pericardium as a simple median tube. Anteriorly it is connected with the pericardium by a funnel, and near the middle communicates with the exterior by means of a lateral urinary duct.

SCAPHOPODA

Dentalium has a pair of symmetrical kidneys, on one each side of the hind-gut. Each nephridium consists of a sac provided with short diverticula. The two nephridia are connected by a tube above the anus, and open into the mantle cavity by two apertures at the sides of the anus. If, as maintained by all authorities, there are no reno-pericardial apertures, the Scaphopoda would be the only group of Molluscs in which these apertures are entirely absent. Apart from the symmetry of the kidneys, a fact to be specially noted is that the genital products pass out of the genital gland into the right kidney (either by the bursting of the wall between the two organs or through an aperture), and only reach the exterior, *i.e.* the mantle cavity, through the right renal aperture. It must further, be noted that near the anus on each side, between it and the renal aperture, a pore, the water-pore, occurs, the function of which is still doubtful. If these pores really lead into the blood lacunar system of the body, as was formerly maintained, and is still held to be possible, this would be the only known case of the direct imbibition of water into the blood.

LAMELLIBRANCHIA

The nephridium (*organ of Bojanus*) is always paired and symmetrical, and lies below the pericardium and in front of the posterior adductor. Each nephridium is tubular or sac-like, opening at one end through a funnel into the pericardium, and at the other into the mantle cavity. This communication of the kidney with the mantle cavity always takes place *above* the cerebrovisceral connective. The lowest Lamellibranchia (*Protobranchia*, *Nucula*, *Leda*, *Solenomya*) are distinguished in two ways. (1) Each nephridium is a simple tube, with a free cavity not traversed by *trabeculae* or *lamellae*. This tube consists of two portions which unite posteriorly at an angle; the anterior end of one of these portions enters the pericardium through the renal funnel, the other end opens into the mantle cavity. (2) The paired genital glands do not open outward directly, but enter the kidneys near their pericardial funnel—a fact which is very important in connection with the arrangement in the *Solenogastridae*, the lower *Prosobranchia* (*i.e.* the *Diotocardia*), and the *Scaphopoda*. In other Lamellibranchia also there is a relation between the genital glands and the kidneys.

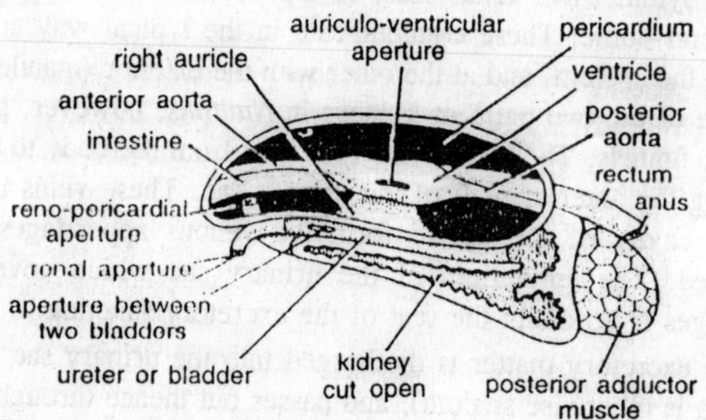

Fig. 12.2. Unio. Excretory organs.

In the *Pectinidae* and the *Anomiidae* the genital gland opens into the kidney, but near its outer aperture. In *Arca*, *Ostroea*, *Cyclas*, and *Montacuta*, the kidney and the genital gland open on each side into the base of a common depression (urogenital cloaca); in all other bivalves the outer nephridial and genital apertures are separate. The simple structure of the Protobranchiate kidney becomes complicated in other Lamellibranchia in the following manner:

1. That portion of the renal tube which opens outward forms and external cavity (vestibular cavity, external sac); this cavity has no excretory epithelium; it encircles the outer side of the pericardial portion of kidney, the renal sac. The latter alone is developed as an excretory organ. Folds or trabeculae, covered with glandular epithelium, project inward from its walls, forming a parenchymatous or spongy structure. The renal sac is connected with the pericardium by means of a nephridial funnel of varying length.
2. The two renal sacs communicate freely in the median plane. The connecting part is widest in the most specialised bivalves (*Pholadacea*, *Myacea*, *Anatinacea*, *Septibranchia*).

In *Anomia*, were all the parts are asymmetrical, the two kidneys, which do not communicate with one another, are also asymmetrical. Venous blood flows through the kidneys on its way to the gills. The afferent renal vessels seen to have walls of their own, but the efferent vessels appear to be lacunar. Open communication between the blood vascular system and the kidneys is nowhere found.

CEPHALOPODA

The Cephalopoda have two (*Dibranchia*) or four (*Tetrabranchia*) spacious symmetrical renal sacs, in the posterior and upper part of the visceral dome. These communicate in the typical way at the one end with the coelom, and at the other with the exterior (mantle cavity). Only one of the two pairs of kidneys in *Nautilus*, however, possesses coelomic funnels. The large veins returning from the body to the heart run along the anterior wall of the urinary sac. These veins bulge out into the cavity of the sac to form the venous appendages already mentioned. The epithelium of the urinary sac which covers these appendages is no doubt the seat of the excretory function.

The excretory matter is discharged into the urinary sac (the wall of which is otherwise smooth), and passes out thence through a ureter of varying length into the mantle cavity. The renal aperture is found on the median side of the base of the gill, and in *Nautilus*, the

Oegopsidoe, and *Sepiotenthis* among the *Myopsidae*, it is simple and slit-like; in the other *Myopsidae* and in the *Octopoda*, however, it lies at the end of a renal papilla which projects freely into the mantle cavity. The two renal sacs in the *Octopoda* are entirely distinct. Near the point where the renal sac passes into the ureter lies the renal funnel, which corresponds with the pericardial aperture of other Molluscs, and which here leads to the coelomic cavity, now reduced to the "water vascular system".

In the *Decapoda*, the two renal sacs communicate with one another in the median plane. In *Sepia*, there are two points of communication, one above and the other below. The lower junction is bulged out to from a large sac, which rises towards the apex of the visceral dome on the anterior side of the paired renal sacs. The veins returning from the body to the heart run in the partition between the unpaired anterior and the paired posterior sacs, and may here bulge out to from venous appendages, not only posteriorly, *i.e.* into the cavities of the two paired renal sacs, but also anteriorly, into that of the unpaired connecting sac. Near the point where each renal sac is produced into the ureter, the reno-pericardial canal springs from it, opening into the secondary body cavity which contains the heart, and corresponds with the pericardium of other Molluscs. The form of the renal sac is at least partly determined by the form and position of the surrounding viscera, the stage of maturity of the genital organs, and the different shape of these organs in the two sexes. All viscera which press against the renal wall from without, bulging it inward, are naturally covered at

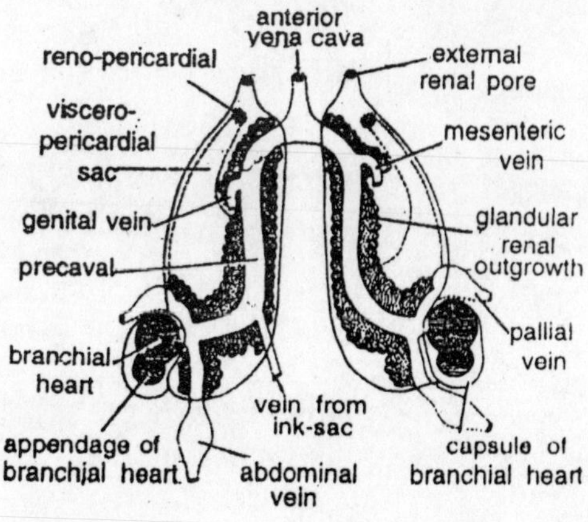

Fig. 12.3. Sepia. Excretory system.

the points of contact with the epithelium of the renal sacs. The same is the case with all organs which, like the stomach, the gastric coecum, and the efferent ducts of the digestive glands in the *Decapoda* (*Sepia*), apparently lie inside the spacious renal sacs.

These organs really lie outside of them, being only suspended into them, like the intestine of an Annelid, which apparently lies within the body cavity, but is entirely separated from it by the peritoneal endothelium. It has been already mentioned that only one of the two pairs of renal sacs of *Nautilus*, viz. the upper pair, has reno-pericardial apertures, The fact was brought forward in support of the view that the two pairs of renal sacs arose by the division of one single pair, corresponding with that of the *Dibranchia*. According to this view, the lower pair of gills, and the two auricles are also to be considered to be new acquisitions. Indeed, the whole question of the original metamerism of the Molluscan body, which has so often been asserted, rests on very weak foundations. It gains no support from the *Chitonidae*, where, in spite of large number of pairs of gills, only two auricles occur, and where no relation exists between the number of the shell plates and that of the gills.

13

Integrated System

Amphineura

The nervous system of the Amphineura is very significant from the point of view of the comparative anatomist. Its most important peculiarities may be briefly described as follows:

1. The ganglionic cells are either not at all or not exclusively localised in definite ganglia.
2. Four nerve cords run through the body from before backward. These contain not only nerve fibres, but ganglion cells distributed along their whole length. They might suitably be called medullary cords, and must be considered as belonging to the central nervous system. One pair of these cords run along the body laterally, these are the lateral or pleurovisceral cords; the second lie ventrally, and are the pedal cords. The visceral and the pedal cords of each side unite anteriorly, and when so united become connected with those on the opposite side by a transverse commissure, which runs in front of and over the oesophagus and contains ganglion cells; this is the cerebral or upper half of the oesophageal ring. The pleurovisceral cords unite posteriorly above the rectum, forming a visceral loop. The pedal cords are connected both *inter se* and with the pleurovisceral cords by anastomoses, so that the whole nervous system strikingly recalls the ladder nervous system of the Turbellaria and Trematoda.

Chitonidae

The scheme just given is founded upon the nervous system of Chiton. The typical ganglia of the central nervous system of the Mollusca are not yet, in Chiton, found as distinct ganglia united by means of commissures and connectives, but the ganglion cells are equally distributed along the commissures and connectives, an arrangement which is probably primitive. The upper oesophageal ring thus corresponds with the cerebral ganglia and the commissures connecting them, and in the same way the pedal cords contain the whole central portion of the pedal nervous system, and the pleurovisceral cords the central portion of the visceral, pallial and branchial nervous systems. Only in one single species of Chiton (*C. rubicundus*) two distinct (cerebral) ganglia occur near each other in the middle line in the upper half of the oesophageal ring. Looking more closely at the nervous system of the Chitonidae, we have to observe: (1) the arrangement of the oesophageal ring and the medullary cords; (2) the peripheral ganglia; (3) the nerves of the ladder-like nervous system; (4) the nerves running from the central nervous system (oesophageal ring and medullary cords).

Form and arrangement of the central nervous system

The visceral cords run back one on each side in the lateral body wall above the branchial groove; these two cords unite above the anus. The pedal cords run in the dorsal part of the pedal musculature somewhat near one another, from before backward, to end without uniting where the rectum commences. The oesophageal ring consists, in the first place, of the semicircular portion mentioned above, which, on account of the peculiar shape of the body of the Chiton, lies in the same plane as the visceral cords. Posteriorly, each limb of this semicircle divides up into the pedal and visceral cords.

At the point where the pedal cord rises from the ring, a cord with a thickened base separates from it and runs inwards; this, uniting below the mouth with a similar cord from the other side, forms the lower half of the oesophageal ring. The upper and lower halves together form the closed oesophageal ring. Besides this central nervous system there are peripheral ganglia connected with it by nerve cords consisting only of nerve fibres.

(*a*) On each side, from the lower half on the oesophageal ring, somewhat further in than the buccal connective, a nerve (the subradular connective) rises and runs forward and inward to the subradular ganglion. This ganglion lies in the subradular organ which is situated on the floor of the buccal cavity. The two subradular ganglia are united by a short commissure.

(*b*) The buccal ganglia together form a horseshoe-shaped ganglionic mass below the oesophagus, which mass is connected on each side by the cerebrobuccal connective with the thickened portion of the lower oesophageal ring. The buccal ganglionic mass in *C. rubicundus* divides into two paired ganglia and one unpaired ganglion joined to one another by connectives. The buccal ganglia innervate the oesophagus as far as the stomach and also the oral aperture.

(c) Two small gastric ganglia, connected by a fine commissure, lie at the anterior end of the stomach, and are joined an each side to the anterior end of the visceral cord by a long connective.

Nerves of the ladder-like nervous system

The two pedal cords are connected by anastomosing commissures along their whole length, but no nerves are given off by these commissures to the pedal musculature. In *Chiton rubicundus* the visceral and pedal cords are united by numerous connectives, which in other Chitonidae, appear either to be wanting or to be reduced to one single anterior or posterior anastomosis.

Nerves running from the central nervous system

(a) *Nerves of the pedal cords*. The pedal cords give off on each side seven or eight nerves outwards to the lateral musculature of the body, and specially numerous nerves run down from it to the pedal musculature (inner and outer pedal nerves). These pedal nerves are richly branched and, anastomosing with one another, form a complete neural network in the foot.

(b) *Nerves of the oesophageal ring*. Numerous nerves rise from the upper or cerebral portion of the oesophageal ring to innervate the cephalic part of the mantle, the snout, the upper and lower lip, the gustatory buds on the lower wall of the oral cavity, and the musculature of the buccal mass. The lower portion of the oesophageal ring, besides the connectives to the buccal and subradular ganglia, sends off from its median portion another pair of nerves, which run along the base of the buccal cavity.

(c) *Nerves of the pleurovisceral cords*. Each of the pleurovisceral cords gives off two nerves to each gill. Besides these they send many nerves to the mantle and, posteriorly, nerves which enter the body cavity, probably running to the kidneys and the heart.

Solenogastres

The central nervous system of the Solenogastres differs from that of the *Chitonidae* principally in a tendency to form distinct ganglia; the pedal and pleurovisceral cords, nevertheless, still retain their outer coating of ganglion cells along their whole length. Fig is a diagrammatic representation of the structure of the nervous system of *Proneomenia Sluiteri*. The fused cerebral ganglia in the middle line are very large. On both the pleurovisceral and the pedal cords ganglionic swellings can be distinguished: (1) three pairs of posterior visceral ganglia; (2) two anterior pedal ganglia. The posterior visceral ganglia are connected by cords, which run transversely over the rectum and correspond, to some extent at least with the loop by which the two visceral strands in *Chiton* are united.

The two anterior pedal ganglia are connected by a strong transverse commissure, which may correspond with the ventral half of the oesophageal ring of *Chiton*. Further, the pleurovisceral cords are joined with the pedal cords, and the latter are also connected *inter se* by transverse connections along their whole length. The pleurovisceral cords likewise are connected by arched transverse commissures. On each side of the cerebral ganglion, a nerve rises, which runs to a ganglion below the pharynx, and behind the radular sheath, this is the sublingual ganglion; this latter is united with the corresponding ganglion on the other side by a short transverse commissure. These sublingual ganglia probably correspond with the buccal ganglia of *Chiton*.

Dondersia is specially noteworthy because distinct ganglionic swellings occur at regular intervals along the pedal cords; this is particularly marked in the anterior part of the body. The equally regularly repeated transverse commissures joining the pedal cords, and the connectives between the pedal and visceral cords, start from these distinct ganglia.

In *Lepidomenia hystrix*, one ganglion occurs posteriorly and one anteriorly in each longitudinal trunk (whether pleurovisceral or pedal), and each is connected with a similar ganglion of the opposite side by a transverse commissure. In *Neomenia* and *Choetoderma*, no connectives between the visceral and pedal cords have been observed, and, so far as is at present known, in *Choetoderma*, the commissures between the pedal cords are also wanting. Further, in *Choetoderma*, the visceral and pedal cords of each side unite together posteriorly to form one single cord, which becomes connected with the similar cord on the other side by a transverse cord which runs over the cloaca.

Gastropoda

The nervous system of the Gastropoda is of great interest to the comparative anatomist on account of the crossing of the pleurovisceral connectives in the *Prosobranchia*, which will be further described in this section. The nervous system of this class consists typically of those parts which we have already mentioned in our scheme of the organisation of the Mollusca, viz.:

1. A simple or complex visceral ganglion lying below the intestine, united to the pleural ganglia by two pleurovisceral connectives.
2. A ganglion, which may be called parietal, almost always occurs in the course of each pleurovisceral connectives. The parietal ganglion divides the connective into two parts, an anterior pleuroparietal and a posterior visceroparietal connective.
3. Two pedal ganglia below the oesophagus, connected with each other by a pedal commissure, and with the cerebral ganglia by two cerebropedal connectives.

   The cerebral and pedal ganglia with the commissures and connectives belonging to them form a ring encircling the oesophagus, which may be compared with the oesophageal ring of the Annulata and Arthropoda.
4. Two cerebral ganglia near or above the oesophagus, which are connected by a cerebral commissure.
5. Two pleural or pallial ganglia (between the cerebral and pedal ganglia), which are connected with the cerebral ganglia by two cerebropleural, and with the pedal ganglia by two pleuropedal connectives.

The cerebral, pedal and pleural ganglia are (with unimportant exceptions) always arranged symmetrically to the median plane in all Gastropoda. The pleurovisceral connectives and their ganglia, however, are only found in such a position in some Gastropoda. In fact, only in the *Opisthobranchia* (including the *Pteropoda* but excepting *Actoeon*) and the *Pulmonata* are they symmetrical, in the sense that the right connective and its ganglion lie entirely on the right, and the left connective and its ganglion entirely on the left side of the body. The *Opisthobranchia* and *Pulmonata* are therefore called euthyneurous Gastropoda. In the *Prosobranchia* and *Actoeon*, the pleurovisceral connectives are asymmetrical, inasmuch as they cross one another, the connective springing from the right pleural ganglion running *over* the intestine to the left before joining the visceral ganglion, while the connective from the left pleural ganglion runs *under* the intestine to the right side of the body.

In consequence of this crossing, the parietal ganglion of the connective which springs from the right pleural ganglion becomes the supraintestinal ganglion, which lies on the left side, and the parietal gaglion of the connective springing from the left pleural ganglion becomes the infra-intestinal ganglion which lies on the right side. The *Prosobranchia* and *Actoeon* are thus streptoneurous Gastropoda.

Areas of Innervation of the various Ganglia

1. The pleural ganglia send nerves chiefly to the mantle, the columellar muscle, and the body walls lying behind the head.
2. The parietal ganglia innervate the ctenidia and osphradium, and also send some nerves to the mantle.
3. The buccal ganglia, which will be described below, innervate the muscles of the pharynx, the salivary glands, the oesophagus, the anterior aorta etc.
4. The cerebral ganglia innervate the eyes, the auditory organs, the tentacles, the snout or proboscis, the lips, the motor muscles of the proboscis and buccal mass, and the body walls lying at the base of the snout. Even when the auditory organs are found in close proximity to the pedal ganglia, or in close contact with them, they receive their nerves from the cerebral and not from the pedal ganglia.
5. The visceral ganglia supply nerves to the viscera. The connectives and commissures also may give off nerves which belong to the areas innervated by the neighbouring ganglia.
6. The pedal ganglia supply nerves to the musculature of the foot, and occasionally to the columellar muscle also (*Patella*).

A comparison of the typical nervous system of the Gastropoda with that of the *Amphineura* reveals the following homologies:

1. The cerebral ganglia of the Gastropoda correspond with the oesophageal ring of *Chiton*, with the exception of the central portion of its lower half; and further with the cerebral ganglia of the *Solenogastres*.
2. The pedal ganglia of the Gastopoda answer to the pedal cords in the *Amphineura*, concentrated each into a single ganglion. The arrangement in the *Diotocardia*, which are the more primitive *Prosobranchia*, is very interesting in this connection; in the Diotocardia the pedal ganglia are continued posteriorly as two true pedal cords which, like those of the Amphineura, and connected by transverse commissures.

It is more difficult to compare the pleural, parietal, and visceral ganglia of the Gastropoda with nerves found in the *Amphineura*. The most satisfactory view seems to be that this whole complex of ganglia, together with its connectives, corresponds with the pleurovisceral cords of *Chitoa*. The areas of innervation coincide, these being the mantle, ctenidia, osphradia (*Chiton*?), and viscera.

3. If this last assumption is correct, the pleural ganglion must be supposed to have arisen by the concentration into one ganglion of that part of the pleurovisceral cord of *Chiton* which contains the pallial ganglionic cells, this concentration having taken place at the anterior end of the cord, where it leaves the oesophageal ring. If, then, the two component portions of each side of the ring, the cerebropedal and the pleural, move further apart, and at the same time the cerebral and pedal ganglia of the ring become more individualised as ganglia, a double cerebropedal connective comes into existence on each side. One of these connectives shows no ganglion in its course, and is the true cerebropedal connective of the Gastropoda. The second, however, has the pleural ganglion in its course, and from this latter spring the visceral cords; this second connective is thus divided into a cerebropleural and a pleuropedal connective.
4. *Chiton* has numerous gills on each side, each of which receives two nerves from the pleurovisceral cord near it. The Gastropoda have at the most two gills, one on the right and one on the left. In correspondence with this reduction, the ganglionic cells of the pleurovisceral cords belonging to the branchial nerves of *Chiton* have become concentrated on each side into a single ganglion belonging to the single gill. The parietal ganglion is thus accounted for. That portion of each pleurovisceral cord which lies between the pleural and the parietal ganglia becomes the pleuroparietal connective, which consists of fibres only without ganglion cells.
5. There is no nerve in *Chiton* homologous with the visceral ganglion or ganglia of the Gastropoda; this is the chief difficulty in the comparison of the two nervous systems. In the *Amphineura*, the pleurovisceral cords unite *above* the intestine; in all other Molluscs the point of junction (which is the visceral ganglion) lies *below* the intestine.

In *Proneomenia* the posterior commissures between the pleurovisceral cords are merely a more strongly developed part of a general commissural system.

ORIGIN OF THE CROSSING OF THE PLEUROVISCERAL CONNECTIVE (CHIASTONEURY)

Several attempts have been made to explain the peculiar crossing of these connectives in the *Prosobranchia*. The one here given is in a high degree probable if not altogether satisfactory. We must start with a supposed racia form which was perfectly symmetrical, even in its nervous system and possessed an organisation somewhat like that of our hypothetical primitive Mollusc. Such an organisation agrees in most important points with that of the extent *Chitonidoe*; only one gill, however, was present on each side. Further, the parietal ganglia innervated the gills and the osphradia, and were thus closely connected with these organs.

The racial form of the *Gastropoda* may have been surrounded by a mantle border which widened posteriorly, *i.e.* covered a somewhat deeper mantle cavity which contained the pallial complex, viz. the median anus, to the right and left of which were the ctenidia and osphradia, and between the ctenidium and anus on each side the nephridial aperture. If we suppose this pallial complex to have changed its position, shifting gradually forward along the right mantle furrow, each ctenidium would drag along with its parietal ganglion. The heart and its auricles which are connected with the ctenidium would also become shifted. As long as the pallial complex had not moved far forward to the right, the pleurovisceral connectives would not cross, but would only be shifted to the right.

We find the *Tectibranchia* among the *Opisthobranchia* apparently at this stage, the only difference being that they have already lost the original left ctenidium and also the original left auricle. If the pallial organs are still further shifted forward along the mantle furrow till they come to lie quite anteriorly, and once more symmetrically, above and behind the neck, the original left ctenidium comes to lie on the right and the original right ctenidium on the left in the anteriorly placed mantle cavity. The original right ctenidium has, however, dragged its parietal ganglion over the intestine to the left side, and the latter becomes the supraintestinal ganglion. The original left ctendium, on the contrary, has dragged its ganglion below the intestine to the right side, and this ganglion becomes the infraintestinal ganglion.

The pleurovisceral connectives, in which these ganglia lie, now cross and give rise to the condition called chiastoneury. The visceral ganglion in which these connectives terminate posteriorly lies as before under the intestine. It is unnecessary to show in detail how this displacement also affects the heart and its auricles, the osphradia, and the nephridial apertures. Although chiastoneury may be satisfactorily explained by this theory of displacement, the cause of the displacement itself has still to be sought.

SPECIAL REMARKS ON THE NERVOUS SYSTEM OF THE GASTROPODA

I. Prosobranchia–Diotocardia

These are the most primitive Gastropoda. The ganglia are not yet very distinct, thus recalling the *Amphineura*. The cerebral ganglia are connected by two long commissures, the cerebral commissure running forward over the pharynx, and the labial commissure running under the oesophagus. The indistinctly separated buccal ganglia together form a horseshoe-shaped figure, and are united on each side by a connective with the thickened root of the labial commissure. The pleural ganglia lies close to the pedal ganglia, so that no distinct pleuropedal connectives can be distinguished. The pedal commissure is very short, and contains ganglion cells. From each pedal ganglion, a long pedal cord runs back into the foot; these two pedal cords contain ganglion cells along their whole length, and are connected by transverse commissures. These cords and commissures thus exhibit the same arrangement as in the *Amphineura*. The pedal cords innervate the musculature of the foot and the epipodium.

There is only one indistinct visceral ganglion, which is joined to the pleural ganglia by two pleurovisceral connectives, crossed in the typical way. In *Fissurella* only does a ganglion occur on the supraintestinal pleurovisceral connective. In no other *Diotocardian* is there a ganglion at the point of departure of the strong branchial nerve from the pleurovisceral connective; this nerve, however, forms the branchial ganglion just below the osphradium at the base of the gill. Where a ctenidium, or merely an osphradium, is found on each side, there is a branchial ganglion close to it; where only the left gill is retained (*Turbinidoe*, *Trochidoe*), only the left branchial ganglion is found. Since, as a rule, the parietal ganglia are wanting in the *Diotocardia*, and the branchial ganglia in the *Monotocardia*, the branchial ganglia of the *Diotocardia* have been considered, with much probability, as intestinal ganglia, which have shifted away from the pleurovisceral connectives and towards the bases of the gills.

As, however, *Fissurella* possesses both a supraintestinal and a left branchial ganglion, it would be necessary to assume that an originally single ganglion had here become divided into two. The symmetrical pallial nerve is always connected by a pallial anastomosis with the asymmetrical pallial nerves on the same side of the body. The symmetrical pallial nerve rises out of the pleural ganglion, the asymmetrical nerves out of the parietal ganglion, or the pleuroparietal connective. The nervous system of the *Neritidoe* and *Helicinidoe* are peculiar, in that the supra-intestinal pleurovisceral connective and its corresponding ganglion are wanting.

Docoglossa

The only essential difference between the nervous system of *Patella* and the typical system of the other *Diotocardia* lies in the fact that the pleural and pedal ganglia are joined by a distinct pleuropedal connective.

Monotocardia

The parietal ganglia are always present. The cerebral commissure is short, and lies behind the behind the pharynx. The labial commissure is wanting (except in the *Paludinidoe* and *Ampullaridoe*). The pedal cords and transverse commissures are wanting (except in the *Architoenioglossa*: *Paludinidoe*, *Cyclophoridoe*, *Cyprocidoc*). The number of visceral ganglia varies from one to three. The progressive development of so-called Zygoneury is noteworthy. In the *Diotocardia*, a pallial anastomosis exists between the symmetrical and asymmetrical pallial nerves on each side. If this anastomosis were to shift along the two pallial nerves of one side to their places of origin, *i.e.* the ganglia from which they spring, it would become a pallial connective uniting the pleural and parietal ganglia of the same side of the body. They would thus arise a new accessory pleurointestinal connective, which would be symmetrical and not twisted, and thus unlike the asymmetrical twisted connective already existing. Zygoneury thus depends on the development of such a pallial connective.

In the large majority of cases in which it occurs it takes place on the right side (a few *Rostrifera*, viz. some of the *Cerithiidoe*, *Ampullariidoe*, *Turitellidoe*, *Xenophoridoe*, *Struthiolariidoe*, *Chenopidoe*, *Strombidoe*, *Calyptroeidoe*, and in all *Proboscidifera siphonostomata* and all *Stenoglossa*). Less frequently, zygoneury takes place on the left side (*Ampullariidoe*, a few *Crepidulidoe*, *Naticidoe*, *Lamellariidoe*, *Cyproeidoe*). In other *Prosobranchia* there is only a pallial anastomosis on each side, as in the *Diotocardia*; the nervous system is then called dialyneurous. The progressive concentration of the central nervous system of the Monotocardia, which keeps pace with the development of zygoneury, must be emphasised. The connectives uniting the various ganglia continually shorten, so that at last anteriorly on the oesophagus there is a collection of ganglia; these are the cerebral, pleural, pedal, infraintestinal, and supraintestinal ganglia, all lying close together, to which must be added my small buccal ganglia. Only the visceral ganglia remain for back in the visceral dome.

In *Natica*, where the anterior part of the foot is strongly developed, and is bent back over the head, a propedal ganglion becomes differentiated from the pedal ganglion. The nervous system of the *Heteropoda* requires fresh investigation. So far as we at present know, they certainly have crossed visceral connective, and are therefore *Prosobranchia* and, as the rest of their organisation shows, Monotocardia. The cerebral ganglia and the pedal ganglia (pleuropedal ganglia?) are far apart, so that the cerebropedal connective are very long.

II. Opisthobranchia

The nervous system of this order, in which the typical Gastropodan ganglia are developed, is further characterised: (1) by the absence of chiastoneury, *i.e.* the pleurovisceral connectives do not cross (except in *Actoeon*); and (2) by a marked tendency to concentration of the ganglia around the posterior end of the pharynx.

Tectibranchia

As a rule only the right parietal ganglion is found (in *Actoeon* the left is also present). A nerve rises from it which innervates the ctenidium, the osphradium, and the mantle, and forms a branchial ganglion at the base of the gill. A delicate lower cerebral commissure is often found, which runs along the pedal commissure below the pharynx, and may be compared with the labial commissure of the *Diotocardia*. As types of the *Tectibranchia* we may take *Bulla* as representative of the *Cephalaspidoe*, and *Aplysia* as representative of the *Anaspidoe* (*Aplysiidoe*).

The nervous system of *Bulla hydatis*; only three points concerning it need be mentioned: (1) The pleural ganglia have shifted till they lie close to the cerebral ganglia, the cerebropleural connectives becoming correspondingly shortened. (In *Actoeon* these ganglia have even fused, and are no longer to be distinguished externally.) (2) There are three visceral ganglia. (3) The commissures are comparatively long. (4) The parapodia are innervated from the pedal ganglia. In many *Cephalaspidoe*, moreover, no distinct right parietal ganglion exists. It seems to have moved up to the right pleural ganglion, or to have fused with it, so that the nerve running to the branchial ganglion rises direct from the right pleural ganglion.

The nervous system of the *Pteropoda thecosomata*, which we derive from *Cephalaspidoe*, bears a general correspondence to that of the

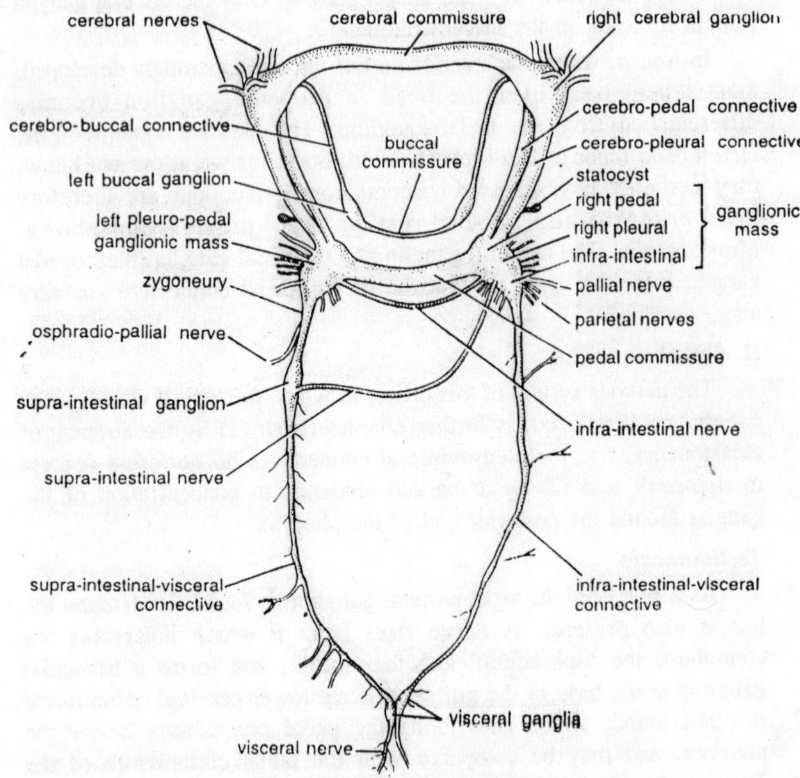

Fig. 13.1. Pila. Nervous system.

latter, especially in the fact that the pleural ganglia shift near to or fuse with the cerebral ganglia. The pleurovisceral connectives are so much shortened that the ganglia occurring in their course lie close to the cerebral and pedal ganglia. There are usually two such ganglia (the right parietal and a visceral ganglion?), less frequently three (two intestinal and one visceral ganglion?). The pedal ganglia also innervate the fins, which correspond with the parapodia of the *Cephalaspidoe*. In the nervous system of *Aplysia*, one of the *Anaspidae*, the two cerebral ganglia have moved close to each other in the middle line. The pleural ganglia here, unlike those of the *Cephalaspidoe*, lie close to the pedal ganglia, so that the pleuropedal connectives are much shortened. The pedal commissure is double, the anterior commissure is, relatively speaking, short and thick, the posterior long and thin. The long pleurovisceral connectives run back from the pleural ganglia and enter two ganglia lying side by side; that to the right represents the right parietal ganglion, innervating chiefly the gill and osphradium, the nerves running to these organs forming a ganglion at the base of each; that to the left is the visceral ganglion. One of the nerves which run from the latter forms a genital ganglion at the base of the accessory glands connected with the genital organs.

In other *Anaspidae*, such as *Notarchus*, the pleurovisceral connectives are so much shortened that the parietal and visceral ganglia lie close to the perioesophageal group of ganglia, which then consists of two cerebral, two pedal, and two pleural ganglia, and further, the right parietal and the visceral ganglia. The two cerebral ganglia are further connected by a thin *lower* commissure. The parapodia are always innervated from the pedal ganglia. The nervous system of the *Pteropoda gymuosomata*, which are nearly related to the *Anaspidae*, corresponds in all essential points with the nervous system of the latter, being of the same type as that of *Notarchus*.

Nudibranchia and Ascoglossa

The nervous system is here characterised by very great concentration of the typical Molluscan ganglia, and by a tendency to the formation of numerous accessory ganglia (at the bases of the tentacles and rhinophores, and at the roots of their nerves, in the course of the genital nerves, etc.). The pleural ganglion has moved close to the cerebral ganglion and may fuse with it. The pedal ganglia have also moved towards the cerebral ganglia so that now the whole oesophageal complex of ganglia lies almost entirely on the dorsal side of the oesophagus. The pedal commissure which runs under the gullet, and is sometimes double, is thus very much lengthened. The pleurovisceral connectives are short, and occasionally enter an unpaired visceral ganglion, which has also been drawn into the oesophageal complex. This single ganglion of the visceral connectives may be wanting ; in that case the two visceral connectives appear like a commissure between the two pleural ganglia running under the oesophagus and parallel with the pedal commissure, sometimes even united with it.

The fusion of all the ganglia belonging to the peri-oesophageal complex is carried very far in such animals as *Tethys*, where the pleural and pedal ganglia of each side may fuse with the cerebral ganglion. The pleuro-cerebropedal ganglion thus formed shifts towards the dorsal middle line close to the similar ganglion of the other side, with which it forms a large supra-oesophageal ganglionic mass. Its composition out of the six typical ganglia can, however, be made out

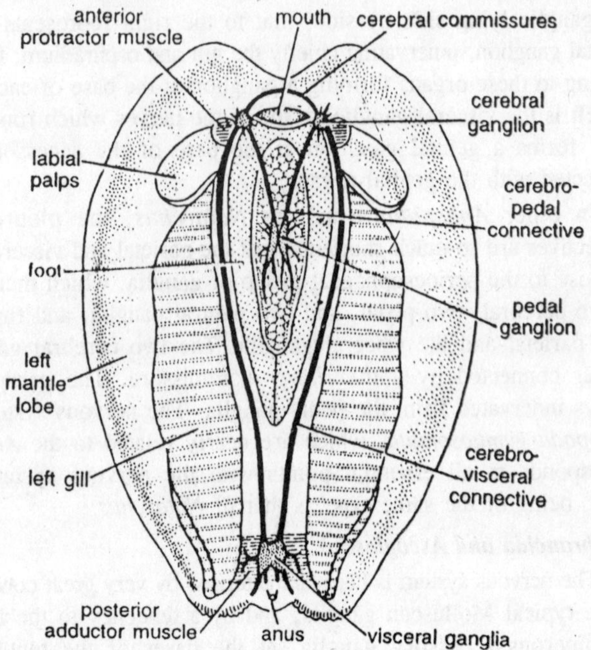

Fig. 13.2. Unio. Nervous system in ventral view.

by the grouping of the ganglion cells and the arrangements of the nerve tracts. A nerve leaves this mass on each side, the two uniting under the gullet. These form the pedal commissure, which when closely examined is found to be double.

A third delicate commissure running under the oesophagus connects the lateral portions of the supra-oesophageal mass, and represents the visceral commissure, in which is found a small visceral ganglion. Among the Nudibranchia the two buccal ganglia are always found on the posterior and lower wall of the pharynx. They are connected with each other by a buccal commissure, and with the brain by two cerebrobuccal connectives, in whose course accessory ganglia may be found. The whole peri-oesophageal complex of ganglia is in the Nudibranchia enclosed in a capsule of connective tissue.

III. Pulmonata

The central nervous system here possesses all the typical ganglia of the Gastropoda. These, grouped together as in so many *Opisthobranchia* and many *Prosobranchia*, immediately behind the pharyngeal bulb, form the peri-oesophageal complex into which even the parietal and visceral ganglia have been drawn. The cerebral ganglia lie close to each other dorsally, and all the other ganglia, which are also close together, lie ventrally. The cerebropedal and cerebropleural connectives are consequently always easily distinguished. In *Testacella* they are even of some length in adaptation, no doubt, to the special shape and the great development of the pharyngeal bulb. All other connectives and commissures, on the contrary, are much shortened, so that the ganglia connected by them lie close together. A visceral ganglion is always found, and usually also in each pleurovisceral connective a parietal ganglion. When an osphradium is present (*Basommatophora*) it is innervated from the parietal ganglion of the same side. In *Pulmonata* with a dextral twist, the osphradium lies on the right, and in those with a sinistral twist on the left; in the former the right parietal ganglion is the larger, and in the latter the left. The smaller parietal ganglion may also fuse with the neighbouring pleural ganglion. Lobes are often formed in the cerebral ganglia, in which certain groups of nerves have their origin. The pedal commissure is often double. Buccal ganglia are always found. They lie posteriorly on the pharynx below the oesophagus, and are joined to one another by the buccal commissure and to the cerebral ganglia by cerebrobuccal connectives.

Scaphopoda

The nervous system of the Scaphopoda is symmetrical; the visceral connectives are not crossed. The two cerebral ganglia lie very near one another in front of (or, if the intestine is regarded as horizontal above) the gullet over the snout; the two pedal ganglia close to one another lie on the anterior side of the foot, more or less at its centre, and are joined to the cerebral ganglia by two long cerebropedal connectives. The two pleural ganglia lie close to and above the cerebral ganglia, so that the cerebropleural connective is very short. The pleuropedal connective at once fuses with the cerebropedal, the two entering the pedal ganglion as one connective.

Posteriorly, to the right and left of the rectum, near the anus, there are two visceral ganglia of the pleurovisceral connectives, joined to one another by a commissure running behind the intestine. There are no special parietal ganglia distinct from the visceral or the pleural ganglia. There are four buccal ganglia, two behind the gullet or below it (if the intestine is supposed to the horizontal), and two lying laterally and anteriorly to (or above) the muscular mass of the radula. The anterior are connected with the posterior, and these to the cerebral ganglia by connectives, and the two posterior and two anterior *inter se* by commissures running behind (under) the oesophagus. Nerves run from the posterior buccal ganglia to the small ganglia of a subradular organ.

LAMELLIBRANCHIA

The nervous system, like the whole organisation of the Lamellibranchia, is perfectly symmetrical, and consists typically of three pairs of ganglia: (1) the cerebropleural; (2) the pedal; and (3) the visceroparietal ganglia. These three pairs of ganglia lie, as a rule, far apart, and the connectives uniting them are therefore long. The two pedal ganglia lie close together, while the two cerebropleural and the two visceroparietal ganglia are connected by distinct commissures beset with ganglion cells.

1. The cerebropleural ganglia are the result of the fusion of the cerebral with the pleural ganglia. In the *Protobranchia*, however, the pleural ganglia are still distinct, and lie immediately behind the cerebral ganglia at the commencement of the visceral connectives. In *Nucula*, the pleuropedal connectives are distinct for some distance and then unite with the cerebropedal connectives. In *Solenomya* they still have separate roots, but are otherwise fused along their whole length with the cerebropedal.

   The cerebropleurat ganglia are supraoesophageal, and are in contact with the anterior adductor muscle, when this is present. They send nerves into the oral lobes, the anterior adductor, and the mantle.
2. The pedal ganglia lie at the base of the foot.
3. The third pair of ganglia, which correspond with the ganglia of the visceral connectives in the *Gastropoda*, lie posteriorly beneath the rectum, behind the foot; and are generally in contact with the posterior adductor muscle; in the *Protobranchia*, however, they lie much further forward. Their area of innervation corresponds with that of the combined parietal and visceral ganglia of the *Gastropoda*, for these visceroparietal ganglia supply with nerves the two ctenidia, the two osphradia, the posterior portion of the mantle, the posterior adductor and the viscera.

The buccal or stomodaeal nervous system is much reduced; this reduction is connected with the absence of a muscular pharynx and of all buccal armature. The anterior portion of the intestine receives nerves from the visceral connectives. Since the fibres of these nerves have been proved to originate in the cerebral ganglia, we may assume that, on the degeneration of the pharynx, the buccal connectives united with the visceral connectives, so that the intestinal nerves now rise from the latter and do not come direct from the brain.

In the *Pholadidoe* and *Teredinidoe* the visceral connectives are united in front of the *visceroparietal* ganglia by a second commissure, which runs under the intestine, and may perhaps be considered as a buccal commissure shifted far back. The mantle is innervated, as is clear from the above, partly from the cerebropleural and partly from the visceroparietal ganglia. The two anterior pallial nerves, which rise from the cerebropleural ganglia, run back along the edges of the mantle to join the two posterior pallial nerves which originate in the visceroparietal ganglia. A nerve thus runs parallel to the edge of the mantle on each side (nerve of the pallial edge), and like a connective, unites the anterior cerebropleural ganglion with the posterior visceroparietal ganglion. This pallial nerve gives off branches to the organs at the edge of the mantle and to the siphons, and is further connected with a rich nerve plexus in the mantle fold, in which certain connecting nerves, further from the edge of the mantle, but running parallel to it, are particularly strongly developed. A varying number of small peripheral ganglia attain development in the pallial plexus and in the siphonal nervous system.

Cephalopoda

The symmetrical nervous system of all Cephalopoda is marked by the great concentration of the typical Molluscan ganglia, including those of the visceral connective. In the following description of the nervous system, we shall consider the body in its physiological, not in its true morphological position, *i.e.* we shall imagine the pharynx and oesophagus to be running horizontally as in other Molluscs. The true morphological position will be given in brackets after the conventionally accepted position.

Tetrabranchia

In the complex of ganglia, which in *Nautilus* surrounds the oesophague behind the great buccal mass, and which is not yet completely enclosed in the cephalic cartilage, the ganglia are not very distinct from the commissures and connectives. The cerebral ganglia are represented by a broad band-like nerve cord running over (morphologically in front of) the oesophagus, and from them run two ganglionic cords, one anterior (lower) and one posterior (upper), which pass just below (behind) the oesophagus. The anterior (3) represents the pedal, and the posterior the combined pleural and visceral ganglia.

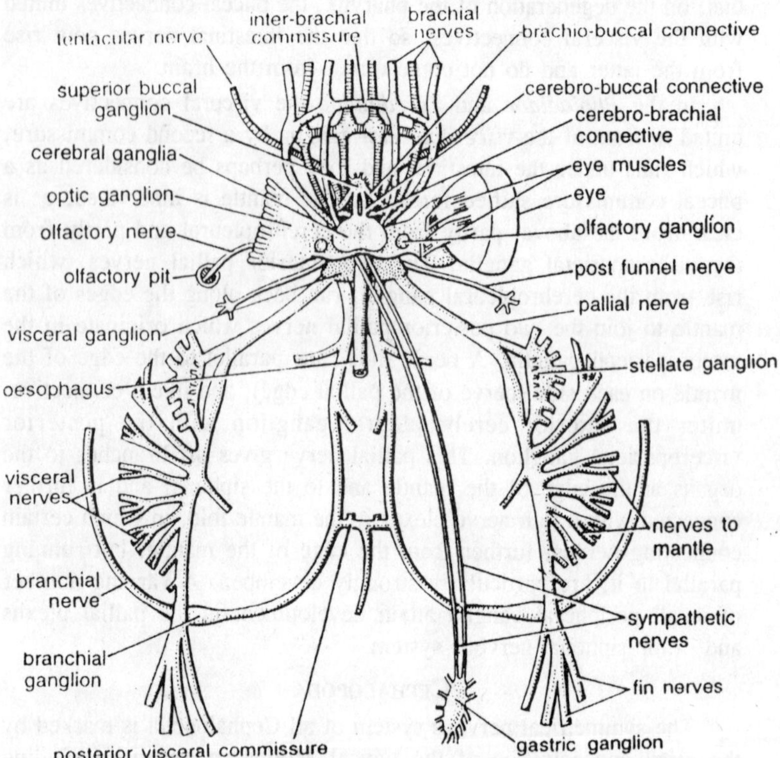

Fig. 13.3. Sepia officinalis. Nervous system in dorsal view.

The cerebral cord gives rise laterally to the large optic nerves (each of which at once swells into an optic ganglion), numerous nerves to the lips, the nerves for the optic tentacles, the auditory and olfactory nerves, and cerebrobuccal connectives. From the pedal cord, nerves run to the tentacles round the mouth and to the funnel. In the female, the nerves for the inner circle of tentacles come from a branchial ganglion, which, however, does not supply all the tentacles this is joined to the pedal ring by a brachiopedal connective.

The plurovisceral cord gives off numerous pallial nerves (there is no stellate ganglion) and two strong visceral nerves which run near the middle line accompanying the vena cava, innervate the gills, the osphradia, and the blood-vessels and form a genital ganglion high up in the visceral dome. The sympathetic nervous system consists of an infra-oesophageal commissure, which rises from the cerebral ganglion and passes close under the oesophagus in the musculature of the buccal

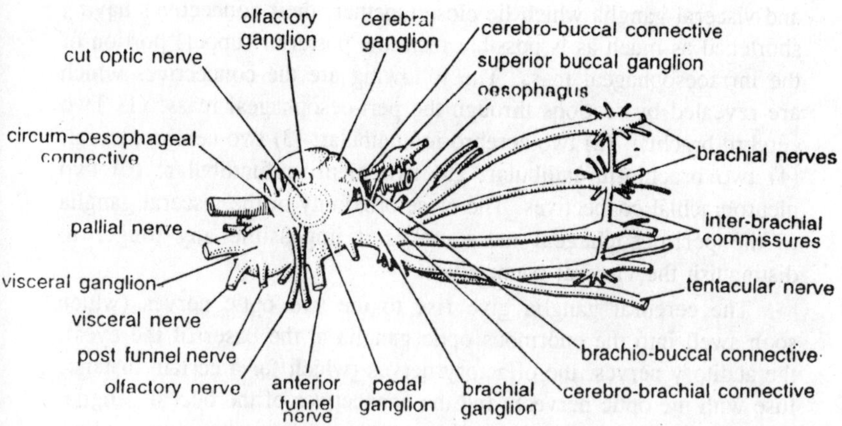

Fig. 13.4. Sepia officinalis. Nervous system in lateral view.

mass; two ganglia, a pharyngeal and a buccal ganglion, are found on each side in its course.

Dibranchia

The peri-oesophageal mass of ganglia comprising the whole of the central nervous system, is entirely enclosed in the cephalic cartilage. The large typical ganglia are so crowded together that it is extremely difficult to distinguish them one from another, and the connectives and commissures are not visible externally. The whole complex has a continuous cortical layer of ganglion cells. The more or less distinct separation of each pedal ganglion into two, one anterior (lower) and one posterior (upper), is characteristic of the Dibranchia. The former of these is the branchial ganglion, and innervates the arms, which must be considered as parts of the foot; and the latter is the infundibular ganglion, and innervates the siphon, which may be regarded as the epipodium. This differentiation of the pedal ganglia can be traced to the great development of that part of the foot (viz. the arms) which surrounds the head. In the same way in *Natica*, where the anterior part of the foot is strongly developed, and is bent back over the head, a propedal ganglion becomes differentiated from the pedal ganglion.

The brachial ganglia become joined in the *Dibranchia* to the cerebral ganglia by cerebrobrachial connectives. In *Eledone* and *Octopus*, they are further connected with one another by a thin supraoesophageal commissure. The pleural ganglia lie laterally in the perioesophageal mass while the ganglia of the visceral connectives, *i.e.* the parietal and visceral ganglia which lie close together, their connectives having shortened as much as is possible form the posterior (upper) portion of the infraoesophageal mass. The following are the connectives which are revealed by sections through the peri-oesophageal mass: (1) Two cerebro-brachial; (2) two cerebro-infundibular; (3) two cerebropleural; (4) two brachio-infundibular; (5) two pleuro-infundibular; (6) two pleurobrachial connectives. The close proximity of the visceral ganglia to the peri-oesophageal mass makes it impossible any longer to distinguish the visceral connectives.

The cerebral ganglia give rise to the two optic nerves (which soon swell into the enormous optic ganglia at the bases of the eyes), the auditory nerves, the olfactory nerves (which for a certain distance fuse with the optic nerves), and the connectives of the buccal ganglia. The branchial ganglia send off separate nerves to the arms, which nerves are connected by a hoop-like commissure round the base of the circle of arms. Running through the arms, the nerves swell into successive ganglia which correspond with the transverse rows of acetabula. The separation of the pedal ganglion into a brachial and an infundibular ganglion can be proved ontogenetically and anatomically.

There is no such separation in the male *Nautilus*, the brachial and infundibular nerves springing from one and the same ganglion. In *Argonauta* the separation is not externally visible, but in *Octopus* we see the first traces of it; in *Sepia*, *Loligo*, and *Sepiola*, it becomes more and more evident, till finally in *Ommatostrephes* the distinct brachial ganglion has moved away from the infundibular ganglion, with which it is joined by a slender externally visible connectives. In this same series, the separation of the so-called upper buccal ganglion from the cerebral ganglion also takes place the buccal remaining united to the brachial ganglion by the brachiobuccal connective. The parietal ganglia give rise to the two large pallial nerves. Each of these runs backward and upward, and enters the stellate ganglion on the inner surface of the mantle.

Numerous nerves radiate into the mantle from this ganglion, one of them, which runs dorsally, looking like the direct continuation of the pallial nerve through the ganglion. The pallial nerve often divides into two branches sooner or later after it has left the parietal ganglion; one of the branches running to and through the stellate ganglion, to unite beyond it with the other branch which runs past the ganglion. The two stellate ganglia are often connected by a transverse commissure. The visceral ganglia give off, near the middle line, two visceral nerves, which innervate the rectum, the ink-bag, the gills, the heart, the genital apparatus, the kidneys, and certain parts of the vascular system. The two genital branches of these nerves are connected by a commissure. The sympathetic nervous system consists of a buccal ganglion lying beneath (behind) the oesophagus in the buccal mass; this ganglion is joined to the upper buccal or pharyngeal ganglion by a buccal connective. Two nerves run up along the oesophagus from the lower buccal ganglion to the gastric ganglion, which lies on the stomach, and innervates the greater of the intestine and the digestive gland (liver).

14

Receptors

Integumental Sensory Organs

In the integument of the Mullusca there are epithelial sensory cells (Flemming's cells), which vary in number and arrangement, and may be scattered over large areas. Two kinds of these cells may be distinguished according to their form. One kind, which is found only in *Lamellibranchs*, consists of large epithelial cells with large terminal plates which form part of the body surface and carry tufts of projecting sensory hairs ("paint-brush cells," Pinsel-Zellen). The second kind of cells are found in all classes of Mollusca. They are long, filiform or spindle-shaped, swelling at one point where the nucleus lies. They sometimes carry a tuft of sensory hairs, sometimes none. Each kind of cells is continued at its base into a nerve fibre, which runs into the nervous system.

A distinct specific function can hardly be attributed to these epithelial cells. They may respond to very various stimuli, chiefly mechanical and chemical, and thus may act in an indefinite way as tactile, olfactory, and gustatory cells. They may become more specialised in function, when crowded together in certain areas of the body, and may then represent special sensory organs. Between the individual cells composing such a sensory organ, however, other epithelial cells (glandular, ciliated and supporting cells) are always found.

Tactile Organs

The tactile function of the integumental sensory cells is likely to assert itself at exposed parts of the body surface, such as the tentacles, epipodial processes, siphons, at the edge of the mantle in the

Lamellibranchia, and at the edge of the foot etc. We cannot, however, assume that even in these places the sensory cells are sensitive *only* to mechanical stimuli.

Olfactory Organs

Osphradium

As has been proved to be the case in the *Prosobranchia*, sensory cells occur scattered among the other epithelial cells throughout the whole epithelial lining of the mantle cavity. Here, as in other parts of the body, three kinds of epithelial cells can be distinguished: (1) undifferentiated cells, which may contain pigment, and are usually ciliated; (2) glandular cells; (3) sensory cells. The proportions in which these three kinds of cells appear varies in different regions of the mantle. If glandular cells prevail on a certain area, that area assumes a glandular character, and may even develop into the sharply localised epithelial gland (*e.g.* the hypobranchial gland). On the gills, undifferentiated ciliated cells predominate. Where sensory cells predominate a sensory character is given to the region; such a region, if sharply circumscribed, the sensory cells continually increasing in number, becomes a pallial sensory organ.

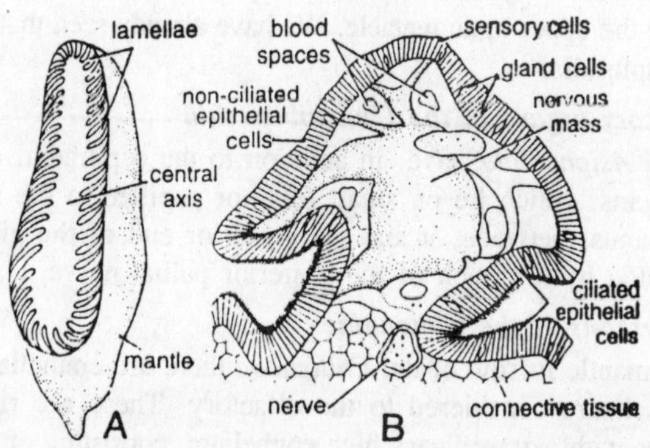

Fig. 14.1. Pila. A—Osphradium; B—Osphradium in T.S.

The gradual development and continuous differentiation of such an organ may be particularly well traced in the *Prosbranchia*, the sensory organ developed being the osphradium. In consequence of its position in the mantle cavity, and especially on account of its proximity to the gill, it has been assumed that its principal function is to test the condition of the respiratory water, or, other words, that it is an olfactory organ.

Olfactory tentacle :

Certain experiments, to which, however, some exception might be taken, seem to show that the large optic tentacles of terrestrial *Pulmonata* are also olfactory. It is also generally accepted, though still not certainly established, that the posterior or dorsal tentacles (rhinophores) of the *Opisthobranchia* are olfactory organs. These rhinophores often show increase of surface, usually in the shape of more or less numerous circular lamellae surrounding the tentacle like a collar. The rhinophores are also often ear-shaped or rolled up conically. Not infrequently they can be retracted into special pits or sheaths. They are innervated from the cerebral ganglion by means of a nerve which forms a ganglion at the base of each.

Olfactory pits of the Cephalopoda

In the *Dibranchia* there is on each side, above the eye, a pit which is considered to be olfactory. Its epithelial base consists of ciliated and cells, and underneath its lies close to the optic ganglion, and olfactory ganglion. The nerves running to this ganglion come form the ganglion opticum, but really originate in the cerebral ganglion. It looks as if these olfactory organs were the remains of the posterior tentacles of the Gastropoda, and were comparable with the rhinophores of the *Opisthobranchia*. In *Nautilus* the place of the Olfactory pit is occupied by the upper optic tentacle. We have already seen that *Nautilus* still true osphradia.

Pallial sensory organs of the Lamellibranchia

Several *Asiphoniata* have, in addition to the osphradia, epithelial sensory organs, which lie on small folds or papillae to the right and left of the anus, between, it and the posterior end of the gill. There are innervated by a branch of the posterior pallial nerve.

Olfactory organs of the Chitonidae

In the mantle furrow of the *Chitonidoe* there are epithelial sensory organs which are considered to the olfactory. These are ridges and prominences with extraordinary high epithelium, consisting of glandular cells and thread-like sensory cells. In *Chiton loevis* and *C. cajetanus* there are, on each side of the mantle furrow, two sensory ridges extending along the whole length of the row of gills; one of these, the parietal ridge, belongs to the outer wall of the furrow, while the paraneural ridge runs along the base of the furrow, above the bases of the gills and under the pleuro-visceral cord. The paraneural ridge is continued a short distance along the inner surface of each gill, so that each gill has an epibranchial sensory prominence. In front of the first pair of gills and near the last the sensory cells in the paraneural ridge become far more numerous in comparison with the glandular cells. *Chiton siculus*, *C. Polii*, and *Acanthochiton* (in which the numerous gills reach for forward) have no parietal and paraneural ridges. The sensory epithelium in these animals is confined to two prominences, paraneural in position, behind the last pair of gills, and connected with a high epithelium covering the pallial wall of the most posterior part of the furrow. All these sensory epithelia seem to be innervated from the pleuro-visceral cords.

"Lateral Organs" of the Diotocardia

At the bases of the epithelial tentacles of *Fissurella* and the *Trochidoe*, and at the base of the lower tentacles of the epipodial ruff of *Haliotis*, and also in other parts near the ruff, sensory organs are found which have been compared with the lateral organs of Annelids. They consist of patches of sensory epithelium, which may form either spherical projections or pit-like depressions. The epithelium of these sensory organs which lie at the lower side of the bases of the epipodial tentacles, consists of sensory cells, each of which is provided with a sensory seta, and pigmented supporting cells. Each of these sensory organs is innervated by the nerve of the tentacle near it, which nerve originates in the pedal cord and forms a ganglion in the base of each epipodial tentacle.

Gustatory Organs

Folds and prominences found in the mouth in some division of the Mollusca have been taken for gustatory, organs, although there are no physiological and hardly any histologically grounds for this opinion. The existence of so-called gustatory pits on a prominence in the buccal cavity has been proved only in a few *Chitonidoe* and *Diotocardia* (*Haliotis*, *Fissurella*, *Trochus*, *Turbo*, and *Patella*). This "gustatory prominence" (which has been best examined in *Chiton*) lies on the floor of the buccal cavity, close behind the lip.

A few gustatory pits are found in its epithelium, sunk somewhat below the surrounding epithelium. They consist of sensory cells with freely projecting sensory cones, and of supporting cells. On each side of the mouth in the *Pulmonata* lies an oral lobe, and under its deep epithelium, which is covered by a thick cuticle, lies a ganglion. Smaller ganglia are found in the small lobes at the upper edge of the mouth. All these ganglia receive nerves which radiate from a branch of the anterior tentacle nerve. These oral lobes (Semper's organ) are considered to be gustatory organs.

Subradular Sensory Organ of Chiton

In the buccal cavity of *Chiton* a subradular organ of unknown physiological significance has been found. It is described as "a prominence lying below and in front of the radula," and in shape resembles two beans with their concave edges turned to one another the end touching; the space between them forms a channel into which a small gland opens. Below this organ lie two ganglia, the subradular or lingual ganglia (*cf.* section on the nervous system). The epithelium of the subradular organ consists of green pigmented ciliated cells are two kinds of sensory cells. A similar organ occurs in *Patella*, but has not been thoroughly examined, and at the same part in various *Diotocardia* there is a prominence, which, however, has no sensory cells. The *Scaphopoda* also possess a subradular organ.

Sensory Organs on the Shell of Chiton

There are numerous organs definitely arranged on the shell of the *Chitonidoe* which have no doubt correctly, been considered as sensory, *i.e.* tactile organs. They are called aesthetes, and lie in pores on the tegmentum; they are club-shaped or cylindrical, and each carries a deep cup-like chitinous cap. Each megalaesthete gives off all round numerous fine branches or micraesthetes, each of which ends in a swelling which carries a small chitinous cap. The body of the aesthetes consists principally of long cells like glandular cells; it is produced into a fibre which runs along the base of the tegmentum, and from here passes together with the fibres of the other aesthetes of the shell-plate, between the tegmentum and articulamentum to the surrounding pallial tissue, or else penetrates of articulamentum.

The sensory nature of the aesthetes is rendered highly probable by the circumstances that in a few species of *Chiton* individual megalaesthetes are transformed into eyes. Each eye is furnished with the pigmented envelope, which is penetrated by the micraesthetes, and outwardly covered by an arched layer of the tegmentum which forms the cornea. Under this is a lens, and under this again a cell layer, which is regarded as a retina, and to which is attached a fibrous strand (optic nerve?) corresponding with the fibrous strands of the ordinary aesthetes.

Auditory Organs

All Mollusca except the *Amphineura* possess auditory organs, which appear very rarely in the embryo. They take the form of two almost closed auditory vesicles (otocysts), whose epithelial walls usually consist of ciliated and sensory cells. The interior of the otocyst is filled with

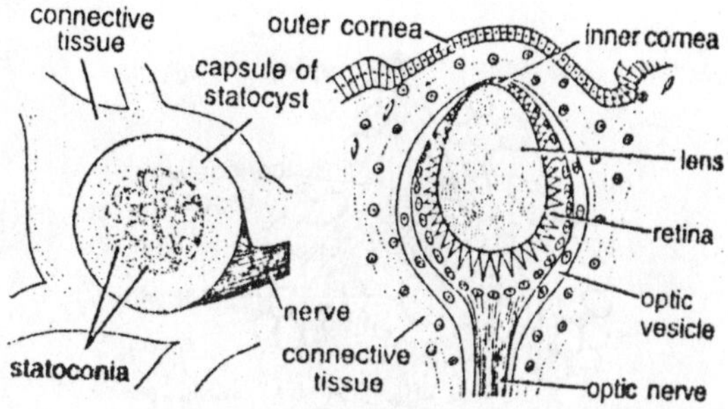

Fig. 14.2. Pila. Statocyst and H.L.S. of eye.

fluid and contains a varying number of otoliths (1 to over 100). These vary in size, form and chemical constitution, and in the living animal oscillate in the fluid in which they are suspended. The otocysts are usually found on or near the pedal ganglia, rarely far from it. It is, however, well established that the auditory nerve does not originate in this ganglion but in the cerebral ganglion, though it often runs along close to and even in contact with the fibres of the cerebropedal connective.

In most cases the otocysts arise as invaginations of the outer epithelium. An interesting discovery has recently been made, that in primitive *Lamellibranchs* (*Nucula*, *Leda*, *Yoldia*) each of the otocysts even in the adult still opens by means of a long canal on the surface of the foot. In such cases the otoliths are particles of sand or other foreign matter taken in from outside. In *Cephalopods*, the remains of the canal of investigation is retained (Kölliker's canal), but it ends blindly. The auditory organs are most highly developed in those Molluscs which are good swimmers, especially in the *Cephalopoda* and *Heteropoda*. Among these, maculae and cristae acusticae are developed.

Heteropoda

The structure of the auditory organ of *Pterotrachea* which has been thoroughly examined, is as follows: The wall of the otocyst consists in the first place of a structureless membrane surrounded by muscle and connective tissue. Inside the vesicle, which is filled with fluid, a calcareous otolith, built up of concentric layers, is suspended. The inner surface of the vesicle is lined by an epithelium, containing three different sorts of cells: auditory, ciliated, and supporting cells. The auditory cells, which carry immobile sensory hairs, are found on the wall of the otocyst at a point (macula acustica) diametrically opposite to the place where the auditory nerve enters. At this spot there is a

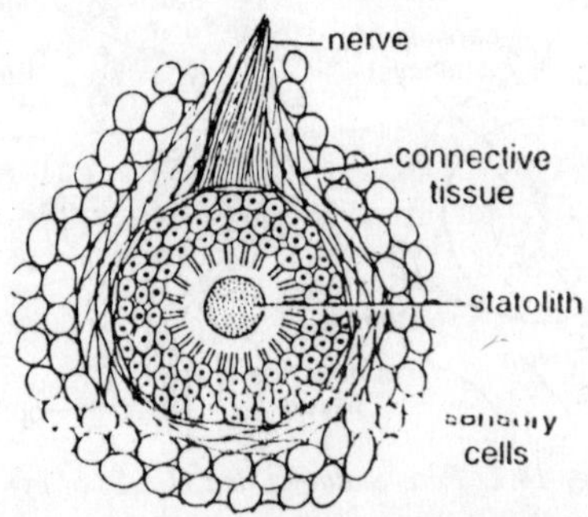

Fig. 14.3. Unio. T.S. Statocyst.

patch formed of numerous auditory cells, and in their midst, separated from the rest by four supporting or isolating cells, one large central auditory cell.

On the larger remaining surface of the wall of the otocyst separated by undifferentiated cells, are found flatter ciliated cells, which carry very long cilia or setae, exhibiting peculiar movements. They sometimes lie flat along the inner wall of the vesicle, and at other times (it is said in response to strong auditory stimuli) stand upright, projecting towards the centre of the vesicle, and supporting the otolith. The auditory nerve, which enters the otocyst at a point exactly opposite the central cell, at once radiates in the form of fibres over the whole wall of the vesicle "as meridians radiate from the pole on a globe," finally innervating the bases of the auditory cells. The two otocysts of the *Cephalopoda* are still more complicated; they lie in two spacious cavities of the cephalic cartilage. The sensory epithelium is here found on a macula acustica and on a kind of ridge, the crista acustica, which projects inwards. Otoliths are only found on the macula acustica.

The auditory nerve divides into two branches, one going to the macula, and the other to the crista acustica. Kölliker's canal, above mentioned, which is internally ciliated and ends blindly, runs out of the otocyst as the remains of the aperture of the original invagination. Experiments made on *Cephalopods* have shown that one of the functions of the otocysts is a to regulate the position of the animal while swimming.

Visual Organs

Optic Pits

These are the simplest form of visual organ. They are cup-shaped depressions of the body epithelium, which at the base of the cup forms the retina. The depression is sometimes very shallow, at other times deep, and like a wide bottle with a short narrow neck. The optic nerve enters at the base of the depression and spreads out over it. The epithelial wall or retina consists, apparently in all *Gastropoda*, of two kinds of long thread-like cells: (1) clear cells without pigment, and (2) pigmented cells. Whether either or possibly both of these kinds can be considered as retinal cells is still a disputed question. In certain it has been proved that the pigment in the second kind lies peripherally; the axis is free from pigment, and many perhaps be considered as the sensitive portion of the cell. On this case, the clear would be undifferentiated supporting cells, or secreting cells. The retina is covered, on that side of its which faces the cavity, by a thick gelatinous cuticle, or the whole cavity is filled by a gelatinous body often called a lens. The clear or secreting cells have been thought to yield this gelatinous mass, but there is a tendency to regard them now rather as retinal cells.

Optic Vesicles or Vesicular Eyes

Optic vesicles are developed from optic pits both ontogenetically and phylogenetically by the approximation of the edges of the pit, which finally fuse. A vesicle is thus formed over which there is a continuous layer of epithelium. The outer epithelium is free from pigment over the eye, and is called the outer cornea, while the immediately subjacent and also unpigmented, epithelial wall of the vesicle forms the inner cornea. The epithelial base of the original depression here again forms the retina; its cells contain distinct rods

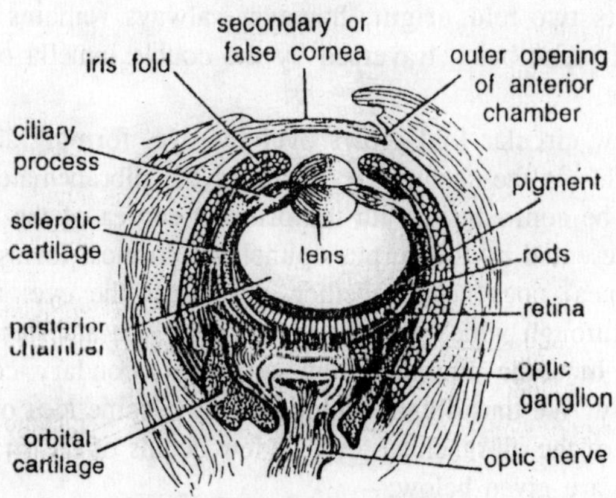

Fig. 14.4. Sepia. Eye in section.

projecting towards the cavity of the vesicle, which is filled with a gelatinous mass. The optic nerve usually swells into a peripheral ganglion opticum before reaching the retina. The tentacular eyes of most *Gastropoda*, except those *Diotocardia* which have cup-like eyes, and of this simple character.

Eye of the Dibranchiate Cephalopoda

This is one of the most highly-developed eyes in the whole animal kingdom. It is further development of the cup-shaped and vesicular eyes. In the *Tetrarbranchiate Nautilus*, as we have seen, the cup-shaped eye persists throughout life. These lower stages (*i.e.* the cup-shaped and vesicular stages) of the eye are passed through ontogenetically. First a cup-like depression is formed (primary optic pit), then this becomes constricted to form a vesicle (primary optic vesicle), the inner wall of which becomes the retina, while the outer (which corresponds with the inner cornea of the vesicular eye) becomes the inner corpus epitheliale. This embryonic optic vesicle then becomes further complicated; the integument over it (the outer cornea of the vesicular eye) rises in the form of a circular rampart, and then grows forward towards the axis of the eye like a diaphragm, which forms the iris, the aperture left in the same being the *pupil*.

The integument which spreads out o ver the circular base of the iris is in close contact with the inner corpus epitheliale, and becomes the outer corpus epitheliale. The inner corpus forms towards the cavity of the primary vesicle an almost hemispherical lens, the outer corpus epitheliale forming a similar lens outwards towards the pupil. The two hemispheres lie in such a way as to form something like a complete sphere; its two-fold origin, however, always remains evident, its equatorial plane being traversed by the double lamella of the corpus epitheliale.

A new circular fold grows over the eye, forming a fresh cavity over it; this is the *secondary cornea* of the dibranchiate eye, which must not be confounded with the primary cornea of the optic vesicle here represented by the corpus epitheliale. In most forms the circular fold (cornea) does not altogether close over the eye; and aperture remains through which the water can enter the anterior chamber of the eye. In some animals, however, the secondary cornea closes completely. We thus obtain, ontogenetically, some idea of the general structure of the dibranchiate eye. A few details of the structure of the adult eye are given below:—

1. The retina consists of two kinds of the cells—(1) *pigmented visual* or *rod cells*, and (2) *limiting cells*. Since the nuclei of the visual cells form, with relation to the centre of the vesicle, an outer, and the nuclei of the limiting cells an inner layer, and since, between these two layers, a limiting membrane traverses the interstices between the retinal cells the retina appears to be laminated, whereas it in reality consists of one layer of cells. The rods of the retinal cells lie on the inner side of the limiting membrane, and are thus turned to the sources of light and at the same time to the cavity of the primary vesicle. The retina is covered on its inner side by a somewhat thick membrane limitans.
2. The eye is surrounded, except on the side turned to the surface of the body, by a *cartilaginous capsule*, which resembles the sclerotica in the vertebrate eye; this cartilage, where it covers the retina, is perforated like a sieve, so that the optic nerves can pass through it.
3. Immediately underneath the cartilaginous floor of the retina lies a very large ganglion opticum, in the form of a massive cerebral lobe. From this rise the nerves which run to the retina through the perforations of the cartilaginous capsule.
4. The two halves of the lens, which are unequal is size (the outer being the smaller), consist of homogeneous concentric laminae.
5. The cavity of the primary vesicle (between the retina and the lens) is filled with perfectly transplant fluid.

It has been proved that, as in the Arthropoda and Vertebrate, the pigment granules of the rod cells, which in the dark lie at the base of the cell, under the influence of light travel towards its free end.

The Dorsal Eyes of Oncidium and the Eyes at the Edge of the Mantle in Pecten and Spondylus

These eyes have been said to resemble vertebrate eyes in structure, because in them the visual rods are turned away from the light, being directed inwards towards the body. They are vesicular eyes, but in them it is the outer wall of the vesicle, that turned to the light, which becomes the retina, while the inner wall (which in other Molluscs forms the retina) is a pigmented epithelium. At the same time the outer or retinal wall is invaginated towards the inner pigmented wall, as is the endoderm towards the ectoderm in the formation of the gastrula. The consequence of this is, that the cavity which in other Mollusca is filled by the gelatinous mass (lens) disappear, and the vesicle becomes a flattened thick-walled plate (*Pecten*) or cup

(*Oncidium*), consisting of a pigment layer and a retina. The body epithelium which passes over the eye is unpigmented and transparent, and here becomes the cornea.

Beneath the cornea, within the optic cup or on the plate, lies a cellular lens, which in the dorsal eyes of *Oncidium* consists of a few (5) large cells but in the pallial eyes of *Pecten* and *Spondylus* of very numerous cells. The development of this lens is unknown; it is perhaps formed by a thickening or invagination of the embryonic ectoderm which covers the eye. In *Oncidium*, the optic nerve penetrates the wall of the optic cup, and in the vertebrate eye, to spread out on the inner surface (with regard to the centre of the vesicle) of the retina, and to innervate the retinal cells. In *Pecten*, the optic nerve which runs to each eye from the nerve for the pallial, edges divides, close to the eye, into the branches. One of these runs to the base of the optic plate, and there breaks up into fibres, which radiate on all sides to the edge of the plate, then bend over towards the retina to innervate some of its cells.

The other branch runs direct to the edge of the plate, there bends round at a right angle and supplies nerves to the rest of the nerve cells. The fibres of this branch are not, however, directly connected with the retinal or red cells, as there is a layer of anastomosing ganglion cells interposed between the two. Between the pigmented epithelium and the rod layer of the retina, a *tapetum lucidum* is found, which gives the eye of the *Pecten* its metallic lustre. Dorsal eyes are found in many species of *Oncidium*. They lie at the tips of the contractile papillae found on the dorsal integument of this curious *Pulmonate*; on each papilla three or four such eyes occur. Besides these, *Oncidium* has the two normal cephalic eyes usually found in *Gastropods*. The pallial eyes of the *Lamellibranchiates*, *Pecten* and *Spondylus*, are found in large number of the edge of the mantle, between the longer tentacles, and on the tips of shorter tentacles. The rods of retina in *Pecten*, when fresh, are of a very evanescent red colour (visual purple?).

Eyes on the Shell of Chiton

These have already been described. Their morphological significance cannot be determined as long as their development is unknown and their histological structure imperfectly investigated.

Compound Eyes of Arca and Pectunculus

These are found in great numbers at the edge of the mantle, and are epithelial organs which do not in any way agrec in structure with the other visual organs found in the Mollusca, but rather resemble certain simple Arthropodan eyes. In form they resemble an externally convex shell. The unilaminar epithelial wall of the shell passes, at its edge, into the surrounding pallial epithelium. In section, its component elements appear to be arranged like a fan ("Fächerauge"). These elements are of three kinds: (1) conical visual cells, with their bases turned outwards; (2) a sheath of six cylindrical pigment shells surrounding each visual cell. Each group, consisting of one visual cell and its surrounding pigment cells, may be considered as a single eye or ommatidium of the simplest structure, in which the retinula is represented by one single visual cell. (3) Slender, almost thread-like interstitial cells which stand between the ommatidia.

Degeneration of the Cephalic Eyes

It is becoming more and more probable that the cephalic eyes of the various Mollusca are homologous structures, and that they primitively occurred in all forms. They may, however, under certain biological conditions become rudimentary, and even disappear, as in boring animals and those living in mud or in the deep sea and in parasitic Molluscs. The *Lomellibranchia* and *Chitonidoe* (?) even have cephalic eyes appearing temporarily during development; they disappear later, when, covered by the shell, they are useless. They may be replaced by secondarily acquired visual organs arising at more suitable parts of the body, and thus we have eyes on the mantle edge in some bivalves and on the shell of some *Chitonidoe*.

15

REPRODUCTIVE SYSTEM

In treating of the genital organs of the Mollusca, we shall have to consider—(1) the *gonads* or germinal glands, those most important organs, in which the reproductive cells (eggs and spermatozoa) are formed; (2) the ducts through which these cells reach the exterior; and (3) the copulatory organs.

GONADS

The gonads or germinal glands have already been recognised as completely or incompletely demarcated portions of the secondary body cavity, and have been described in their relation of the other divisions of that cavity. The gonads are paired and symmetrical in the *Lamellibranchia* and *Solenogastres*, occurring in one pair. In all other Mollusca, only one unpaired gonad is found. In very rare cases, such as that of some hermaphrodite *Lamellibranchs*, which will be described later, there are two pairs of gonads; one female and one male. The sexes are separate, among the *Amphineura*, in the *Chitonidoe* and *Choetoderma*, in many *Lamellibranchs*, in the *Scaphopoda*, among the *Gastropoda* in the *Prosobranchia* (excepting a few *Marseniadoe* and the *Valvata*), and in all *Cephalopoda*. Hermaphroditism prevails among the *Amphineura* in *Proneomenia*, *Neomenia*, and allied forms; in many *Lamellibranchs*, among the *Gastropoda* in the *Pulmonata*, *Opisthobranchia*, and in the *Prosobranchiate* family of the *Marseniadoe*.

In hermaphrodite animals, it is the rule that the same gland, the *hermaphrodite gland*, produces both eggs and spermatozoa, but in exceptional cases there are in the same individual distinct male and female gonads (testes and ovaries). This is the case, as already mentioned, in certain bivalves, viz. the *Anatinacea* and the *Septibranchia*, which possess two testes and two ovaries.

Position of the Gonads

The long tubular hermaphrodite glands of the *Solenogastres*, which are separated from one another by a median septum, lie in the anterior prolongation of the pericardium, over the intestine. In the *Chitonidoe*, the gonads are found in a similar position, but are not in open communication with the pericardium. In the *Gastropoda* they lie in the visceral dome, usually in its uppermost part, between the lobes of the digestive gland. Where the visceral dome has disappeared, the gonad with the intestine and the digestive gland shift back into the primary body cavity above the foot. The gonads in the *Scaphopoda* occupy a position similar to that of the Gastropodan gonads, lying dorsally in the high visceral dome, above the anus and the kidneys. The same is the case in the *Cephalopoda*.

The paired much-lobed genital glands of the *Lamellibranchia* lie in the typical position in the primary body cavity, above the muscular part of the foot, between the coils of the intestine. They may lie behind the "liver," or else, passing between its lobes, spread out at the sides of and below the kidney. The epithelium which lines the gonads is, morphologically, the endothelium of the secondary body

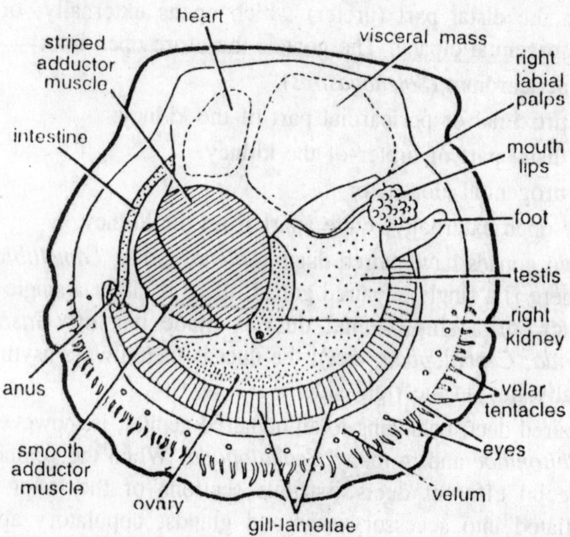

Fig. 15.1. Pecten. Internal organ after removal of right shell value.

cavity. The reproductive cells may either be produced from any part of the epithelium of the gonad, or from definite areas of this epithelium (*Cephalopoda*), which areas may then be called germinal epithelium or germinal layers. It may then appear as if the germinal gland lay in or on a special sac, whereas this sac is, in reality, the gonad itself, and the germinal gland is only the much-developed germinal layer of the gonad. The ripe reproductive cells become detached from their place of formation, and fall into the cavity of the gonad, *i.e.* into a part of the secondary body cavity, from which they pass out in various ways.

Ducts

The gonads either have separate ducts (*Chitonidoe*, *Monotocardia*, *Pulmonata*, *Opisthobranchia*, *Cephalopoda*, many *Lamellibranchia*) or they utilise the nephridia as ducts. In the latter case the genital products either pass direct into the kidney, and reach the exterior through the nephridial aperture (all *Diotocardia*, the *Scaphopoda*, and many *Lamellibranchia*), or they first pass into the pericardium, and then are ejected through the nephridia (*Solenogastres*). Where the gonads open into the kidneys, their apertures may lie in various part of these organs; either in the proximal part, which communicates with the pericardium by means of the renal funnel, and is usually widened into the renal sac or in the distal part (ureter) which opens externally, or into a shallow urogenital cloaca. The gonads therefore open into:

(a) The pericardium (*Solenogastres*).

(b) The proximal or pericardial part of the kidney.

(c) The distal part of ureter of the kidney.

(d) The urogenital cloaca, or

(e) They open externally, quite apart from the kidney.

Paired gonads have paired ducts (*Solenogastres*, *Lamellibranchia*). Where there is a single unpaired gonad, there is either a single efferent renal duct, or a single renal duct is made use of (*Gastropoda*, *Scaphopoda*, *Cephalopoda*, etc.); the duct is then always asymmetrical and usually lies on the right side.

A paired duct, belonging to an unpaired genital, is, however, found in the *Chitonidoe* and in many *Cephalopoda*. When the genital glands have special efferent ducts, various sections of the latter may be differentiated into accessory sacs and glands, copulatory apparatus, etc., which, especially in the *Pulmonata*, *Opisthobranchia*, and *Cephalopoda*, transform the ducts into a very complicated apparatus.

In males, this complication arises through the development of copulatory organs, and of special glands which form the capsules of the spermatophores, and of seminal vesicles, etc.; in females, through the development of albuminous glands, shell glands, receptacula seminis, vagina, etc. Since, in hermaphrodite Molluscs, both kinds of complication occur simultaneously in the same genital apparatus, the most complicated arrangement is found in the (hermaphrodite) *Pulmonata* and *Opisthobranchia*.

Copulatory Organs

Copulatory Organs are wanting in many Molluscs, such as the *Amphineura*, nearly all Diotocardia, the *Scaphopoda*, and all *Lamellibranchia*. They are present in the *Monotocardia*, the *Pulmonata*, *Opisthobranchia*, and *Cephalopoda*. In the *Gastropoda*, in the nuchal region, to the right, there is a male apparatus, consisting sometimes of a freely projecting muscular penis, sometimes of an organ which can be protruded or evaginated through the genital aperture. In the *Cephalopoda*, this is a definite arm in the male, which is specially modified (*hectocotilised*), sometimes in a very remarkable manner, and which plays a more or less important part in copulation.

Gonads in Different Groups

Amphineura

The long hermaphrodite gland of *Proneomenia* and allied forms has been called paired. As a matter of fact it is divided into two more or less distinct lateral tubes, by a median much-folded septum. In the lower portion of each tube, that which lies next the intestine, the germinal epithelium produces spermatozoa, in the upper portion eggs. Posteriorly, these tubes separate for a certain distance, and open as a pair of distinct ducts into the anterior end of the pericardium. The male or female gonad of the *Chitonidae* lies as a long unpaired sac on the dorsal side of the intestine, in front of and partly under the pericardium.

In the ovary, numerous pear-shaped tubes project from the epithelia wall into the cavity. Each of these tubes is a stalked follicle, with egg cells surrounded by follicular cells. These follicles are found in all sizes and all stages of development. Each egg is at first a simple ovarial epithelial cell, which is distinguished by its size from the surrounding epithelial cells. As it grows and becomes more and more rich in yolk, it sinks down under the ovarial epithelium, bulging out this latter towards the ovarial cavity, and thus forming a young follicle.

The wall of the pear-shaped testicle also rises into its cavity in the form of numerous folds, in which the epithelium becomes multilaminar, and produces the mother cells of the spermatozoa. The fact that the gonad of *Chiton* has two ducts makes it probable that it was originally paired. The two ducts, *i.e.* the two seminal ducts in the male and the two ovarial ducts in the female, open into the mantle furrow on each side, somewhat in front of the renal aperture.

Gastropoda

The gonads of the *Prosobranchia* offer but few points of interest to the comparative anatomist. In the *Pulmonata* and *Opisthobranchia*, the germinal gland is a hermaphrodite gland, in which spermatozoa and eggs are produced simultaneously. This gland is much lobed, or else consists of numerous converging diverticula; the spermatozoa and eggs arise intermingled on the walls, become detached at one of the stages of their development, and then lie free in the cavity of the gonad. The same applies to the large hermaphrodite gland of the *Tectibranchia*, which varies much in its outer form. It lies in the posterior part of the body, on the digestive gland, penetrating at times between its lobes; it is itself more or less lobed, its lobes consisting of secondary lobes (vesicles or acini). In all these acini, spermatozoa and eggs are simultaneously produced. It is only in the *Pleurobranchoea*

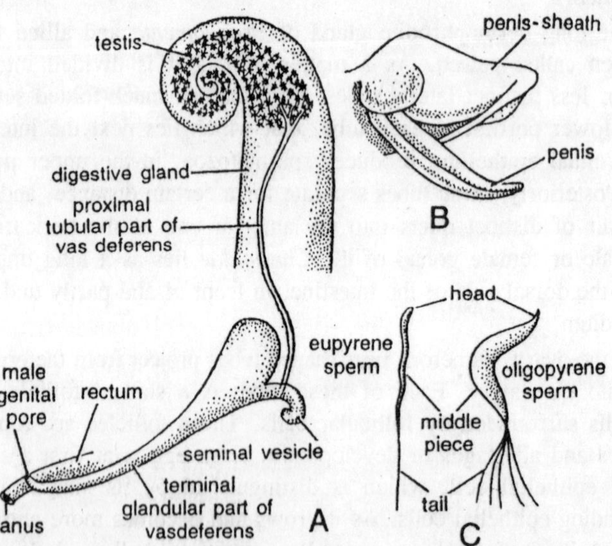

Fig. 15.2. Pila. A—Male reproductive organs. B—Male copulatory organs. C—Sperms.

and allied forms that the parts of the gland which produce spermatozoa and those which produce eggs are localised; this arrangement resembles that in the *Nudibranchia*, which will presently be described.

The constituent lobes or vesicles are either male or female, the former producing only spermatozoa, the latter only eggs. This is arrangement found also in some *Nudibranchia* (*Amphorina*, *Capellinia*), but in most *Nudibranchs* the male and the female germinal regions become separated in such a way that the terminal acini yield eggs only, but open in groups into lobes of the gland which produce only spermatozoa. Each lobe has its duct; these ducts, uniting together, finally form the duct of the hermaphrodite gland. This gland thus forms an extensive organ spread out in the larger posterior part of the primary body cavity; where there is a compact digestive gland it covers this organ.

Phyllirhoe has 2 to 6 (usually 3) separate globular acini whose long and thin ducts combine to form a hermaphrodite duct. The hermaphrodite gland of the *Pteropoda* (*Tectibranchia natantia*) always lies in the upper (dorsal) portion of the visceral dome; it is sometimes acinose and sometimes consists of converging tubular follicles or of laminae closely crowded together. The eggs are always produced at the peripheral part of the acini, tubes, or lamellae, while the spermatozoa arise in the central parts, near the ducts. These two

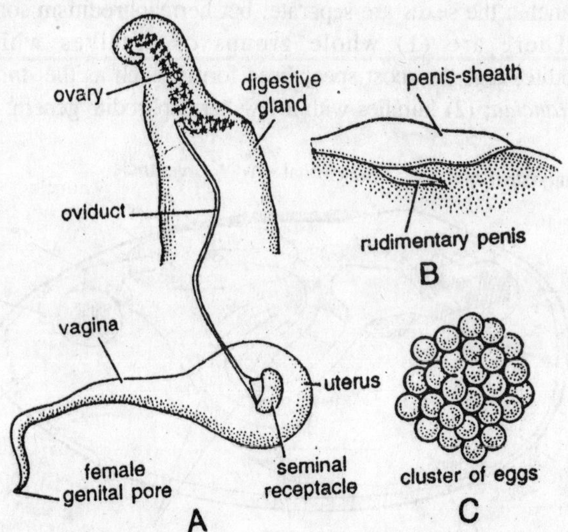

Fig. 15.3. Pila. A—Female reproductive organs. B—Rudimentary penis. C—Eggs.

parts are generally separated by a membrane, which the eggs have to break through to reach the hermaphrodite duct. The *Pteropoda* are protandrously hermaphrodite, *i.e.* the spermatozoa are produced before the eggs, an arrangement found in many hermaphrodite Molluscs.

Scaphopoda

The gonad (testis, ovary) in these animals is a long spacious sac, provided with lateral diverticula; it lies above the anus, rising high up into the visceral dome along the posterior side of the body. In the *Solenopoda* (*Siphonodentalium*, etc.) a large part of the gonad stretches into the mantle. In young animals, the gonad is closed on all sides, but in adults its wall appears to fuse with the right kidney, and in the partition wall so formed an aperture arises which establishes communication between the gonad and the right nephridium.

Lamellibranchia

The gonads are here found in the form of much-branched tubular or lobate masses lying on each side in the primary body cavity, surrounding and partly penetrating between the other internal organs. In some case (*Anomiidoe*, *Mytilidoe*), the gonad on each side stretches into the mantle. In others (*Axinus*, *Montacuta*), it bulges out the body wall in such a way that branched outgrowths, containing the germinal tubes, project from the body into the mantle cavity. In most Lamellibranchia the sexes are separate, but hermaphroditism sometimes occurs. There are (1) whole groups of bivalves which are hermaphrodite; *e.g.* the most specialised forms, such as the *Anatinacea* and *Septibranchia*; (2) families with a few hermaphrodite genera: *Cyclas*,

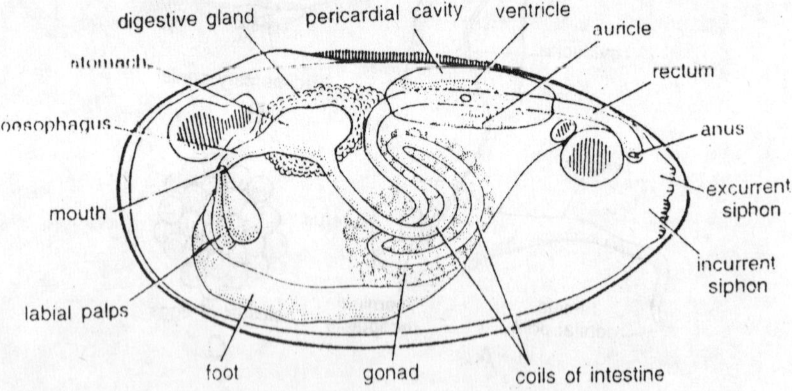

Fig. 15.4. Unio. Alimentary canal and digestive gland.

Pisidium, *Entovalva*; (3) genera (*Ostroea*, *Pecten Cardium*) with a few hermaphrodite species; (4) occasional cases of hermaphroditism in species the sexes of which are usually separate: *Anodonta*.

The hermaphroditism of the Lamellibranchia is, however, always incomplete in the sense that the spermatozoa and the eggs do not ripen simultaneously. In the *Anatinacea* and *Septibranchia*, there are on each side entirely separate male and female gonads, whereas all other hermaphrodite Lamellibranchs have a hermaphrodite gland on each side.

Cephalopoda

The sexes are always separate in this class. It has already been mentioned that the germinal sacs form a part of the secondary body cavity, with which they are in open communication. One single unpaired gonad is always found, lying in the uppermost part of the visceral dome. It is a variously-formed sac (peritoneal sac or genital capsule), lined on all sides by an epithelium often to a great extent ciliated, which is in reality the peritoneal epithelium of secondary body cavity. The whole of the epithelium covering the wall of the gonad is not, however, germinal, but only that on its anterior side (that turned to the shell).

The germinal layer here forms what may be called, in the narrower sense, the ovary of the testis, which is then said to be contained in a peritoneal sac or an ovarial or testicular capsule, or else to project

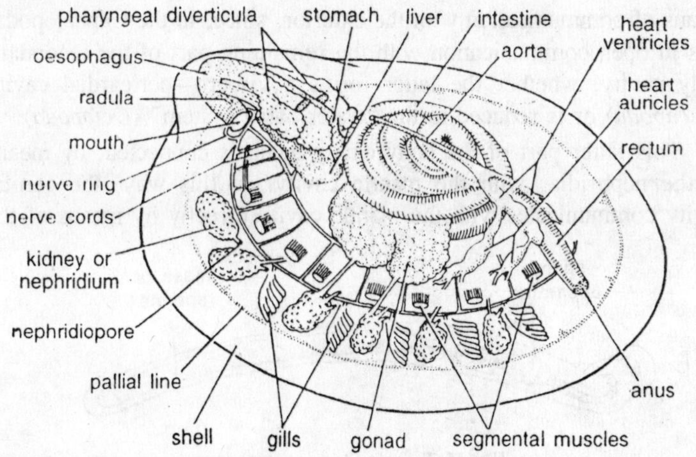

Fig. 15.5. Neopilina. Internal anatomy in side view.

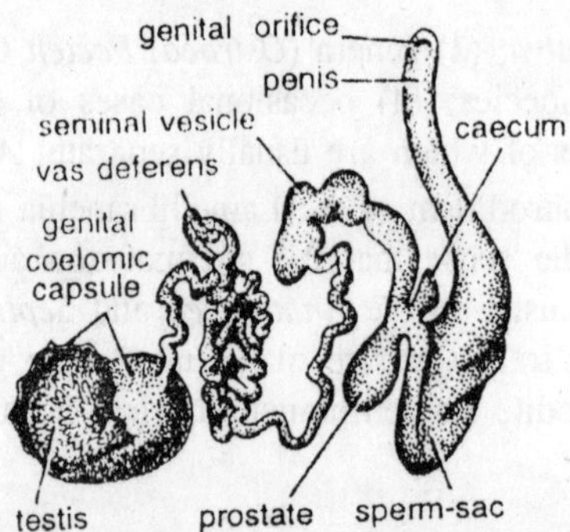

Fig. 15.6. Sepia. Male reproductive system.

into or be suspended in such sac or capsule. The whole apparatus is really a gonad, in which the places of formation of the reproductive cells are localised on the anterior wall. From this it is clear why the tested and ovaries do not appear to possess efferent ducts of their own, but to empty their products into their respective capsules, these products passing out into the mantle cavity through the ducts of these capsules (oviducts and seminal ducts).

Since, however, the entire germinal sac corresponds with the genital gland of a *Gastropoda* or a *Lamellibranch*, the reproductive products in reality merely fall into the cavity of this gland (the testicular and ovarial capsules), and pass out through the ovarial and seminal ducts, which exactly correspond with the same ducts in the *Gastropoda*, *Lamellibranchia*, and *Chitonidoe*. The genital cavity has also another means of communication with the exterior, since, in the Cephalopoda, it is in open communication with the remaining part of the secondary body cavity, whether the latter forms a viscero—pericardial cavity (*Decapoda*) or is reduced to the "water canal system" (*Octopoda*).

This latter part of the body cavity again is connected, by means of the nephridia, with the mantle cavity. In this way, the genital cavity communicates with the mantle cavity directly by means of the

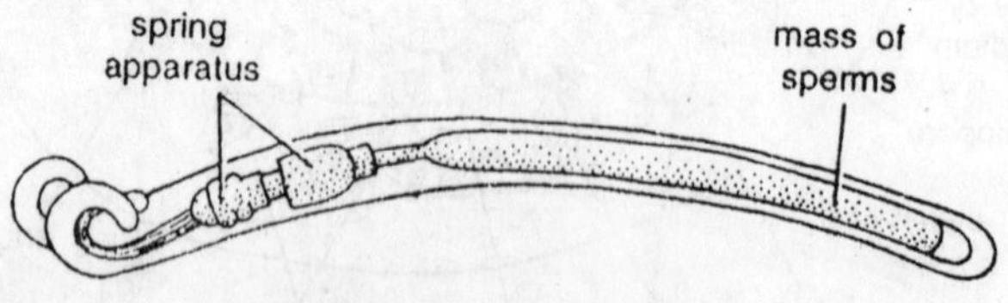

Fig. 15.7. Sepia. A spermatophore.

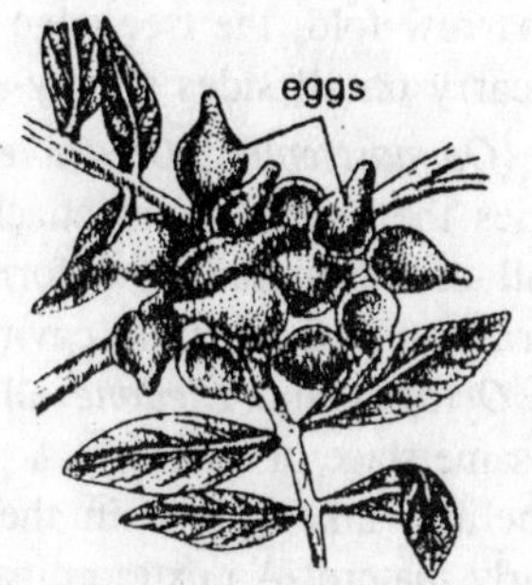

Fig. 15.8. Sepia. A cluster of eggs.

oviduct or seminal duct, and indirectly through (1) the viscero-pericardial cavity or the "water canal system," and (2) the nephridia. This second way of communication, however, is never used for discharging the genital products.

The female germinal layer or ovarial layer (the ovary in the narrower sense) is always found on the anterior wall of the gonad, and varies considerably in structure. We can always distinguish (1) the eggs, and (2) the ovigerous wall. The former are stalked, and project from the wall into the cavity of the gonad (the cavity of the ovarial capsule). The largest and oldest eggs are covered by a follicular epithelium, and this latter by the general epithelium of the wall of the gonad, which also covers the stalk. Each egg has a separate stalk. The youngest eggs are mere prominences on the wall, which in the process of growth acquire a stalk, by means of which they remain connected with the wall from which they project. This arrangement is exactly like that in the *Chiton*. When the eggs are mature, the follicle bursts, they fall into the genital cavity, and thence reach the exterior through the oviduct. In *Nautilus* and *Eledone* the whole wall of the gonad, with the exception of the posterior surface, can produce eggs; these stand out from it all over on simple stalks.

In *Argonauta* and *Tremoctopus* also, the whole ovarial capsule except the posterior wall produces eggs, but the egg-bearing region (to obtain increase of surface) projects into the genital cavity in the form of numerous dendriform processes, the eggs being attached by simple stalks to the stems and branches. In *Parasira* (*Tremoctopus*) *catenulata* there is a central region containing more than twenty large "egg trees" surrounded by a circle of smaller "trees." On the anterior wall of the gonad in *Octopus* there is a single but very richly-branched "egg tree" (*C*). In *Sepia*, *Sepiola*, and *Rossia* the egg-bearing surface bulges out in the shape of a ridge on the anterior wall of the gonad. This ridge, in *Loligo*, becomes a narrow fold, the free edge of which is produced into filaments, which carry on all sides simply-stalked eggs.

In the *Oegopsidoe* (*Ommastrephes*, *Onychoteuthis*, *Thysanoteuthis*) the region which carries the eggs is only attached by its upper and lower ends to the wall of the gonad, and forms an otherwise free spindle-shaped body traversing the genital cavity, and beset all over with stalked eggs. In *Octopus* and *Eledone* all the eggs in a given ovary are found at the same stage of maturity. A peculiar transformation of the follicular epithelium takes place in the ovarial eggs of the Cephalopoda when nearly mature. An extraordinary increase of surface occurs in the shape of numerous folds, which run longitudinally along the egg, either reticulating or remaining parallel to one another, and projecting for into the yolk of the egg which they surround. This arrangement may be connected with the nutrition of the egg.

The *male* germinal layer (germinal body, or testis in the narrower sense) is a variously-shaped (often globular or oviform) compact organ, which usually lies free in the genital cavity, suspended to its anterior wall by a thin ligament (*mesorchium*) in which the genital artery runs. The germinal body is everywhere covered with epithelium, which is continued over the mesorchium into the epithelium of the wall of the gonad (endothelium of the testicular capsule). On the surface of the germinal body which is turned away from the mesorchium, there is a funnel-shaped depression towards this, from all sides, the tubular testicular canals which form the male germinal body converge, in order to open into it. In these testicular canals, between which there is a slight framework of connective tissue, the spermatozoa are produced, and are passed on to the genital cavity through the depression into which all the canals open; they reach the exterior by means of the seminal duct.

The testicular canals originally possess a multilaminar germinal epithelium, which yields the spermatozoa, and which passes at the common aperture into the outer epithelium of the germinal body, and so into the epithelium of the germinal sac. This description applies to the male germinal body of most Cephalopoda. In *Loligo*, however, the funnel-shaped depression into which all the testicular canals open is replaced by a longitudinal furrow, into which these converging canals open.

In *Sepia*, the germinal body has no ligament, but lies immediately in front of the anterior wall of the gonad, and is thus outside the genital cavity. The germinal body here has a central channel towards which the radially arranged seminal canals converge from all sides, and which they enter. This channel, again, opens through an efferent duct into the genital cavity, from which the spermatozoa are conducted to the exterior by the seminal duct. The *Spermatozoa* of the Mollusca are of the common pin shape.

In many species of *Prosobranchia* two different forms of spermatozoa, the hair-shaped and the vermiform, occur in one and the same individual. This phenomenon has by some been taken as an indication of developing hermaphroditism, and by others as pointing to a former hermaphrodite condition; in the first case the vermiform spermatozoa would be the eggs beginning to form, in the second the rudiments of eggs. There is, however, no solid foundation for either of these views. With regard to the question whether the hermaphrodite or the dioecious condition is the original condition, the latter alternative may be considered as the more probable. Of the five classes of the Mollusca, two, the *Scaphopoda* and *Cephalopoda*, are altogether dioecious. Among the *Amphineura*, the *Chitonidoe*, which most recent observers hold to be less specialised than the *Solenogastres*, are sexually separate. Among the *Lamellibranchia*, the sexes are separate in the *Protobranchia*, which are rightly considered as primitive forms; and most other bivalves are also dioecious. Among the *Gastropoda*, the sexes are separate in the *Prosobranchia*, especially in the *Diotocardia*, which are universally considered to be the lowest and least specialised *Gastropods*.

Duct in Different Groups

The manner in which the sexual products are conducted to the exterior in the *Amphineura*, *Scaphopoda*, and *Lamellibranchia* need not again be discussed, as it has already been described in the general part of this section, and in the section on the nephridial system. We thus have now only to treat of the very complicated ducts of the *Gastropods* and the *Cephalopoda*.

Gastropoda

It has been seen that in all *Diotocardia* (*Haliotis*, *Fissurella*, *Patella*, etc.) the genital products are ejected through the right kidney. In the *Monotocardia*, the right kidney has atrophied as such, but, according to the most recent investigations, its duct persists as genital duct. In the *Pulmonata* and *Opisthobranchia*, the genital aperture is no longer in the mantle cavity, but has shifted for forward along the right side of the neck, probably in connection with the development of the copulatory apparatus. The position of this aperture is thus not necessarily affected by any further displacement of the pallial complex, or indeed of the whole visceral dome, which explains the fact that, in *Daudebardia* and *Testacella*, the common genital aperture, and in *Oncidium*, the male aperture, lies for forward on the right side of the body, although the pallial complex has shifted completely to the posterior end of the body. In the *Opisthobranchia* also, the single or (secondarily) double genital aperture lies to the right in front of the anus and even in front of the kidney. This position seems inexplicable except by the supposition of a shifting back of the pallial complex in which the genital aperture, emancipated from the complex, took no part, thus coming to lie in front of the shifted anal and renal apertures.

Monotocardia

Unlike the *Diotocardia*, which , with the exception of the *Neritidae*. have no copulatory organs, the Monotocardia possess a penis, which, however, does not lie in the mantle cavity where the genital aperture originally lay. It would be unable to function in this position, and is therefore placed on the right side of the head or neck, and forms a freely projecting, extensible, muscular appendage, which often attains a considerable size.

The male genital aperture, however, in very many perhaps in most, Monotocardia, remains in its original position in the mantle cavity, to the right, near the rectum. In such cases, a ciliated furrow runs forward on the floor of the respiratory cavity, along the right side of the neck, to the base of the penis, to the tip of which it is continued as a deep groove. This furrow conducts the semen to the penis from the genital aperture. In some cases the furrow closes, and forms a canal; the penis then becomes tabular, and the seminal duct enters into it. The genital aperture is thus shifted for forward from its original position. The seminal duct, which arises from the testis, usually forms coil as it runs along the columellar side of the shell. The vas deferens has no special appendages, although it may widen into a vesicle at some point in its course.

In the female, the genital aperture remains in the mantle cavity, lying to the right near the rectum, behind the anus. The duct remains, as a rule, more or less simple; it is divided into the following consecutive sections: (1) an *oviduct*, rising from the ovary, which may bulge out to form one or more *receptacula seminis*; (2) the *uterus*, a wider section with thick glandular walls, in which the eggs are provided with albumen and a shell; (3) a muscular sheath, the *vagina*, which leads to the outer genital aperture. In *Paludina*, there is a special *albuminous gland* opening into the oviduct. In hermaphrodite *Prosobranchia* (*Valvata*, a few *Marseniadoe*, *e.g. Marsenina*, *Onchidiopsis*) a hermaphrodite gland is found. This gland gives rise either to one duct, which divides later into a vas deferens and an oviduct, or to a vas deferens and an oviduct which are from the first distinct. The vas deferens runs to the penis as in the males of dioecious *Prosobranchiates*; the oviduct runs to the female genital aperture. Both these ducts are, owing to the occurrence of accessory glands, etc., more complicated than in other *Prosobranchiates*.

Opisthobranchia and Pulmonata

The ducts in these orders are extremely complicated, both by division into many consecutive sections and by the development of various accessory organs. In the following descriptions of several types of genital ducts only the most important points can be mentioned. We give first the type of duct commonly found in the *Cephalaspidoe* (*Tectibranchia*).

1st Type—The hermaphrodite gland has a single undivided efferent duct, opening out through a single genital aperture. From this aperture the fertilised eggs pass out direct, but the spermatozoa pass into a ciliated seminal furrow which runs along in the mantle cavity, and by which they are conducted to the penis. This lies more or less far forward in front of the genital aperture, near the right tentacle. If we imagine the testis of a male Monotocardian transformed into a hermaphrodite gland, and the vas deferens into a hermaphrodite duct, the above conditions would be realised.

Gastropteron may be chosen as a good example of this arrangement, which is further found in other *Cephalaspidoe* (*Doridium*, *Philine*, *Scaphander*, *Bulla*) and all *Pteropoda*. The hermaphrodite gland or ovotestis, which lies between the lobes of the liver inthe posterior part of the body, gives rise to a hermaphrodite duct, which, after a long coiled course, enters a short but much widened terminal section known as the uterus or genital cloaca. This cloaca opens outward in front of the base of the gills through the genital aperture. Into the cloaca open: (1) the common efferent duct of two glands, one of which, the albuminous gland, supplies the egg with albumen, while the other, the nidamental or shell gland, yields its outer protective envelope; (2) the duct of a globular vesicle (receptaculum seminis, Schwammerdam's vesicle), which receives the spermatozoa during copulation. From the genital aperture, which has a more or less median position on the right side of the body, the seminal furrow runs forward to the penis. The latter is enclosed in a special sheath, out of which it can be protruded, and into which it is withdrawn by means of a retractor muscle. A gland called the prostata opens into the penis. The penis itself lies on the right anteriorly, on the boundary between the head and the foot. When it is at rest its sheath lies in the cephalic cavity, near the buccal mass. The very complicated ducts of *Aplysia* and *Acera* do not essentially differ from that above described. The hermaphrodite duct, on reaching the region of the albuminous gland, coils back upon itself, the ascending and descending portions of this coil surrounding the albumen gland with their spiral coils. The penis has no prostata.

2nd Type— The hermaphrodite gland gives rise to a hermaphrodite duct, which soon divides into two parts the vas deferens or seminal duct, and the oviduct. The former runs to the male copulatory apparatus, the latter to the female genital aperture. The male aperture and the penis lie in front of the female, far forward on the head or neck; the two apertures are quite distinct, and both lie on the right. This second type may be deduced from the first, if we assume not only that the common duct of the hermaphrodite gland divided into a male and a female duct, but also that the seminal furrow closed to from a canal in continuation of the male duct. When the duct of this second type split into a male and a female duct, the accessory organs also so divided that the male opened into the vas deferens, the female into the oviduct. To this type belong, among the *Pulmonata*, the *Basommatophora*, a few species of *Daudebardia* (*D. Sauleyi*, in which the two apertures lie close together), the *Oncidia*, and *Vajinulidoe*.

In both these latter groups, the female aperture has followed that part of the pallial complex which shifted to the posterior end of the body, and lies near the anus. The male aperture has, however, retained its anterior position on the head, behind the right cephalic tentacle. The two apertures thus lie at the opposite ends of the body. Among the *Opisthobranchia*, this second type is exemplified in *Oscinius* (*Tectibranchia*). Taking *Limnaea stagnalis* and *Oncidium* as examples, we find in the former that the hermaphrodite gland which lies embedded in the "liver," high up in the visceral dome, gives rise to a thin hermaphrodite duct; this soon divides into a male and a female duct.

The male duct first widens into a flattened sac, then into a large pear-shaped glandular vesicle (prostata). From this vesicle it runs as a long thin vas deferens through part of the pedal musculature, and finally enters the male copulatory apparatus, which is, in fact, merely the widened muscular and protrusible end of the vas deferens. A small penis tube is first formed by the vas deferens, and this projects on a papilla into a subsequent larger tube (the penis sheath), which is evaginated during copulation. Protractors are attacjed to the sheath, and retractors to the small tube; the latter alone with its papilla enters the vulva during copulation. An albuminous gland opens into the female duct immediately after its separation from the male duct. It then forms a uterus consisting of wavy folds, and is continued into a large pear-shaped body as oviduct, the narrow end of which is the vagina and leads to the female genital aperture. The oviduct receives a lateral accessory gland called the nidamental gland, and the vagina the efferent duct of the globular receptaculum seminis.

In *Oncidium celticum* the hermaphrodite gland and female accessory glands lie in the posterior part of the body, between the lobes of the liver and the coils of the intestine. From the gland rises a hermaphrodite duct, which at one point carries a small lateral caecum, and opens into an irregularly-shaped organ, the uterus. Within the uterus two projecting folds border a channel; if these folds become apposed, the channel becomes a tube. This channel runs from the point of entrance of the hermaphrodite duct to the point where the seminal duct leaves the uterus, and serves for conducting the semen. The remaining wider portion of the uterus serves as oviduct and egg-reservoir, and carries a large caecal appendage; the ducts of the two much-lobed albuminous glands also enter the uterus. A comparison of *Limnoea* and *Oncidium* shows that in the latter the male and female ducts separate from one another further back than in the former.

The vas deferens in *Oncidium* is only incompletely separated as a groove in the uterus. Its differentiation into a separate duct takes place here, as in terrestrial *Pulmonates*, at the distal end of the uterus. The thin seminal duct (vas deferens) passes into the body wall to the right, and runs forward along the longitudinal furrow between the foot and back, passing again at the anterior end of the body into the primary body cavity, where it forms numerous coils, and finally enters the copulatory apparatus. This apparatus, in *Limnoea*, consists of a large evaginable terminal widening, into which the vas deferens projects in the form of a papilla. Blood pressure causes the penis sheath or praeputium to be evaginated through the genital aperture, into which it is again withdrawn by means of a retractor.

In other species of *Oncidium*, the copulatory apparatus is complicated by the occurrence of accessory penis glands and variously-shaped cartilaginous armature. The oviduct which separates from the vas deferens at the end of the uterus is also a vagina. It is a simple tube which opens outward to the right near the anus through the genital aperture. Near the middle of its course it is joined by the stalk-like duct of a globular vesicle, the receptaculum seminis (bursa copulatrix), and by a long glandular caecal appendage.

3rd Type—We find this in the *Stylommatophora* among the *Pulmonata*, and also in all *Nudibranchia* and a few *Tectibranchia* (e.g. *Pleurobranchoea*). The hermaphrodite gland gives rise to a hermaphrodite duct, which, as in the second type, sooner or later divides into a male and a female duct. These, however, do not open out through distinct apertures, but again unite to form a common atrium genital or a genital cloaca. This third type may be deduced from the second by supposing that the male and female apertures became approximated, and finally opened together. *Helix pomatia* and *Pleurobranchoea Meckelii* afford good examples of this arrangement.

Helix pomatia

From the hermaphrodite gland a hermaphrodite duct, in zigzag coils, passes into the long folded uterus. The straight band which passes along the folds of the uterus is that portion of it which belongs to the seminal duct; the folds belonging to the female ducts. The seminal channel, however, is merely a furrow within the uterus, divided from the cavity of the latter by two projecting folds, the edges of which become superimposed. A longitudinal glandular band, which is regarded as a prostata, accompanies this duct. At the point where the hermaphrodite duct passes into the uterus, the large linguiform albuminous gland opens into it. At the end of the uterus, the male and female ducts become entirely distinct. The thin vas deferens runs is coils to the copulatory apparatus, which again opens into the genital cloaca. The copulatory apparatus consists of a protrusible penis; at the point where the vas deferens enters this organ, the latter carries a long hollow appendage, the flagellum, the glandular epithelium of which perhaps yields the substance of the spermatophoral capsules. At the same point a retractor muscle is attached to the penis.

The short oviduct widens before opening into the genital cloaca. The widened portion has the following appendages: (1) a long stalked pear-shaped receptaculum seminis, lying close to the uterus,—the stalk has a lateral bulging, which is sometimes rudimentary; (2) two tassel-shaped organs, the digitate glands, the milky secretion of which contains calcareous concretions, and no doubt assists in the formation of the outer envelope of the egg; (3) the dart sac, which lies close to the cloaca, and contains a pointed calcareous rod, the *spiculum amoris*, which is thrust by each individual into the tissue of the other as an excitant during copulation. The common outer genital aperture lies in the nuchal region behind the right optic tentacle.

Pleurobranchaea mockelii

The hermaphrodite duct, which rises from the gland, forms a long amulla or widening, and then divides into a male and a female duct. The vas deferens runs in coils to the penis sheath, which it enters, coiling up in almost like a watch-spring, and then forms the evaginable widened end portion which is called the penis, and which can be invaginated by a retractor muscle. The oviduct has a shorter course, and receives the short efferent duct of a globular receptaculum seminis. The widened terminal portion of the oviduct (the vagina), which enters the genital cloaca with the penis, receives the ducts of the albuminous and nidamental glands (shell and slime glands); the second of these may be regarded as the homologue of the digitate gland of *Helix*. There is a general agreement between the ducts of the *Nudibranchia* and those just described; in details, however, extraordinary variety prevails.

The male and female ducts nearly always unite in the base of a genital cloaca, which often lies anteriorly on the right, on a papilla. The male and female apertures are rarely separate; when they are so, they lie close together. The penis is often armed in various ways. The important subject of the mutual relations of the three types of genital ducts in hermaphrodite Gastropoda has been much discussed, but no satisfactory conclusion has been reached. Ontogenetic research has been appealed to so far in vain. It is thus not at present known whether the single hermaphrodite duct has arisen by the fusing of separate male and female ducts, or whether the separate ducts have come into existence by the splitting of an originally single hermaphrodite duct.

The difficulty is increased by the fact that the genetic significance of the hermaphrodite gland is uncertain. Fertilisation is mutual in hermaphrodite Gastropods. It is, however, certain that, in the *Pulmonata* at least, when copulation does not take place, self-fertilisation can occur. The hermaphrodite duct not infrequently carries one or two lateral caeca or vesiculae seminales, in which an animal can store up its own sperm to be used in fertilising its own eggs if cross-fertilisation does not take place. The eggs and the sperm are often not ripe at the same time.

Cephalopoda

Although the gonad in all extant Cephalopoda is unpaired, the ducts are originally paired in both sexes. In *Nautilus*, the *Ocgopsidoe*, and the *Octopoda*, there is one pair of ducts in the female; but in the males a paired seminal duct occurs only in *Nautilus* and *Philonexis carenoe* (*Tremoctopus*). In *Nautilus*, in which both sexes possess paired ducts, the left duct is in both cases rudimentary and no longer functions. It is the so-called pear-shaped vesicle, which is attached on one side to the heart and the lower and of the gonad, and on the other opens into the mantle cavity at the base of the lower gills. Where only one duct is retained, it is, in both sexes, the one on the left, as in *Loligo*, *Sepia*, *Sepiola*, *Rossia*, *Sepioteuthis*, *Chroteuthis*, *Cirrhoteuthis*, etc. The genital ducts rise on the wall of that part of the secondary body cavity which is known as the genital cavity (peritoneal) sac, genital capsule), and open into the mantle cavity at the sides of the anus, between the nephridial aperture and the base of the gills.

Male ducts

In more complicated form of the male duct, such as that of *Sepia*, four principal divisions may be distinguished. From the testicular capsule rises a vas deferens, which runs along in close coils, and then widens into a vesicula seminalis, the highly developed and much folded epithelium of which plays an important part in the formation of the spermatophores. The vesicula seminalis is continued as a thin vas deferens to the last division, the spermatophoral pouch (Needham's pouch), which serves as a reservoir for the spermatophores. This pouch is flask-shaped and projects freely, with the end which corresponds to the neck of the flask, at which the male genital aperture lies, into the mantle cavity. The vas efferens receives (1) the short duct of an oviform gland, the *prostata*, and (2) a simple, lateral, non-glandular *caecum*. The prostata takes part, like vesicula seminalis, in the formation of the spermatophores. The prostata, caecum, and vesicula seminalis, in their nature position, form a coil, which lies in a special division of the secondary body cavity, the peritoneal sac.

It is remarkable that the vas deferens is in open communication with this peritioneal sac by means of a narrow tube. The male efferent apparatus of *Octopus*, as compared with that of *Sepia*, is distinguished chiefly by the absence of a separate vas efferens. The long vesicula seminalis opens into the large prostata near the point where the latter enters the spermatophoral pouch. This point lies, not in the base, but in the neck of the pouch, where the latter is produced into the long fleshy penis, the point of which projects into the mantle cavity. The penis is provided with a lateral caecum. It has already been mentioned that, as far as we know at present, only two living Cephalopods, *Nautilus* and *Philonexis carenoe*, have paired male duicts. In *Nautilus*, the left duct is rudimentary. Whether the two ducts of *Philonexis carenoe* correspond with the two ducts which we may assume that the Cephalopoda originally possessed is very doubtful. The two vasa deferentia of *Philonexis*, which arise out of the testicular capsule, and differ considerably in structure, unite together later, and both lie on the left side. It is also remarkable that the spermatophoral pouch has two apertures, and that there are thus two genital apertures.

Female ducts

The complicated female efferent apparatus consists of two entirely distinct parts, opening separately into the mantle cavity: (1) an unpaired oviduct (to the left), the position and aperture of which correspond with those of the seminal duct in the male; and (2) the *nidamental glands*. The two large nidamental glands are pear-shaped organs, lying just beneath the integument in the posterior part of the visceral dome, symmetrically, at the side of and anterior to the descending efferent duct of the ink-bag. They open into the mantle cavity at their ventral ends. Each gland appears symmetrically divided by a series of glandular lamellae, traversing it from side to side.

The spaces between the lamellae open into the central slit-like duct; this structure is to be seen even on the exterior of the gland. Besides these two nidamental glands there is an *accessory nidamental gland* lying below and in front of the former. It is of a brick-red colour, and consists of a central part and two lateral lobes. It consists of numerous coiled glandular canalicules, which open into a glandular area in the mantle cavity. This glandular area forms a depression between the central and lateral lobes. As the aperture of the large nidamental gland also lies in this depression, the secretions of the two glands here mingle.

The oviduct which rises from the ovarial sac is, during the reproductive season, so full of eggs, that it becomes much distended, especially at the part which opens into the ovarial sac. Before this duct opens outward into the mantle cavity at the same point and in a similar manner as the seminal duct in the male, it becomes connected by means of a freely projecting portion with a doubly-lobed or heart-shaped accessory gland, the *gland of the oviduct*, which repeats the structure of the nidamental gland. The terminal portion also (from the point of entrance of this gland to the aperture of the oviduct) is glandular, two symmetrical rows of perpendicular glandular leaflets projecting from its wall into its lumen. The secretions of the nidamental glands, accessory nidamental glands, and the glands of the oviducts yield the outer envelopes of the ovarial eggs. Nidamental glands occur, among the Cephalopoda, (1) in the *Tetrabranchia* (*Nautilus*); (2) in the *Dibranchia*, among the *Decapoda*, in the *Myposidoe* (*Sepia*, *Sepiola*, *Rossia*, *Loligo*, *Sepioteuthis*, etc.); in a few *Oegopsidoe* (*Ommastrephes*, *Onycotecuthis*, *Thysanoteuthis*). They are wanting inthe *Octopoda* and in some *Oegopsidoe* (*Enoploteuthis*, *Chiroteuthis*, *Owenia*).

Nautilus is distinguished from all other living Cephalopoda (1) by the possession of only one nidamental gland, and (2) by the fact that this gland does not lie in the visceral dome but in the mantle. Accessory nidamental glands are found only in the *Myopsidoe*. The two glands are either separate (*Rossia*, *Loligo*, *Sepioteuthis*) or fused together (*Sepia*, *Sepiola*). Glands of the oviduct occur in all Cephalopoda, but vary in position and in structure. Outgrowths of the oviduct, which function as receptacula seminis, occasionally occur (*Tremoctopus*, *Parasira*). In all Cephalopoda, certain quantities of spermatozoa are collected in extremely complicated envelopes, the *spermatophores*. The substance of these large filamentous spermatophores is yielded by the prostata and the vesicula seminalis, abut the mechanism by which so complicated a case is produced is still unknown. When touched, or when they reach water, the spermatophores burst at definite points, and scatter their contents. At the reproductive season the spermatophoral pouch is entirely filled with spermatophores. In *Philonexis carenoe*, however, only one very long spermatophore is produced.

Copulatory Apparatus in Different Groups

Hectocotylisation in the Cephalopoda

The copulatory apparatus of the *Gastropoda*, and the penis which projects into the mantle cavity in certain Cephalopoda have already been described. One of the most remarkable and enigmatical phenomena is connection with the Cephalopoda is their hectocotylisation. This consists in the transformation of one of the oral arms of the male into a copulatory organ and spermatophore-carrier. This arm is said to be hectocotylised; during copulation it becomes detached, and finds its way into the mantle cavity of the female. Typical hectocotylisation is found only in the Octopodan genera *Argonauta*, *Philonexis*, and *Tremoctopus*. In *Tremoctopus* and *Philonexis* (*Parasira*) the third arm on the right is the one transformed, in *Argonauta* the third on the left.

The arm is at first enclosed in an outwardly pigmented sac, when this bursts, the arm becomes free, and then its special form can be recognised (B). The folds which formed the sac bend back so as to form a new sac, which receives the spermatophores and is now inwardly pigmented. An aperture leads from this sac into a seminal vesicle inside the hectocotylised arm this vesicle is continued into a long thin efferent duct, which runs the whole length of the arm and opens outwardly as its end.

The end of the arm is transformed into a long filamentous penis, which at first is also enclosed in a special sac. When the penis is evaginated the sac remains as an appendage as its base. The spermatophores then pass from the pigmented sac into the seminal vesicle, and are ejected through the efferent duct which opens at the tip of the penis. It is probable that Cephalopods grasp one another, during copulation, with their arms, in such a way that their mouths face each other. In this position the hectocotylised arm of the male becomes detached, and in some way or other forces its way into the mantle cavity of the female. Detached arms are often found in the mantle cavity of the female, as many as four have been found at one time. We still do not know (1) how the hectocotylised arm fertilises the eggs of the female, or (2) how the spermatophores reach the hectocotylised arm.

The males and females in the above-mentioned genera differ from one another, apart from the sexual dimorphism caused by the development of the hectocotylised arm. The males are much smaller, and in *Argonauta* the female only has a shell. It is very probable that the detached hectocotylised arm can be replaced by a new one. Although a true hectocotylised arm which can be detached, is only developed in the three general above- mentioned, it has been proved that in all other Cephalopoda (even *Nautilus*), a certain arm or portion of the head in the male is in some way modified, differing in some (often unimportant) manner from the other arms. Such as arm is said to be hectocotylised, and it is assumed that it plays some part in copulation, although its exact function is unknown. In *Sepia* and *Nautilus* it is even difficult to imagine what part is can take in copulation.

The constant occurrence of a hectocotylised arm is the more remarkable as it is by no means always the same arm that is thus transformed. In the *Octopoda*, as a rule, it is the third on the right side, but in the Octopodan subgenus *Scoeurgus* and in *Argonauta* it is third on the left. In the *Decapoda* the hectocylised arm is generally the fourth on the left, but in the genus *Enoploteuthis* it may be the fourth on the right, or even in one and the same species of *Ommastrephes*, it is sometimes the fourth on the left and sometimes the fourth on the right.

In *Sepiola* and *Rossia*, it is the first arm which is hectocotylised. Finally, both the arms of one pair may be thus transformed; in *Idiosepion* and *Spirula* this is the case with the fourth pair, in *Rossia* with the first. The difference in size between the male and the female, which has been mentioned as occurring in those forms which have true hectocolylised arms, is also found, though not to the same degree, in many other Cephalopoda, in which the male is slightly smaller than the female.

16

EMBRYONIC DEVELOPMENT

AMPHINEURA

Ontogeny of Chiton Polii

The egg possesses little nutritive yolk. The segmentation is total and somewhat unequal; a coelogastrula is formed by invagination.

(*a*) The blastopore of the gastrula larva marks its posterior end. A pair of endoderm cells near the dorsal edge of the blastopore are specially large. A longitudinal section shows two ventral ectodermal cells with larger nuclei; these belong to a double row of cells on which is developed the *preoral ciliated ring* which, in Molluscs, is called the velum.

(*b*) At a later stage, the blastopore appears shifted somewhat towards the ventral side, and an inward growth of ectodermal cells begins at its edge; this is the commencement of the formation of the ectodermal *stomadaeum*. At the posterior and upper edge of the blastopore, there is a cell lying between the endoderm and the ectoderm; this is, not doubt, a *mesodermal cell*.

(*c*) *The larva elongates*; a distinct stomodaeum (embryonic oesophagus), leading through the blastopore into the archenteron, is formed by the continuous growth inward of the ectodermal cells; this organ becomes shifted still further forward along the ventral surface.

(*d*) The larva is an oblique section from an anterior upper to a posterior lower point through a slightly older larva, which shows the stomodaeum, and the sides of the blastopore, the *first mesoderm cells*. These are probably derived from the endoderm, and are symmetrically placed at the two side of the blastopore.

(*e*) A median section through the next stage shows no mesoderm cells as yet in the median plane. The mouth, however, appears shifted forward along the ventral side as far as the ciliated ring or velum, the double row of cells in the latter being very clear.

(*f*) Transverse section of an older stage. The mesoderm cells have increased in number, and are arranged in two groups at the sides of the stomodaeum, between the ectoderm and the endoderm.

(*g*) At a later stage, a longitudinal section of the principal feature is a stronger development of the mesoderm, in which a space, the *body cavity*, now appears. A bulging backward of the stomodaeum forms the first rudiment of the *radular sac*. Behind the mouth, a sac—like depression is formed, evidently by the ectoderm; this has been called the *pedal gland*, although it has not yet been discovered what becomes of it in the adult animal.

(*h*) When the body cavity forms, the cells of the mesoderm become divided into two layers, in inner *visceral layer* becoming applied to the intestine, and the outer *parietal layer* to the ectoderm. In the transverse section, we see, deep down in the ectoderm, the first rudiments of the *pleurovisceral cords*. The *pedal cords* arise in the same way, and anteriorly, in the cephalic area, which is encircled by the preoral ciliated ring, the rudiments of the *supra-oesophageal central nervous system* form as a neural plate, *i,e,* as a thickening of the ectoderm, which carries at tuft of long cilia.

(*i*) At later stages, the central nervous system with the pleurovisceral and pedal cords become detached from the ectoderm and take up their mesodermal position. The rudiments of seven shell-plates appear on the back as cuticular formations the eighth only appears later. A posterior invagination of the ectoderm represents of rudiment of the proctodeum (the embryonic hind-gut with the anus). The first teeth appear in the radular sac. The whole of the cephalic area and the region of the foot become covered with cilia. On the dorsal ectoderm, on the parts that are not covered by the shell-plates, the first calcareous spines appear. In the posterior part of the body, a great accumulation of mesodermal elements evidently marks the position of a formative mesodermal zone.

At this stage, the larva leaves the egg envelope, and swims about freely, and, on the degeneration f the ciliated ring, sinks to the bottom transformed into a young *Chiton*. During this last transformation two lateral larval eyes appear on the anterior ventral side of the body. The development of the circulatory system, the nephridia, the genital organs and the etenidia has not been followed.

Solenogastres

The ontogeny of this order is as yet only known through a very incomplete account of the development of *Dondersia banyulensis*. The segmentation is unequal and total, and takes place through the formation of micromeres. The process of gastrulation seems to occur in a manner half way between epibole and invagination. The blastopore marks the posterior end of the larval body, which is divided by two circular furrows into three consecutive regions. The anterior region consists of two circles of cells, and evidently corresponds with the pretrophal area. It is partially ciliated, and carries in the middle a group of longer cilia, one which is sometimes to be distinguished from the rest as a flagellum. The second region, which consists of a single row of cells, carries a circle of long cilia, and evidently represents the velum. The third region consists of two rows of cells carrying short cilia the second row edges the blastopore.

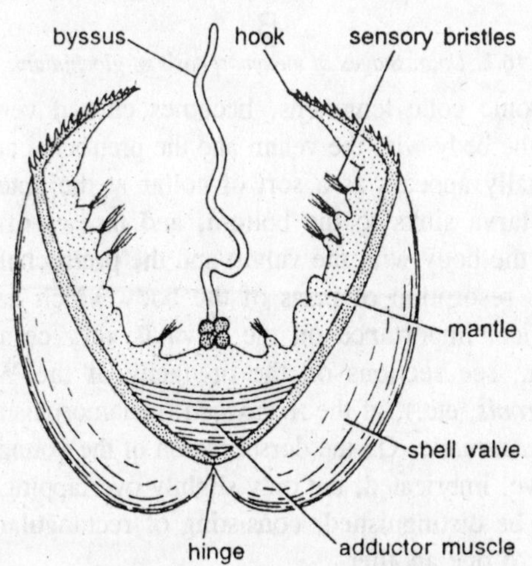

Fig. 16.1. Unio. A glochidium.

At an older stage, the posterior part of the larva appears to be withdrawn into a invagination of the anterior part. The whole or by far the greater part of the body of *Dondersia* is said to be produced from this posterior part (the "*embryonic cone*") alone. On this embryonic cone, there appear, first of all, on the two sides of the middle line, three pairs of consecutive imbricated spiculae, still retained in their formative cells. They soon break through these cells, and their number is increased by the appearance of new once anteriorly.

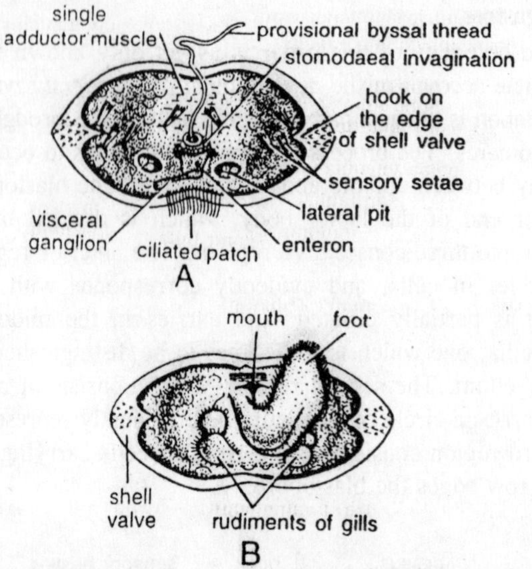

Fig. 16.2. Unio. Stages of metamorphosis of glochidium.

The embryonic cone lengthens, becomes curved ventrally. The anterior part of the body with the velum and the pretrochal area becomes reduced and finally appears as a sort of collar at the anterior end of the body. The larva sinks to the bottom, and throws off the whole anterior part of the body with the velum and the pretrochal area. Such throwing off or resorption of parts of the body which have been of great physiological importance on the larva is very common in the animal kingdom; see sections on the ontogeny of the Worms (e.g. *Nemertina*, *Phoronis*, etc.), of the *Arthopoda* (metamorphism of Insects) and of the *Echinodermata*. On the dorsal region of the young *Dondersia*, seven consecutive, imbricated, but only slightly overlapping, calcareous plates can bow be distinguished, consisting of rectangular rods lying close alongside of one another.

This fact is very significant with regard to the shell of the *Chiton*, which in the adult consists of eight, but in the larva of only seven plates. If it could be proved that the ***Solenogastridae*** pass through a *Chiton* stage, the view that the are more specialised animals than the *Polyplacophora*, and are to be derived from Chiton-like forms, would receive almost decisive support. Besides the seven dorsal calcareous plates, the young *Dondersia* has numerous circular calcareous spicules, covering it laterally; the ventral side is, however, naked. A mouth is still wanting, the endodermal mass is not yet hollow, and on each

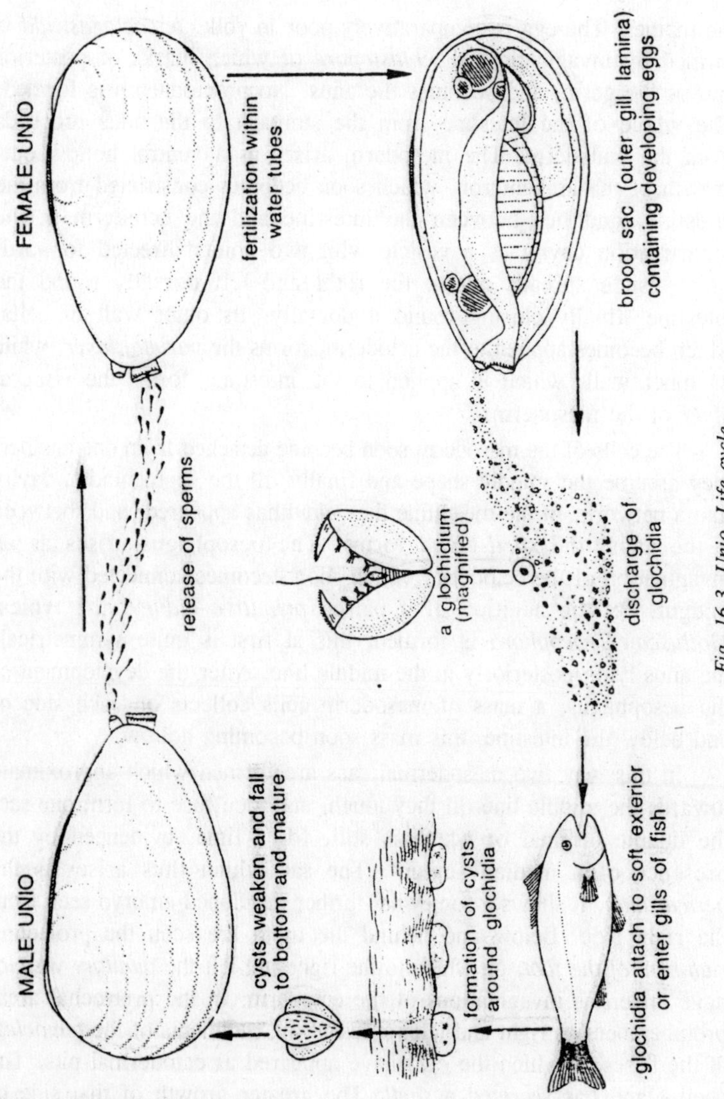

Fig. 16.3. Unio. Life cycle.

side, between the endoderm and the integument, there is a solid mesodermal streak.

Gastropoda

As a type of the development of the Gastropods, we may take *Paludina vivipara*, the ontogeny of which has recently been again very carefully investigated. Development here takes place with the body of the mother. The egg is comparatively poor in yolk. A *coelogastrula* is formed by invagination, the *blastopore* of which marks of posterior end of the germ, and becomes the anus. No proctodaeum is formed. The whole of the intestine from the stomach to the anus proceeds from the endoderm. The mesoderm arises as a ventral hollow outgrowth of the archenteron, which soon becomes constricted from the intestine, and lies between the intestine and the ectoderm in the segmentation cavity as a vesicle with two points directed forward. This vesicle spreads out to the right and left dorsally round the intestine, finally closing round it dorsally. Its outer wall of cells, which becomes applied to the ectoderm, forms the *parietal layer*, while its inner wall, which is applied to the intestine, forms the *visceral layer* of the mesoderm.

The cells of the mesoderm soon become detached from one another; they assume the spindle shape and finally fill the segmentation cavity like a network. In the meantime the *velum* has appeared, and, between it the anus, the *shell-gland* forms. The oesophagus arises as an invagination of the ectoderm, which soon becomes connected with the midgut. By the addition of a paired *primitive kidney*, the typical *Molluscan Trocophora* is formed; this at first is quite symmetrical, the anus lying posteriorly in the middle line. After the development of the oesophagus, a mass of mesoderm cells collects on each side of and below the intestine, this mass soon becoming hollow.

In this way two mesodermal sacs are formed which approximate towards the middle line till they touch, and then fuse to form one sac, the double original of which is still, for a time, evidenced by the presence of a median septum. The sac which thus arises is the *pericardium*. It shows a somewhat further developed embryo seen from the right side. Below and behind the moth are seen the projecting *rudiment of the foot*, on which to the right and left the *auditory vesicles* have arisen as invaginations of the ectoderm. In the pretrochal area, protuberances to right and left represent the rudiments of the *tentacles*, at the bases of which the *eyes* have appeared as ectodermal pits. The shell gland has secreted a *shell*. The greater growth of that side of the body which is covered by the shell has caused a bending by which the anus is shifted towards the ventral side.

Immediately behind the anus, the ectoderm bulges out to form the rudiment of the mantle fold, so that the anus comes to lie in a shallow depression, the rudiment of the *pallial* or *respiratory cavity*. It is important to note that at this outwardly symmetrical stage, the mantle cavity and the anus lie posteriorly. The fore-gut (oesophagus) has greatly lengthened. The *digestive gland* has grown out from the stomach ventrally in the form of a wide sac, but is still connected with the latter by a wide aperture. The *pericardium*, in which the septum is still visible, has already somewhat shifted from below the stomach to its right side. The rudiments of the definite nephridia next form in the following way. In each division of the pericardium (the left division being smaller than the right) the wall bulges out; the right outgrowth becomes the *secreting portion of the permanent kidney*; the left outgrowth must be regarded as a temporarily appearing *rudiment of the* (original) *left kidney*.

The mantle cavity, which lies beneath the pericardium, presses into it to the right and left in the form of two projections. The right projection, continuing to grow, becomes connected with the rudiment of the right kidney and forms its *efferent duct*. The left projection does not grow further, nor does it become connected with the rudiment of the left kidney. A further stage is depicted from the right side. The following are the most important alterations. The optic pit has become constricted into an optic vesicle. The *mantle fold* has grown further forward, and has become *deeper to the right*.

The undivided pericardium has shiftd altogether to the right of the stomach, and lies above the rectum, which bends forward and downward. The body is already asymmetrical. At the following stage the posterior and dorsa region of the body rise distinctly from the rest as a visceral dome; the shell covering this part of the body has increased considerably in size. The mantle fold has becomes much broader, and the mantle cavity much deeper; the latter now lies chiefly on the right side of the body. The looping of the intestine is far more marked.

On the posterior and dorsal side of the pericardium, the pericardial wall sinks in the form of a channel, which soon closes and forms a tube; this is the *rudiment of the heart*. The two apertures of the tube, where the wall of the heart passes into that of the pericardium, communicate with the body cavity. The heart tube becomes constricted in the middle, the anterior division forming the auricle and the beginning of the branchial vein, the posterior, the ventricle and the rudiment of the body aorta. It shows a somewhat older embryo which already resembles in form the adult animal. The velum is reduced, and a ventral bulging of the anterior division of the oesophagus represents the rudiment of the *radular* sac. The *ventricle* and the *auricle* are distinct. An ectodermal depression on the foot forms the *operculum*.

The mantle cavity which lies on the right side, and into which the rectum opens, now also stretches to the left on the anterior and dorsal side of the sharply demarcated visceral dome. The gill appears in the form of a protuberance on the inner surface of the mantle cavity, and the osphradium at the left of the gill as an ectodermal protuberance. Finally shows us an embryo in which the mantle cavity has assumed the anterior position on the visceral dome. The ctenidium and osphradium have developed further. The velum is very much reduced and can only be seen in sections. This stage is important on account of the appearance of the rudiment of the *genital organs*, which is identical in the two sexes.

A depression of the (mesodermal) pericardial wall, which becomes separated from the pericardium, forms the rudiment of the gonad, while an ingrowth from the base of the mantle cavity runs towards this, and is the (ectodermal) rudiment of the genital duct. The latter arises on one side of the anus, just as the efferent duct of the permanent kidney rises on one side; this ontogenetic fact confirms what was stated above that the genital duct of the *Monotocardia* corresponds with a part of the right (with originally, and in the young embryo, is the left) kidney of the *Diotocardia* (apparently wanting in the *Monotocardia*). The vascular system arises very early in the form of spaces between the mesoderm and ectoderm or entroderm, round which the mesoderm cells grow, and which become secondarily connected with the heart.

All the ganglia of the nervous system, the cerebral, pleural, pedal, parietal, and visceral ganglia arise separately as ectodermal thickenings, which become constricted off from the ectoderm by delamination. They only secondarily become connected with one another through the growing out of the nerve fibres. The parietal ganglia arise to right and left in the middle region of the body, but soon become shifted by the displacement of the organs of the visceral dome, one above the intestine and the other below it. The rudiment of the visceral ganglion is said to appear dorsally to the hind-gut and to move later to its position beneath the same. The observations on the development of *Paludina vivipara*, here briefly described, are in many ways of great importance, and confirm in the most unmistakable manner the results arrived at by comparative anatomy. The following are specially noteworthy:

1. The manner in which the pericardium originates favours the opinion that it is a secondary body cavity. It is important to note that the pericardium is at first paired, being divided into two lateral halves by a septum, which afterwards disappears.

2. The fact that the gonad arises as an outgrowth of the pericardium, confirms the view arrived at by comparative anatomy, that the genital cavity also is a secondary body cavity.
3. The anus and the mantle cavity originally lie symmetrically at the posterior end of the body, but, through asymmetrical growth, come to lie first on the right side of the visceral dome, and finally on its anterior side.

The development of other Gastropods cannot here be described in detail. We refer the reader to a bibligraphy at the end. As a rule, nutritive yolk is present in larger quantities than in the viviparous *Paludina*, in which the small provision of yolk is evidently connected with the favourable conditions of nutrition of the embryo. The blastopore generally corresponds in position with the future mouth; it often, perhaps usually, remains open; notwithstanding this, the oesophagus arises by the sinking in of ectoderm cells. *Paludina* is, as far as is known, the only Mollusc in which the mesoderm originates as an outgrowth of the archenteron. This fact is no doubt connected with its poverty in nutritive yolk.

In other Gastropods, the mesoderm arises in the manner already described for other Molluscs, as two large symmetrical primitive cells, at the posterior edge of the blastopore; these cells look more like endodermal than ectodermal cells, and soon pass into the segmentation cavity. A *Veliger* larva, *i.e.* a Trochophora with Molluscan characteristics, always forms (1) a dorsal shell gland with the embryonic shell, and (2) a ventral rudiment of the foot. The outward appearance of the Veliger larva, however, varies much in different groups, the variations being connected with the manner of life and of feeding of the embryo.

In the marine Gastropods, *i.e.* in the majority of the *Prosobranchia* (including the *Heteropoda*), the Pulmonate genus *Oncidium*, and all *Opisthobranchia*, the embryo leaves the egg envelope early, as a free-swimming Veliger larva. In all these forms, the preoral ciliated ring is well developed. The ectodermal floor of the ciliated ring usually bulges out anteriorly, so that the cilia appear to be carried by a distinct circular ridge. This ridge even grows out laterally to form a lobe of varying size, which carries at its edge long and strong cilia, and is occasionally itself produced into an upper and a lower lobe. This is the true velum of the free-swimming Gastropod larva, and is its only organ of locomotion. It is internally traversed from wall to wall by contractile mesoderm cells (muscle cells), which make it highly contractile. In the older larvae, the head with the velum can be withdrawn into the shell.

It is probable that the velum of the larva also serves for respiration, and perhaps for bringing about a circulation of the body fluid by means of its contractility. The embryos of fresh-water and terrestrial Gastropods, where these animals are not viviparous, remain longer in the egg, and leave it only after their transformation into young Gastropods, the larval organs (the velum, the primitive kidney, the cephalic vesicle, and the pedal vesicle or podocyst) having degenerated within the egg envelope.

Even in these forms, the mass of nutritive yolk contained in the egg is not very great, but there is a large quantity of albumen stored up within the egg capsule, which serves as food for the developing embryo; this is either absorbed through the body wall or swallowed. The egg capsules are always large, in some cases (in tropical terrestrial Gastropods) as large as the egg of a small bird; but their size is not, as in the *Cephalopoda*, determined by that of the egg contained, but by the quantity of albumen in which the small egg is embedded. The mature egg capsule contains a young Gastropod of considerable size with a well-developed shell.

In terrestrial and fresh-water, the velum is not needed as a locomotory organ, and is therefore reduced to a single ring of cilia or to two lateral ciliated streaks. It is entirely wanting in the embryos of a few terrestrial *Gastropod* snails. The respiratory and circulatory functions, which were originally merely accessory functions of the velum, here become of greater importance. The nuchal region becomes much bulged forward, and forms a *cephalic vesicle*, which is sometimes very large, and undergoes regular pulsations. The posterior division of the foot, in the same way, often widened into a pulsating *pedal vesicle* or *podocyst*.

Towards the end of larval life the cephalic and pedal vesicles and other similar "larval hearts" degenerate. The embryonic shell is either retained throughout life or is thrown off at an early stage, and replaced by the rudiment of the definitive shell. Even a *second temporary shell* occasionally attains development. It must again be noted that shell-less Gastropods, to whatever natural division they belong, pass through a typical Veliger stage, and at the older Veliger stage have a distinctly demarcated coiled visceral dome, with a corresponding shell, and usually an operculum on the metapodium. In the larva of the gymnosomatous *Pteropoda* three postoral accessory ciliated rings are developed on the body.

SCAPHOPODA

Ontogeny of Dentalium

Segmentation, in these animals, leads to the formation of a coeloblastula arise by invagination. The blastopore at first lies posteriorly on the ventral side, but gradually shifts, as in *Chilton*, more and more forward along the ventral side. The stomodaeum arise as an ingrowth of the ectoderm, the blastopore nevertheless remaining open. A typical Molluscan Trocophora is developed, although no primitive kidney has been found. The velum is a thick ridge round the body of the long oviform larva. This ridge consists of three rings of very large ectoderm cells, each row carrying a circle of long cilia. The shell gland spreads out at an early stage, its lateral edge soon growing out ventrally and posteriorly as the mantle fold. The free edges of the two folds fuse at a later stage below the body. The anus forms very late. The development of the cerebral and pedal ganglia and of the auditory organ has been specially carefully observed.

On the ventral side of the pretrochal area, in front of the velum and behind the tuft of cilia, two symmetrical invaginations of the ectoderm form the cephalic sacs or tubes. These become constricted from the ectoderm at a later stage, their lumen gradually narrows and finally disappears, while their walls become thick and multilaminar by the continuous growth of the cells. The two cells masses which thus arise become connected in the middle line above and below the oesophagus, and form the cerebral ganglion. The otocysts arise at the base of the pedal rudiment on each side as ectodermal epithelial pits, which soon become detached from the ectoderm in the form of epithelial vesicles. Immediately beneath these auditory vesicles, certain ectoderm cells sink below the surface, and form on each side an ectodermal cell mass, which becomes detached form the rest of the ectoderm, sinks into the mesoderm of the foot, and fuses with the similar mass on the other side to form the pedal ganglion.

LAMELLIBRANCHIA

Development of Teredo

Segmentation is here total and unequal. The gastrula, formed by epibole consists of (1) two large endoderm cells (macromeres), a thick cap of ectoderm cells (micromeres) closely covering these, and two symmetrical primitive mesoderm cells of medium size at the posterior edge of the blastopore. The blastopore closes from behind forward, the ectoderm cells by continual division growing entirely round the endoderm cells; during this process the two mesoderm cells become covered by the ectoderm and come to lie between the latter and the endoderm. Somewhat anteriorly on the ventral side, a depression of the ectoderm forms a pit, the *stomodaeum*. The ectoderm separates off from the two-called mesoderm, thus giving rise to a segmentation cavity, or primary body cavity. A double *preoral ciliated band* is formed. The two large endoderm cells, by fission, produce other smaller cells. Cilia appear over the whole surface of the germ.

With the exception of the posterior dorsal surface, where the ectoderm cells, which have become cylindrical, sink in to form the shell gland. The latter secretes the first rudiment of the shell in the form of a simple cuticular membrane. The endoderm cells begin to collect to form the intestinal wall. After the formation of the first rudiment of the shell, the shell gland flattens and spreads out; its edge can still be found as a ridge running under the edge of the shell. The endoderm now forms a large globular hollow mid-gut, into which the oesophagus breaks through. Each of the primitive mesoderm cells has given rise to two or three smaller cells. The thin cuticular shell becomes bivalvular by the appearance of a mediodorsal boundary line.

A further stage is distinguished first by the appearance of a small posterior ectodermal invagination, the *proctodaeum*, which produces the rectum and anus. An ectodermal thickening, the neural plate, appears in the pretrochal area, carrying three flagella. Some of the mesoderm cells become muscle cells. The next stage may be called that of the Trochophora larva. This larva differs from a typical Annelidan Trochophora only by possessing a shell, which now covers the greater part of the body, and by a mantle which appears, at first posteriorly, and then at the sides, as a fold, and continues to grow from behind forward. The region of the body which lies behind the cephalic area has spread out on each side to form a broad fold, which becomes outwardly applied to the shell. The neural plate has become multilaminar, and the proctodaeum has broken through into the mid-gut. The primitive mesoderm cells have given rise to a short mesoderm streak on each side. At the anterior end of each mesoderm streak a somewhat long body, the *primitive kidney*, has formed; this contains a channel which opens externally, and whose lumen is ciliated at a later stage. The rudiment of the digestive gland appears in the mid-gut as a paired semi-spherical outgrowth.

The body is no longer ciliated all over; cilia are retained only on the neural plate and in the anal region. The double preoral ciliated band now becomes very distinct, and a *postoral band* is now added. The region between the two ciliated bands also carries cilia and forms an *adoral ciliated zone*. A further stage of development is depicted. The rudiment of the pedal ganglion can be recognised as an ectodermal thickening on the ventral side, and that of the gill as a thick epithelial ridge. The stomach has formed a coecum posteriorly, and the narrow mid-gut has formed a loop. The two auditory vesicles containing otoliths have arisen between the mouth and anus as ingrowths of the ectoderm which have become detached. The mesoderm consists of branched muscle cells, branched cells of connective tissue, the primitive kidney and the still undifferentiated cells of the mesoderm streaks.

The ectodermal thickening, which represents the rudiment of the pedal ganglion at a later stage, becomes rounded off and detached from the ectoderm, at the same time becoming surrounded by the cells of the mesoderm streak, which have rapidly multiplied, and which unite in front of it to form a median mass of cells. This median mass of mesoderm cells increases greatly by rapid division, bulging forward the ectoderm in the anterior ventral region, and thus forming the rudiment of the foot. In the growing branchial fold slits occur, a single slit appearing first, and another soon following in front of the first. The further development of this larva is unknown.

The development of other marine bivalves runs very much the same course as that of *Teredo*, the same larva being formed. The ciliated band is very strongly developed in all marine bivalves (*Teredo*, *Ostrea*, *Modiolaria*, *Cardium*, *Montacuta*, etc.), and is generally carried by a collar-like expansion of the integument, or velum, which is often divided into two lateral lobes. The velum, which on account of the band of strong cilia it carries is the locomotory organ of the free-swimming larvae of these Lamellibranchs, can be protruded out of and withdrawn into the shell.

Among fresh-water Lamellibranchs there is one form, *Dreissensia polymorpha*, whose larva is free-swimming and carries a well-developed velum. This form is said to have migrated from salt to fresh water in (geologically speaking) recent times. Special arrangements are found among the other fresh-water forms. The eggs of *Pisidium* and *Cyclas*, for instance, develop in special brood capsules in the gills of the mother animal, and leave these as young bivalves. The Trochophora stage is, nevertheless, passed through, but the velum, not being used for locomotion, remains rudimentary.

Ontogeny of Cyclas Cornea

We shall here only mention the points in which the development of *Cyclas* differs from that *Teredo*, and describe such observations as complete those made on the latter. The blastula consists of a cap of small cells (ectoderm cells) and a floor made of three large cells, one very large primitive endoderm cell and two symmetrical primitive mesoderm cells. The primitive endoderm cell yields through fission a disc of endodermal cells. The two primitive mesoderm cells become overgrown by the ectoderm cells, and thus reach the segmentation cavity. The endoderm invaginates in such a way that a slit-like blastopore arises, which reaches from the region of the future mouth to that of the future anus. This blastopore closes completely. The oesophagus arises as an ingrowth of the ectoderm.

A Molluscan Trochophora is formed with a shell gland, a rudimentary foot, a mid-gut, a stomach, anus, primitive kidney, and a neural plate. The velum is reduced to a ciliated area lying at the sides of the mouth; this reduction is connected with the fact that the Trochophora of *Cyclas* is not a free-swimming larva, for the *eggs* of *Cyclas* pass through the whole course of their development within the gills of the mother animal. Above the neural plate the ectoderm cells are large and flat, and form a projecting *cephalic vesicle*. The *mesoderm* consists of (1) *scattered cells*, which lie under the ectoderm of the cephalic cavity, in the foot and on the intestine (especially on the oesophagus, where they are already changed into muscle cells); and (2) two *mesoderm streaks* lying at the sides of the intestine. The *pedal ganglia* arise together with the paired rudiment of the *byssus gland*, as thickenings of the ectoderm at the posterior end of the foot. The *auditory vesicles* originate as ingrowths of the ectoderm. The *mantle* forms by degrees from behind forward as a ridge, which grows more and more ventrally downwards. At the same time the shell gland, which at its edge secretes the delicate shell membrane, spreads out and becomes flattened. Beneath the shall membrane the rudiments of the permanent shell valves are produced from two small round areas lying to the right and left of the dorsal middle line.

The digestive gland (liver) develops from two lateral globular outgrowths of the wall of the stomach. The gonads arise from cells of the mesoderm streaks, which are larger than the rest and also in other way differentiated, so that they can very early be distinguished. In the anterior and dorsal part of each mesoderm streak a group of cells surrounds a cavity, which at first is very small, but becomes continually larger. The two vesicles thus formed, the cavities of which represent the *secondary body cavity*, from the pericardium. Behind these the mesoderm cells collect in such a ways as to form on each side a strand, which becomes hollow; this is the rudiment of the *nephridium*, which at once becomes connected with the pericardial vesicle, and, growing further towards the ectoderm, soon opens outward.

The two pericardial vesicles lengthen posteriorly and upward, each becoming divided into two parts, one lying behind the other, by a constriction, the parts still communicating dorsally with one another. The two double vesicles grow towards one another above the rectum, and finally fuse in the dorsal middle line. In a similar manner they fuse below the rectum. The inner wall of the pericardial vesicle becomes the wall of the ventricle, and its lateral wall becomes that of the auricle. At the points where the lateral vesicles were constricted lie the slits through which the auricles communicate with the ventricle, and the atrioventricular valves. The visceral ganglion arises at the posterior end of the mantle furrow from an ectodermal thickening. The pleurovisceral connectives form, in all probability, throughout their whole length, through constriction from the ectoderm. The gill arises on each side as a fold on the dorsal edge of the inner surface of the mantle. It develops from behind forward. In the contrary direction furrows form on the branchial fold, commencing from below upwards; these are found on the inner as well as the outer surface, and exactly correspond. The inner furrows join the outer right through the gill, and thus give rise to the branchial slits.

Development of the Unioniae

The development of the Unioniae (*Anodonta*, *Unio*) is much influenced by the *parasitic manner of life* of the larva. The fertilised eggs reach the outer leaf of the gill of the female, and there run through the first stages of their development.

Segmentation leads to the formation of a coeloblastula, in which the rudiment of the *shell gland* appears very early as an incurved plate of large and high cells of the blastula wall. The archenteron forms by invagination at a very late stage; this is no doubt connected with the later parasitism of the larva. Before this invagination occurs the mesoderm has begun to form; its two primitive cells lie in the blastocoel at the part where, later, the enteric invagination appears. The embryo known as *Glochidium parasiticum* has, in the last stage of its development, which is passed through in the gills of the mother animal but within the egg-shell, the following structure. It is bilaterally symmetrical, and has a *bivalve shell*. Each valve has, at its ventral edge, a triangular process, the exterior of which is base with short spines and thorns. Between the two valves, which are markedly concave, lies the soft body which lines the shell internally in such a way that its ventral epithelial layer might, incorrectly, be called a mantle. It may be called the *false mantle*. If this false mantle is examined from below, when the shell is open, it is seen to have on each side four sensory cells furnished with long sensory hairs; three of these cells lie near the shell process, and the fourth near the middle line.

Between the two more median sensory cells a long *adhesive filament* projects from the opening of the gland which secretes it. Behind this gland are found—(1) the *oral sinus*; (2) a small prominence, the *pedal swelling*; (3) the *ciliated lateral pits*, one on each side; and (4) furthest back of all, the *ciliated shield* or *patch*. Between the mantle and shell the *embryonic adductor* runs across from the one valve to the other. Besides these are only found a few isolated muscle fibres, and the *rudiment of the mid-gut*, the latter as an epithelial vesicle, which becomes entirely separated from the ectoderm, and in no way communicates with the exterior.

The embryo at this stage leaves the gills, at the same time emerging from the egg shell. Its adhesive filament floats in the water. If a passing fish comes in contact with such an embryo, the latter can, by closing its shell, attach itself by means of the triangular processes mentioned above, to its integument, into which the spines on these processes penetrate. The embryo of *Anodonta* attaches itself chiefly to the fins, that of *Unio* to the gills of the fish. The epithelium of the part of the flash attacked grows very rapidly in such a way as in a few hours to surround the parasite completely. The embryonic false mantle grows out from each valve of the shell as a fungus-like body to penetrate the tissues of the host, and probably serves for nourishing the embryo. During this endoparasitic life, which lasts for several weeks, the transformation of the embryo into the young Mussel is completed.

In the course of the is process of transformation some larval organs are resorbed, and also serve for nutrition; first the sensory cells disappear in this way, then the gland of the adhesive filament with the remains of the filament itself, then the adductor, and finally the false mantle. The rudiment of the definitive mantle and shell then appear. The vesicular mid-gut joins the oral sinus; the pedal swelling grows into the linguiform foot, and, in this, the rudimentary byssus gland appears as an ingrowth of the epithelium. The rudiments of the inner branchial leaves, the digestive gland, the nephridium, the heart, the cerebral, pedal, and visceral ganglia, and the auditory vesicle appear during the parasitic stage, in the same way as in other Lamellibranchs. During to last week of parasitic life the capsule formed by a growth of the tissue of the host which surrounds the embryo becomes thinner; the parasite breaks through it, and falls to the bottom of the water as a young Mussel. The only organs still wanting are the genital organs, the outer leaves of the gills, and the oral lobes.

Cephalopoda

Tetrabranchia

Nothing is known of the development of *Nautilus*.

Dibranchia

The egg is usually very large, and contains, like that of the sharks, reptiles, and birds, a great quantity of nutritive yolk. It belongs to the telolecithal meroblastic type, and is enclosed in a capsule. A number of such capsules may become cemented together to form strings. The partial segmentation takes place at the animal pole of the egg, and leads to the formation at that point of a germinal disc (blastoderm).

Ontogeny of Sepia

The blastoderm grows so very slowly round the yolk, that long after all the outer organs of the embryo are quite recognisable in the region of the original germinal disc, the opposite pole is still occupied by the yolk. The germ lies in such a way that the centre of the germinal disc or animal pole is placed dorsally, and corresponds with the uppermost point of the visceral dome of the adult animal, while the mass of nutritive yolk lies ventrally.

1st Stage

In the centre of the germinal disc there appears an oval-rhombic bulging; this is the rudiment of the *visceral dome* and the *mantle*. On each sides of his there arises a bean-shaped prominence, the rudiment of the *eye*. Behind the eye, on each side, a long narrow ridge runs backward in a curve; about half way down this ridge a small prominence, the rudiment of the *funnel cartilage*, forms close to its outer side. The part of the ridge lying in front of this prominence becomes the *muscle* which runs *from the funnel to the nuchal cartilage*; the posterior part (which lies behind the rudiment of the visceral dome and mantle) forms the paired rudiment of the funnel itself. Between the two rudiments of the funnel two other prominences rise symmetrically—the rudiments of the *gills*. A pit in the centre of the rudiment of the visceral dome has been indicated as the rudiment of a shell gland (?).

2nd Stage

The rudiments just described stand out more prominently. On the outer and posterior sides of the rudiments of the funnel the rudiments of the two posterior *pairs of arms* first appear as prominences, then those of the third and fourth pairs. The first indications of the *head* are seen in the form of a large double swelling on each side, the outer and anterior part of which carries on each side the rudiment of the eye. The embryo becomes covered with cilia. At the extreme anterior end the *mouth* appears in the middle line, forming the opening of the oesophagus, which begins to sink inwards.

3rd Stage

The whole embryo has become more arched dorsally, and more marked off from the yolk. On the latter, the blastoderm, which consists of two layers, an external ectoderm and an internal yolk membrane, has spread out further towards the ventral (vegetative) pole of the egg. At the posterior edge of the rudiment of the visceral dome, the *mantle fold* has grown out forward in such a way as to form a small *mantle cavity*, which already partly covers the rudiments of the gills. In the space between the rudiments of the funnel and the gills the *proctodaeum* has formed by invagination, and its aperture, the *anus*, can be seen. The rudiment of the fifth pair of arms appears.

4th Stage

The visceral dome projects further, and has a free mantle edge all round its base. The gills have shifted further into the mantle cavity, which is now larger, and lies posteriorly. The rudiments of the funnel also now lie close to the mantle, and are so approximated posteriorly as nearly to touch. *The rudiments of the arms has shifted from behind further forward round the rudiments of the head.* As the whole embryo projects more distinctly from the yolk, the rudiments of the arms shift nearer to one another and under the rudiments of the head. The anus is already covered by the mantle fold.

5th Stage

The arms shift still nearer to one another (*i.e.* towards the axis of the embryo), grouping below the rudiments of the head (which have become fused), and form a somewhat narrow circle on the ventral side in such a way that, when the embryo is seen from the dorsal side, some of them are hidden by the head. As a consequence of this the embryo, which is already recognisable as a young *Sepia*, now becomes sharply constricted from the yolk beneath it. The free edges of the rudiments of the funnel fuse and move to a position within the mantle cavity.

6th Stage

The rudiments of the head and arms have now assumed the typical position to form the "head" (Kopffuss). The embryo is now altogether distinct from the yolk, to which it merely hangs instead of, as before, lying upon it. The blastoderm finally grows round the yolk and so forms a yolk sac. At first this sac is four or five times the size of the embryo, but in proportion as the latter grows at the expense of the yolk and develops further, the sac becomes smaller, so that when the embryo is hatched the size of the yolk-sac is only one-third of that of the young animal.

It must further be mentioned with regard to the yolk sac that it is at no time in communication with the intestine. As the embryo becomes constricted from the yolk the latter divides into two parts—an inner part, lying inside the embryo, and an outer part, filling the sac. These two parts are connected by means of the stalk of the yolk sac, which projects downward from the 'head." The yolk within the embryo is divided into three unequal parts, the largest of which fills the visceral dome; another mass of considerable size fills the "head," and these two masses are connected with a smaller portion lying in the nuchal region.

Loligo and Argonauta have a smaller yolk sac, round which the blastopore grows at an earlier stage than in *Sepia*. The yolk sac of *Argonauta* is entirely taken into the body before the latter has completely closed ventrally. The quantity of nutritive yolk is still less in a *Cephalopod* (*Ommatostrephes* ?), the spawn of which floats in the sea. Segmentation is in this case also partial and discoidal, but the blastoderm almost completely encloses the yolk before any organ develops on the germ, and no external yolk sac is formed. The results of the investigations hitherto made with regard to the germinal layers, the development of the inner organs, and the inner differentiations of the outwardly visible organs are so contradictory and in many cases so incomplete, that no description of them is here attempted.

Further investigation is much needed. The development of the eye has already been described, and that of the hind-gut and ink-bag was illustrated. Two important facts in the ontogeny of the *Dibranchia* should be noted. (1) In considering the arms as part of the foot, it is important to notice that they arise behind the rudiments of the head, and only secondarily come to lie round and below the latter. The mouth, at quite a late stage, lies at the anterior end of the circle of arms. (2) The funnel consists of two separate lateral rudiments, the free edges of which fuse secondarily.

This point is important in connection with the separation of the two lobes of the funnel, which lasts throughout life in *Nautilus*. The fact that the velum is wanting in the Cephalopod embryo must also be noted. The absence of this organ is explained by the direct development of the *Cephalopoda* within the egg capsule at the expense of a large quantity of nutritive yolk. Investigations as to the development of the shell, and as to the nature of the organ which has been called the shell gland, are much needed.

PHYLOGENY

No actual points of connection between the Molluscan phylum and any other division of the animal kingdom have as yet been found; the origin of the *Mollusca* is therefore merely a matter of speculation. The present writer favours the view that the *Mollusca* descended from animals like the *Turbellaria*, which had become differentiated from the modern *Platodes* by the acquisition of a hind-gut and a heart, and the (at least partial) transformation of the genital cavity into a secondary and primitively paired body cavity. There is a striking agreement in the nervous system of the lower *Molluscs* (*Chiton*, *Solenogastres*, and in some respects the *Diotocardia*) and that of the *Platodes*; in both there is a ladder-like nervous system with the principal trunks beset along their whole length with ganglion cells; the pleurovisceral cords answer tot he lateral trunks of the *Platodes*, and the pedal cords to the ventral longitudinal trunks of the latter. If such a hypothetical racial form were to secrete a dorsal shell, perhaps at first in the form of a thick cuticle containing calcareous particles, a typical Molluscan organisation would be produced.

The development of a shell would deprive the greater part of the surface of the body of its original respiratory function, and would lead to the formation of localised gills. By means of the development of a mantle fold these delicate-skinned organs could be brought under the protection of the shell. The musculature on the dorsal side, which the shell covered, would disappear, and with it the dorsal longitudinal nerve trunks. The musculature on the ventral side, which was already strongly developed in the *Planaria*, would become strengthened in the development of the foot with its sole for creeping. A part of the dorsoventral musculalture would be changed into the shell muscle. In this derivation of the *Mollusca* their characteristic larval form might be explained, without any need for traçing it to the Annelidan Trocophora, in the following way. It would correspond to a Tubellarian larva (Müller's Polyclade larva, etc.), on to which certain Molluscan characteristics such as the shell gland, the shell, the anus, and the foot had been shifted back. The preoral ciliated band (the velum) of the Molluscan larva would correspond with the same structure in the Tubellarian larva. The primitive kidney of the former would answer to a simplified water vascular system, while the permanent nephridia as ovarial and seminal ducts might be homologised morphologically with the ducts of the genital products in the *Turbellaria*.

17

LARVAL FORMS IN MOLLUSCA

In molluscs the development may be direct or indirect. In the direct there being no larval stage where the young ones hatching from the eggs resemble their parents in general appearance, except the size. The direct development is seen in cephalopods. In indirect development a larval stage is seen during the development in majority of mollusc classes. *Paludina* is a viviparous form. In some cases like *Lottorina*, *Cenia* etc. the metamorphosis takes place inside the egg case. However, three main larval forms are met with in Mollusca. There are:

1. Trochophore
2. Veliger
3. Glochindium

TROCHOPHORE

The ciliated trochophore in its typical state appears in *Patella* and *Trochus* and in these it is formed at a very early period, even before the formation of the mesoderm, and becomes free at once. It is pear shaped and measures about 0.5 mm in length. In some cases that are free where as they are enclosed in egg. A circlet of preoral cilia, the *prototroch* or *velum* divides the body into two unequal parts, the upper on consist of prostomium where as the lower part bearing mouth and anus. The preoual part is large and convex with an *apical plate* bearing the long cilia called the *apical cilia* as seen in the larvae of *Patella*, *Dentalium* etc.

Near the apical cilia there are two ciliated elevations each consisting of a single cell. At the posterior end there are large anal cells bearing a bunch of cilia called *telotroch*. Between the ciliary girdle and anus, on the dorsal side, the area of ectoderm constitute

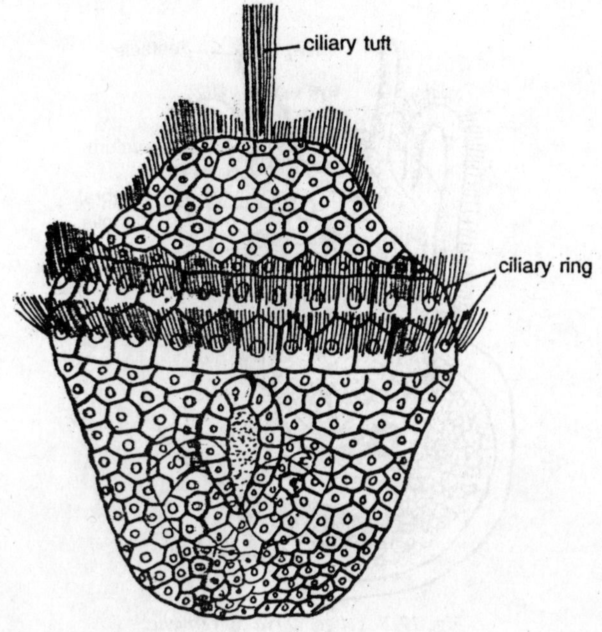

Fig. 17.1. Trochophore of Patella.

the *shell glands* which secretes a thin, saucer-shaped larval shell. The trochophores of Lamellidae consists of an additional shell inside the first one which is cast off later. A slight ridge in the neighbourhood of shell represents the border of mantle. Alimentary canal is made up of single layered epithelial cells in the form of a tube. It comprises mouth, stomodaeum, oesophegus, stomach and intestine (mesenteron). A single large cell gives rise to the mesoderm. A rudiment of redula is present in oesophegus. The trochophore contains two cerebral ganglia, a pair of eyes from the underside of apical plate. The statolith sacs appear as depression of ectoderm at the sides of the mouth. Soon the trochophore develops into a yet more efficient locomotor form, the *veliger*.

Veliger Larva

The preoral ciliate area or *velum* begin to protrude on both sides as a bilobed flap. The velum are very delicate and extensive, and a very delicate organs of locomotion. Between the bases of velar lobes the anterior end of the larva is provided with eyes and tentacles. The larva has a shell into which the velar apparatus can be withdrawn and which encloses the partly developed viscera. Alimentary canal is complete and the anus is shifted to anterior side. A foot usually bearing

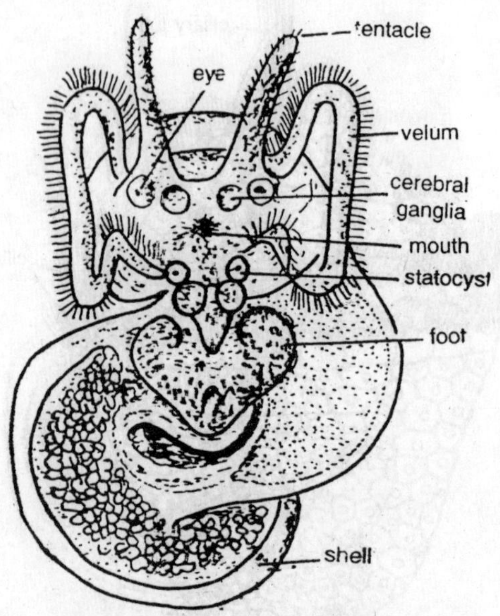

Fig. 17.2. Veliger larva of Vermetus.

an operculum in present. In addition to velar apparatus, there are some organs like larval heart and kidney also present. The larval kidneys are paired and generally, symmetrical organs situated at the anterior end of the body immediately behind the velum. Statocyst and gill-rudiments are also present. Some gastropods have feeding veligers (planktotrophic) with a larval life that may last as long as three months. Others have a brief yolk-laden nonfeeding veliger (lacithotrophic). The long cilia of the velum function not only in locomotion but also suspension feeding. During the course of the veliger stage, torsion occur and the shell and visceral mas & twist 180° in relation to the head and foot.

In most of the pulmonates, the young are hatched out with the characters of the adult. In the species in which veliger stage is passed in thc cgg, the velar lobes are reduced and the larva rotates in the egg with the help of velar cilia. With the exception of *Dreissena* and *Nausiforia*, which have a free-swimming veliger larva, fresh water bivalves exhibit modifies development.

Glochidium Larva

In the worldwide Unionacea, the African Mutilidae, and the South American Mycetopodidae the veliger, called as *glochidium*, *haustorium*

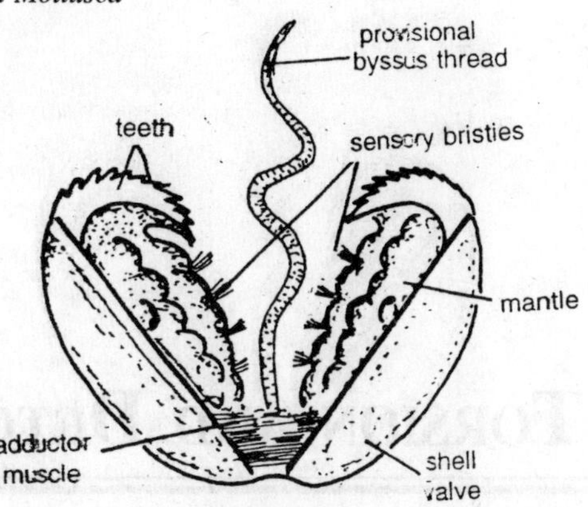

Fig. 17.3. Glochidium larva of Unio.

and *lasidium* respectively, has become highly modified for a parasitic existence on fish. The glochidium larva is enclosed by two valves, each edge of which bears a hook. The shell valves cover a larval mantle, which bears four groups of sensory bristles. A rudimentary foot is present, to which is attached a long adhesive thread, the *byssal thread*. There is neither mouth nor anus. When mature in measure from 0.1 mm to 0.5 mm depending upon the species. In *Unio* and *Anodonta*, the glochidium leave the gills through the suprabranchial cavity and exhalant aperture.

In *Lampsilis*, the glochidia emerge directly from gills through temporary openings. The glochidia of *Anodonta* immediately clamp into the fins and other parts of the body surface of the fish. The glochidia require certain species of fish as host; others can tolerate a wide range of host species and the byssal thread aids in initial contact and adhesion (*Wood,* 1974). The larval mantle contains phagocytic cells that feed on the tissue of host and obtain nutrition for the developing clam. The parasitic period lasts from 10–30 days. In the mean time the parasite is surrounded by the over growth of skin of fish forming a cyst. Some of the larger freshwater mussels may produce as many as 3,000,000 glochidia.

18

TORSION AND DETORSION

Chiastoneury, *i.e.* the crossing of the two pleuro-visceral connectives in the *Prosobranchia*, may be explained on the three following assumptions : (1) The ancestors of the *Prosobranchia* were symmetrical animals; the mantle cavity lay behind the visceral dome and in it the pallial complex, that is, the ctenidia, osphradia, nephridial apertures, genital apertures, and, in the centre, the median anus. (2) The visceral commissure or ganglion lay beneath the intestine. (3) The pallial complex shifted gradually from behind forward, along the right side of the body. The position of the pallial complex in the *Tectibranchia*, among the *Opisthobranchia* on the right side, can also be thus explained. The pallial complex in its forward movement in these animals has either not yet reached the anterior position or, having reached it, has shifted back again. The visceral connectives are therefore not crossed.

The above assumptions do not, however, explain—

1. The relation existing between the manner in which the visceral dome and shell are coiled, on the one hand and the special asymmetry of the asymmetrical organs (ctenidia, osphradia, nephridia, anus, genital organs) on the other.
2. The asymmetry which is brought about in some Gastropoda by the disappearance of one ctenidium, one osphradium, and one renal aperture.
3. The coiling of the visceral dome and shell, especially the dextral or sinistral spiral twist.
4. The cause of the shifting forward of the pallial complex.

It is unnecessary to discuss the first of the above assumptions. viz. that the ancestors of the Gastropoda were symmetrical animals, since all Molluscs except the Gastropoda are symmetrical, *i.e.* the *Amphineura*, the *Lamellibranchia*, the *Scaphopoda*, and the *Cephalopoda*. The assumption that the pallial complex originally lay posteriorly is also well founded.

In all symmetrical Molluscs, the anus lies as the centre of the complex posteriorly in the middle line, and further, in all symmetrical Molluscs, the nephridial and genital apertures lie posteriorly at the sides of the anus. When the ctenidia and osphradia have been retained in symmetrical Molluscs, they lie symmetrically on the posterior side of the visceral dome. This is the case in the *Cephalopoda*, and in the most primitive *Lamellibranchia*, the *Protobranchia* (*Nucula*, *Leda*, *Solenomya*), and even in some *Chitonidoe*, and those *Solenogastres* which still have rudiments of gills. In keeping with the posterior position of the pallial complex, the mantle fold which hangs down round the base of the visceral dome is, in symmetrical Molluscs, widest posteriorly where it has to cover the complex; at this part the mantle furrow deepens into a mantle cavity.

In connection with the second assumption, it still remains unexplained why in the *Amphineura* the commissure between the pleuro-visceral cords runs over the intestine; whereas on the other hand, *in all other symmetrical Molluscs*, the visceral ganglion lies, as in the Gastropoda, below the intestine. The third assumption, that the pallial complex has shifted forward, requires separate discussion. If the pallial complex did thus shift forward, chiastoneury must necessarily have taken place; the original left half of the complex must necessarily have become the present right half, and *vice versa*. Further, the right pleuro-visceral connective would have to become the supra-intestinal connective and the left the infra-intestinal connective; the original right parietal ganglion the supra-intestinal ganglion, and the original left parietal the infra-intestinal ganglion. But *why* did such a shifting take place? We shall here attempt to answer this question.

Cause of the Shifting Forward of the Pallial Complex

We have assumed the symmetrical racial form of the Gastropoda (with posterior mantle cavity and symmetrical pallial complex) to be a dorso-ventrally flattened animal with a broad creeping sole, a snout-like head with tentacles and eyes, and a somewhat flat cup-shaped shell covering the dorsal side of the body. It therefore resembled in outward appearance a *Fissurella*, a *Patella*, or a *Chiton*, if we assume the imbricated shell of the last to be replaced by a single shell. The body of such a racial form was only protected dorsally by the shell.

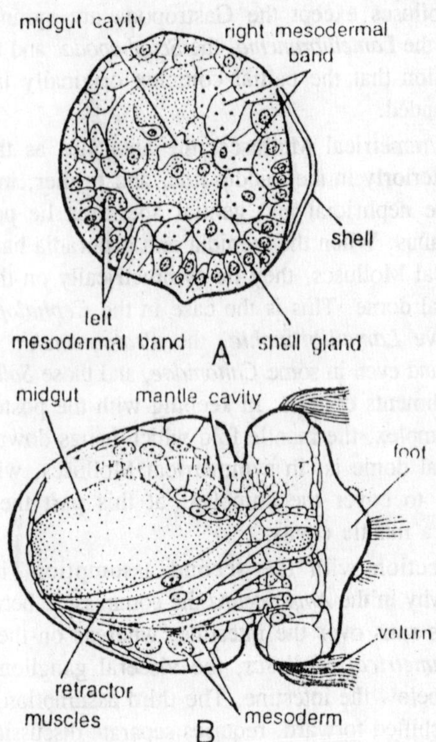

Fig. 18.1. Mechanism of torsion. A—T.S. early veliger of Haliotis showing disproportionate growth of right mesodermal cells. B—48 hour larva of Patella vulgata showing a symmetrical retractor muscle.

The hard surface along which the animal slowly crept served to protect its lower side, the dorsal shell being pressed firmly against the substratum, when necessary, by the contraction of a powerful shell muscle.

When the shell was thus pressed down, communication between the pallial cavity and the exterior (for the purpose of inhaling and exhaling the respiratory water, and ejecting the exereta, excrement, and genital products) was rendered possible by means of a cleft in the posterior edges of the mantle and shell. Unlike their racial form, all known Gastropoda (except those whose body form has been secondarily modified, generally in connection with the rudimentation of the shell) are distinguished by the fact that the viscera with their dorsal integumental covering protrude hernia-like in the form of a high spire-like visceral dome, with which the shell corresponds in shape. The

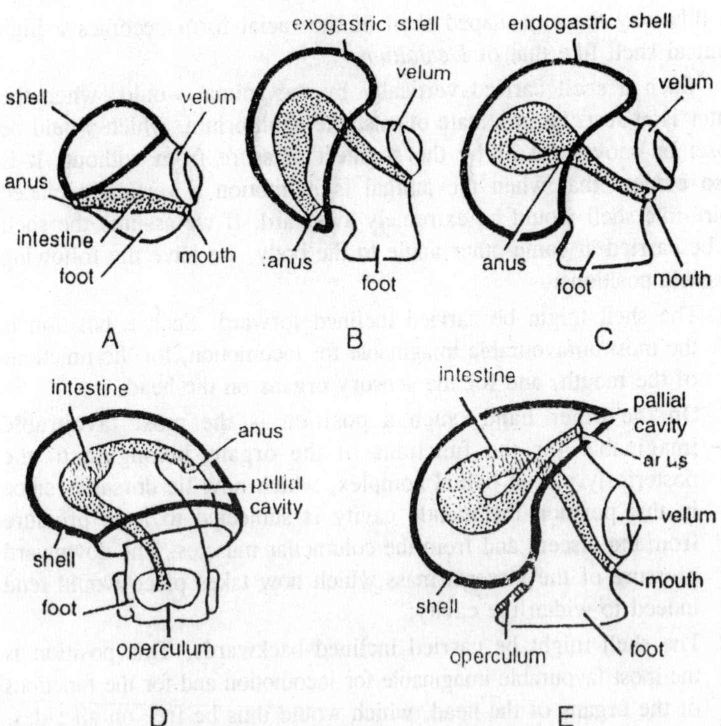

Fig. 18.2. Five successive stages in the development of a gastropod to show occurrence of torsion. A—Early veliger larva or pretorsional stage in lateral view. B—Larva with ventral flexure and an exogastric shell in lateral view. C—Stage showing 90° of lateral anticlockwise torsion. Shell becomes endogastric. Mantle cavity and anus move on to right side. D—90° torsion stage in posterior view. E—Adult stage with complete or 180° torsion in lateral view.

uncoiled shell of every snail is as a matter of fact spire-shaped. The development of such a shell and some has already been recognised as due to the increased protection needed by the body when the capacity for creeping becomes developed. The whole of the softer part of the body can be withdrawn into such a shell, and, further to increase the protection, an operculum is often developed on the foot for closing the aperture of the shell, when the animal has retired into it. The shell muscle of the racial form no longer serves for pressing the shell against the surface on which it rests, but for withdrawing the head and foot into the shell. It becomes the columellar muscle. Taking in turn the different stages in the development of the Gastropod shell, we have as the first and most important its dorsal spire-like prolongation.

In this way the cup-shaped shell of the racial form becomes a high conical shell like that of *Dentalium*.

Such a shell carried vertically by the animal would, when the latter is at rest, be in a state of unstable equilibrium, which would be upset by movement or by the slightest pressure from without. It is also evident that when the animal is in motion a vertically placed spire-like shell would be extremely awkward. If we assume the shell to be carried at some other angle to the body, we have the following possible positions:—

1. The shell might be carried inclined forward. Such a position is the most unfavourable imaginable for locomotion, for the functions of the mouth, and for the sensory organs on the head.

   On the other hand, such a position is the most favourable imaginable for the functions of the organs belonging to the posteriorly placed pallial complex, which now lie dorsally, since in this position the mantle cavity is subjected to least pressure from the viscera and from the columellar muscles. The downward pressure of the visceral mass which now takes place would tend indeed to widen the cavity.
2. The shell might be carried inclined backwards. This position is the most favourable imaginable for locomotion and for the functions of the organs of the head, which would thus be free on all sides. It is, however, the most unfavourable imaginable for the functions of the organs of the pallial complex, which now lie *beneath* the visceral dome. The mantle cavity has to bear the whole pressure of the visceral mass, and especially that of the columellar muscle; it would be squeezed together, so that the circulation of the respiratory water would be prevented or at least rendered more difficult, as would also the ejection of the excreta, excrement, and sexual products.
3. Finally, the shell may be carried inclined to the right or left. This is neither the most favourable nor the most unfavourable position for locomotion, for the head and for the pallial complex. It is an imaginable intermediate position.

In this position there is no dead point, as shifting of the parts would always be possible, and the shell be enabled to take up the position most suitable for locomotion and for the functions of the cephalic organs, and the mantle cavity the best suited for the exercise of the functions of the pallial complex lying within it. Assuming that the shell is inclined to the left, the pressure brought to bear on the mantle cavity would vary in amount in different areas of that cavity. It would be greatest on the left side, and would continually decrease towards the right. On the left there would be a pressure from the front which would, so to speak, squeeze out the pallial complex backwards over to the right.

It must further be noted that the point subjected to least lateral pressure and to the greatest downward pull lies on the right, which has become the upper side of the visceral dome. At this point the mantle furrow will most easily deepen, and become more spacious. Into such a deepening the organs of the pallial complex which are being pressed from the left have room to move forward to the right. Here we have the first step in the shifting forward of the pallial complex along the right mantle furrow. Further, as soon as the least

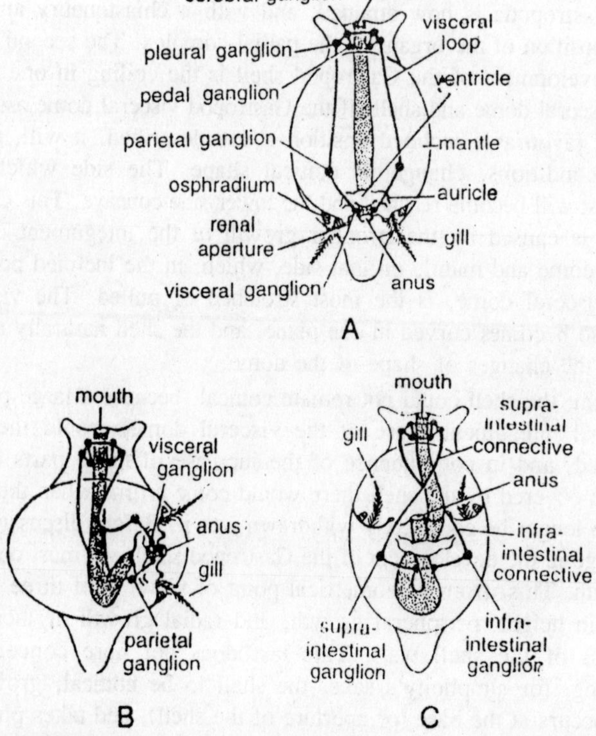

Fig. 18.3. Effects of torsion upon position of gills, digestive tract and nervous system. A—Hypothetical primitive stage before torsion. B—Intermediate stage showing 90° torsion with mantle cavity and pallial complex displaced to the right side of the body. C—Final stage showing 180° torsion.

shifting of this sort has taken place, the shell and visceral dome can move slightly from their present position on the left, towards that backward position which we have seen to be the most favourable imaginable for locomotion and for the functions of the cephalic organs. If we suppose this process gradually to be completed, the shell and visceral dome finally gain the most favourable backward position, and the pallial complex is gradually shifted forwards along the right mantle furrow.

The pallial complex this lies anteriorly on the upper side of the visceral dome, which now points backwards. This anterior position is that of the least upward pressure, or rather of the greatest downward pull, *i.e.* it is the point at which the mantle cavity can most easily deepen and widen, and where the pallial organs can best fulfil their functions. The position of the shell and the pallial complex characteristic of the Gastropoda is now attained, and with it chiastoneury and the inverse position of the organs of the pallial complex. The second stage in the development of the Gastropod shell is the coiling in one plane of the visceral dome and shell. If the Gastropod visceral dome assumes the most favourable inclined position above described, it will, under normal conditions, change its conical shape. The side which lies uppermost will become reached and the lower side concave. This change of form is caused by the stronger growth of the integument of the visceral dome and mantle on that side, which, in the inclined position of the visceral dome, is the most stretched or pulled. The visceral dome also becomes curved in one plane, and the shell naturally adapts itself to the changes of shape of the dome.

Again, the shell could not remain conical, because a large part of the dorsal integument (base of the visceral dome) would them be uncovered, and in consequence of the increase of those parts of the body not covered by the shell there would come a time when the body could no longer be completely withdrawn into it. Before discussing the third stage in the development of the Gastropod shell, we must consider its growth. This, from a geometrical point of view, is of three kinds: growth in height, peripheral growth, and radial growth or increased thickness of the shell wall. This last does not here concern us. Supposing, for simplicity's sake, the shell to be conical, growth in height occurs at the base (or aperture of the shell), and takes place by means of continual deposits of bands of new material at the edge of the aperture, by the growing edge of the mantle. Peripheral growth is the enlargement of the circumference of the base or aperture of the shell.

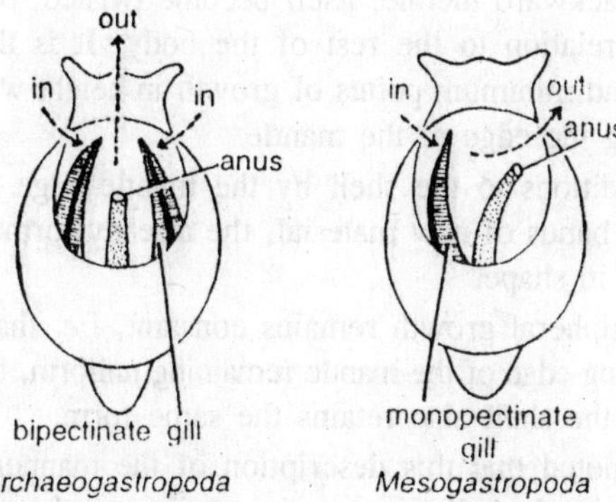

Fig. 18.4. Diagrams to illustrate the loss of symmetry and atrophy due to torsion. The arrows indicate the course of respiratory water currents.

If the height and the peripheral growth remain uniform round the whole aperture of the cone (which is assumed to be round), the cone increases without altering its shape. If, however, the growth in height is not uniform, but steadily and symmetrically increases along each side from an imaginary minimum point to a diametrically opposite maximum point, the peripheral growth, however, remaining uniform, a spirally twisted hollow cone is produced. If the minimum and maximum points in this growth continue throughout in one and the same plane, a symmetrical shell coiled in this plane of symmetry results. If, however, as growth increases, the maximum point shifts from the symmetrical plane, say to the left (the minimum point shifting in the opposite direction to the right), the maximum and minimum points no longer trace on the spirally coiled shell straight but spirally twisted lines, and the conical shell is then not coiled symmetrically in one plane, but asymmetrically in a screw-like spiral. We then have what conchologists call a dextrally twisted shell.

The growth of the Gastropod shell actually takes place in this last manner. This, the dextral (or sinistral) coiling of the Gastropod shell, is the last stage to be discussed. If the visceral dome and shell which are twisted in one plane pass, in growth, from an incline to the left to a backward incline, this is equivalent to the continual shifting of the point of maximum growth to the left and that of minimum growth to the right; the necessary consequence being a dextral screw-like spiral twist. It must be borne in mind—

1. That the growing edge of the mantle, which secretes the shell substance, does not, in the course of the gradual change from the

left to the backward incline, itself become twisted, but retains its position in relation to the rest of the body. It is thus only the maximum and minimum points of growth in height which become shifted along the edge of the mantle.

2. That the additions to the shell by the mantle edge are made in the form of bands of new material, the already formed firm shell not altering in shape.
3. That the peripheral growth remains constant, *i.e.* that the outline of the growing edge of the mantle remaining uniform, the increasing aperture of the shell also retains the same form.
4. It must be noted that this description of the manner in which a dextrally twisted shell arose only applies to that stage in the ontogenetic or phylogenetic development of the shell during which its displacement in a backward direction and the shifting forward of the pallial complex occur. When once the result most favourable to the animal, *i.e.* the anterior position of the mantle cavity and the backward direction of the shell, are attained, further displacement, which would be disadvantageous, does not take place. It is, then not at first sight evident why, when the need for displacement ceases, its action still continues, *i.e.* why, though displacement ceases, the viseral dome and shell continue to grow in a dextral twist and not symmetrically. This point will be explained below.

For the sake of clearness we have treated separately the three important factors in the development of the Gastropod shell, viz. (1) the formation of a tall conical shell, (2) the spiral coiling of the same, and (3) the special manner of coiling in a dextra twist. In reality these three factors do not denote special stages, but all operate simultaneously. The continually increasing protrusion of the visceral dome was accompanied by the dextral twist, as a consequence of the

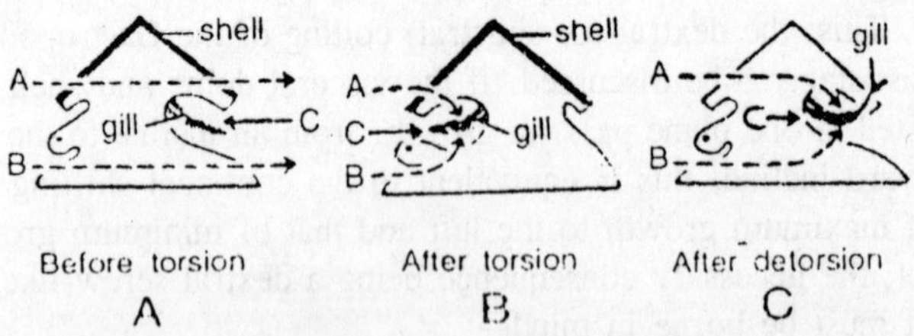

Fig. 18.5. Diagrams illustrating the possible advantage of torsion. A—Current due to flow of water. B—Current due to movement of gastropod. C—Respiratory current.

twisting of the visceral dome from its incline to the left to the most favourable backward incline, by which the pallial complex was shifted forward. The results of ontogenetic research favour the theory here advanced.

We have first to note the fact that the anus (the centre of the pallial complex) and the mantle fold originally lie posteriorly. They come to lie anteriorly in the embryo not by active shifting, but by the cessation of growth on the right side between the mouth and anus, and its continuation on the left side. There is, however, no difficulty in harmonising this ontogenetic method of gaining the object with the phylogenetic method. So far we have placed mechanical and geometrical considerations in the fore-ground. But these necessarily coincide with utilitarian considerations. Every alteration in the direction we have been considering means an improvement in the organisation of the animal, an advantage to enable it the better to maintain the struggle for existence.

The formation of a spire-like shell, which has been recognised as the starting-point in the development of the asymmetry of reptant Gastropods, was the only method by which complete protection of the whole body could be attained, and must therefore be considered to have been advantageous under the circumstances. We might further conclude this from the fact that the possession of such a shell actually distinguishes the Gastropoda from the primitive Mollusca, which the *Chitonidae* are rightly consider most nearly to represent. One apparently important objection to the theory here set forth must be mentioned. If the first factor in the asymmetry of the Gastropod body is the development of a high spire-like shell, and if the arrangement of the nervous system is necessarily connected with the coiling of the shell in a definite direction, how can we account for forms such as *Fissurella*? This Diotocardian genus actually belongs to the most primitive Gastropods, because the symmetry of the pallial complex is still retained. But it possesses an asymmetrical nervous system and the typical chiastoneury of the *Prosobranchia*, and nevertheless a flat cup-shaped symmetrical shell. We thus here have secondary characteristics of the inner organisation combined with an apparently primitive shell.

The latter is, however, only *apparently* primitive, as can be proved systematically and ontogenetically. The forms most nearly related to *Fissurella*, such as the primitive genus *Pleurodomaria*, *Polytremaria* and *Scissurella*, have spacious spirally coiled dextrally twisted shells. In *Haliotis*, the shell becomes flat and the coiling indistinct, as is also the case to some extent in *Emarginula*, till finally in *Fissurella*, it again secondarily becomes flattened or cup-shaped and symmetrical. *Fissurella* even passes ontogenetically through an *Emarginula* stage, in which the shell is distinctly spirally coiled. We may therefore conclude, with as much certainty as is possible in morphological questions, that the outwardly symmetrical *Fissurella* descends from forms with high spirally coiled shells. It return to a flat symmetrical shell may have been determined, as in the *Patellidoe*, *Copulidoe*, etc. by adaptation to certain biological conditions. The explanation given above seems to throw new light on many as yet unsolved problems in the morphology of the Mollusca, such as the asymmetry of the pallial complex in most Gastropoda.

Many *Diotocardia*, all *Monotocardia*, all *Opisthobranchia*, and all *Pulmonata* shows marked asymmetry in the pallial complex. The asymmetry consists principally in the absence of one gill, one osphradium, and one nephridial aperture. The inner organisation also shows reflections of this asymmetry in the nervous system, and the absence of one kidney and one auricle. On closer inspection, it is found that it is the original left half of the pallial complex (which in a *Prosobranch* would lie to the right in the mantle cavity near the anus) which is wanting. The anus is no longer the centre of the pallial groups of organs, but lies outermost on one side. While in the *Prosobranchia*, for example, the original left half of the pallial complex (which would now lie on the right) has disappeared, those organs of the complex (the original right) which are retained, shift from the left to occupy the empty space. Consequently, we find the anus no longer anteriorly in the middle line, but on the right side, close to the extreme right of the mantle cavity.

But what is the reason of the disappearance of the left half of the pallial complex in the *Monotocardia*, *Opisthobranchia*, and *Pulmonata*? In answering this question we must refer back to paragraph 3, where it was seen that if the spire-like shell assumes the only possible lateral inclination, the mantle cavity and the pallial complex within it are subjected to unequal pressure. If the shell is inclined to the left, the side of the posterior mantle cavity subjected to the greatest pressure is the left, and the pressure continually decreases towards the right. These variations of pressure are also retained during the whole time in which the backward displacement of the shell and the forward displacement of the pallial complex takes place. In other words, *i.e.* described in terms of our theory, from the very commencement of the development of the Gastropod organisation, the original left organs of the pallial complex were subjected to unfavourable conditions. In this left-sided compression of the mantle cavity the ctenidium especially would necessarily be reduced in size and become rudimentary, and might entirely disappear. As a matter of fact, the original left half of the pallial complex (which would now lie on the right) has entirely disappeared in many *Diotocardia* (the so-called *Azygobranchia*), in all *Monotocardia*, and in the *Opisthobranchia*. The fact that the original right gill, the only one remaining, has also disappeared in the *Pulmonata* is accounted by the change to aerial respiration.

It is an interesting fact that in the *Basommatophora* the original right osphradium is retained. If, however, the original left gill did not quite disappear, but only became smaller, we should have to expect that in such *Diotocardia* as still possess two gills, the original left (now the right) would be the smaller. This would be the case at least in the more primitive forms with shells still twisted. *Haliotis* and *Fissurella* are the only Molluscs to which this applies. In *Haliotis*, whose shell is still twisted, the right (originally left) gill is in reality the smaller. But in *Fissurella* and *Subemarginula*, were the asymmetry of the mantle cavity has been secondarily lost, the inequality in the size of the gills has also disappeared.

Another unsolved problem remains. Why does the shell continue to grow asymmetrically coiled with a dextral twist, after the cause of this asymmetry, viz. the change from the incline to the left to the backward incline of the shell, simultaneously with the shifting forward of the mantle cavity and pallial complex, has ceased to act, *i.e.* when the shell has definitely assumed the posterior, and the pallial complex the anterior, position ?

The explanation of this lies in the asymmetry so early apparent in the mantle cavity, which from the beginning is more spacious to the right (now left) than to the left, the consequence being that the left half of the pallial complex atrophied. This asymmetry of the pallial complex and mantle cavity remained after the displacements of shell and pallial complex had been definitely accomplished in the *Prosobranchia*, i.e. the asymmetrical growth, and therefore the continuous coiling of the visceral dome and shell in a spiral twist, continued. In altogether exceptional conditions, which rendered a flat cup-shaped shell useful, the return to symmetry in the pallial complex and mantle cavity or fold would be advantageous, since then symmetrical growth of the sheil could take place.

If the difference between the maximum and minimum growth in height is but slight the shell would be but slightly coiled, and if the peripheral growth is pronounced, while the growth in height is insignificant, a flat cup-shaped shell would result (*Haliotis*, *Emarginula*, *Fissurella*, *Patella*, etc.). Chiastoneury only takes place when the original right half of the pallial complex crosses over to the left of the median line anteriorly. This crossing of the line of symmetry has actually taken place in the *Prosobranchia*. The original right gill in them lies quite to the left of the mantle cavity.

In the *Azygobranchia* and *Monotocardia* the hind-gut with the anus has at the same time become displaced into the right (original left) narrower gill-less half of the mantle cavity, which, however, is still spacious enough to contain the rectum. The *Prosobranchia* are *streptoneurous*. In the *Tectibranchia* and *Opisthobranchia* the pallial complex is found on the right side of the body, and has nowhere crossed the median line anteriorly. There is therefore no chiastoneury among the *Opisthobranchia*, i.e. their visceral connectives are never crossed. In the *Pulmonata* the pallial complex has shifted far forward, but it has not passed the middle line with any organ which, drawing the parietal ganglion and the visceral connective with it, could have brought about chiastoneury. For the left (original right) gill, the only one elsewhere retained, disappeared (apparently very early) in the *Pulmonata*.

The osphradium, which is retained in aquatic *Pulmonata*, is the original right, and still lies on the right side. In considering the arrangement of the nervous system, it is really immaterial whether we assume that the hind-gut has shifted back to the right secondarily, and the osphradium moved to near the respiratory aperture, or that the hind-gut never reached the median line, and that the osphradium never passed over it. The *Pulmonata* are thus *euthyneurous*. We saw, that with a strongly developed visceral dome and posteriorly placed pallial complex, a shell inclined forward or coiled forward is an impossibility for a reptant Gastropod. But such a shell is not an impossibility for an animal which does not creep. For example, in a swimming animal, whose shell, partly filled with air, serves as a hydrostatic apparatus, there is no reason why a much developed visceral dome and shell should not become coiled forward, the original posterior position of the pallial complex being retained as the most favourable under such circumstances.

As an example of this we have the *Nautilus*, all *Nautiloidea* and *Ammonitidea*, with their *exogastrically* (anteriorly) coiled shells and posteriorly placed pallial complexes. The coiling of the shell of *Spirula* forms an exception to that of all other *Mollusca*, being *endogastric*. With regard to this we have to consider first, that the shell of *Spirulla* is internal and rudimentary, and that the backward coiling does not in any way affect the posteriorly placed mantle cavity; and second, that only the *modern* genus *Spirula* has such a shell.

The Miocene genus *Spirulirostra* has its phragmacone endogastrically bent but not coiled, and the older *Belemnitidoe* never have either curved or coiled shells. Moreover, the shell of this whole group, being internal and, as far as the original purpose of a shell, protection of the body, is concerned, rudimentary, does not come under consideration in the present discussion. In an animal living in mud, like a limicolous bivalve, there appears no reason why the shell should not simply become elongated, and why the mantle cavity and pallial complex should not retain the posterior position. *Dentalium* is distinctly in this condition, being the symmetrical primitive Gastropod adapted to life in mud, and provided with a turret-like shell and posterior pallial complex. The perforation at the upper end of the shell, which freely projects from the mud, is of great morphological importance, corresponding physiologically with the siphons of the limicolous *Lamellibranchia*.

A comparison between *Dentalium* and a *Fissurella* with its pallial complex twisted back, and with a long and turret-like shell, is from our point of view, very appropriate. A *Fissurella*, so transformed, would almost exactly resemble the hypothetical symmetrical racial form of the Gastropoda, in which, however, we should have to assume a mantel- and shell-cleft reaching to their edges.

The anatomy of the *Protobranchia*, which has recently been more closely studied, and especially the posterior position of the two gills, the flat sole for creeping, and the presence of the pleural ganglia, justify us in deriving the *Lamellibranchia* also from the racial form of the Gastropoda, in which the cleft edge of the mantle would correspond with the posterior or siphonal edge of the mantle in the former. This edge of the mantle, having a similar physiological function, often possesses tentacles, papillae, etc., in both groups. *Dentalium* further fits in with our theory, for the forward curve and the position of the columellar muscle on the anterior side of the visceral dome which would be disadvantageous to a freely reptant, is not so to a limicolous, animal.

Dextral and Sinistral Twists

Most Gastropods have the visceral dome and shell twisted dextrally. The direction of the twist has been determined by the fact that the visceral dome and shell originally inclined to the left, and them more and more backward, thus pushing the pallial complex along the right mantle furrow. It cannot be determined why the incline to the left was originally chosen. The shell might just as well have inclined to the right at first, and then more and more backward, pushing the pallial complex along the left mantle furrow. The consequent asymmetry would then have been exactly reversed.

To take a concrete example; in a *Monotocardian*, with visceral dome and shell twisted sinistrally, the original left parietal ganglion would become the supra-intestinal ganglion on the right. The original right half of the pallial complex would disappear, and the left half which persisted would lie to the right of the anus or rectum which would take up its position to the left of the median line. Gastropoda with sinistrally twisted shells are actually known, many of them having the asymmetrical organs in the inverse position which corresponds with this twist. Such are, among the *Prosobranchia*, *Neptunea contraria*, *Triforis*, and occasional specimens of *Buccinum*; among the *Pulmonata*, *Physa*, *Clausilia*, *Helicter*, *Amphidromus*, and occasional specimens of *Helix* and *Limmaea*. In *Bulimus perversus*, individual specimens with either sort of shell are found with the special asymmetry of the organs belonging to it.

There are, however snails whose shells are dextrally twisted, but which possess the organisation of animals with sinistrally twisted shells. This is the case among the *Prosobranchia* inthe sinistrally twisted sub-genus *Lanistes* of the genus *Ampullaria*; among the *Pulmonata*, in *Choanomphalus Maacki* and *Pompholyx solida*; among the *Opisthobranchia*, in those *Pteropoda* which, whether as adults (*Limacinidoe*) or larvae (*Cymbuliidoe*), have a twisted shell. This fact is entirely against our theory in explanation of the asymmetry of the Gastropoda, for this theory points to a causal connection between the spiral coiling of the visceral dome and shell on the one hand and the special asymmetry of the asymmetrical organs on the other.

The above-mentioned exceptions to the rule can, however, be explained as follows. The spiral of a dextrally twisted shell can by degrees become flattened in such a way that the shell may be simply coiled in one plane or may nearly approach that condition. In this case the spiral might again assert itself, but on the side opposite to that on which the umbilicus originally lay, and in this way a false spiral might form on the umbilical side and a false umbilicus on the spiral side. The transition from a dextrally twisted to a falsely sinistrally twisted shell, which latter was, however, genetically dextrally twisted, by means of the shells of seven species of the genus *Ampullaria*, *Ampullaria Swainsoni* Ph ? (G) and *A. Geveana* Sam (F) are dextrally twisted with distinctly projecting spiral. In *A. crocastoma* Ph (E) the spiral is flat, in *A. (Ceratodes) rotula* Mss. (D) and *A. (Ceratodes) chiquitensis* d'Orb (C) the spiral is already pushed through or sunk, yet we find a true umbilicus on the umbilical side.

In *A. (Lanistes) Bolteniana* Chemn. (B), and still more in *A. purpurea* Jon. (A), the false spiral appears on the umbilical side, and on the spiral side a false umbilicus is found. However plausible this explanation may appear, it can only be proved to be correct if it is found that where a spiral operculum occurs, the direction of its spiral is opposite to that of the spiral of the shell and the commencement of the spiral is always turned to the umbilical side of the shell. *Lanistes* has not a spirally twisted operculum, but such occur in the *Pteropoda*. In those *Pteropods* which combine a sinistrally twisted shell with the organisation belonging to a dextrally twisted Gastropod, the operculum exactly corresponds with that of a dextrally twisted shell.

In *Peraclis*, in the larvae of the *Cymbuliidoe* and in *Limacina retroversa* Flemming, the operculum (the free surface of which must be viewed) is sinistrally twisted, and the starting point of the twist faces the (false) spiral, which in these falsely sinistrally twisted Gastropods lies in the place of the original umbilicus. This apparent exception is thus shown to be quite in keeping with the rule above established.

19

PARASITISM

Thyca ectoconcha is a Prosobranchiate Gastropod which is parasitic on the Star-fish *Linckia multiforis*. The chief points in its organisation are longitudinal section in which, however, several organs which lie laterally to the section are also represented. The organisation of the Gastropod is as yet little influenced by its parasitic manner of life. It possesses a shell, shaped somewhat like a Phrygian cap. In the mantle cavity lies the gill. The alimentary and nervous systems also are in no way remarkable. It has eyes and auditory organs, and a short powerful snout, and muscular oesophageal bulb, which penetrates the Star-fish between the calcareous parts of its integument into the tissues. There is no radula.

The base of the snout is surrounded by a muscular disc consisting of an anterior and a posterior part. This disc, the so-called false foot, is the grasping organ by which the animal attaches itself to the integument of its host so firmly that it cannot be torn away without injury. The rudiment of a foot occurs without an operculum. The Gastropodan organisation is somewhat more strongly modified in *Stilifer Linckiae*, which is parasitic on the male *Linckia*. The whole body of this parasite penetrates into the calcareous layer of the integument of the host, on which it raises pathological globular swellings, and further causes the peritoneum to bulge inwards towards the body cavity.

The parasite communicates with the outer world only by means of a small aperture at the tip of the swelling. The parasite, thus established in the integument of its host, is surrounded on all sides by a fleshy envelope. This envelope is only broken through by an aperture at the point where the apex of the dextrally twisted shell lies; this aperture

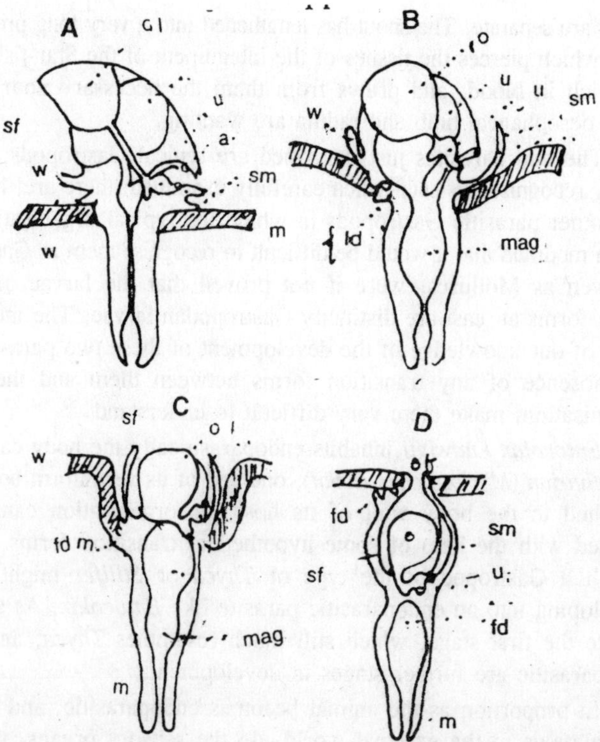

Fig. 19.1. A, B, C, D, Hypothetical transition stages between Thyca and Stillifer on the one side and Entocolax on the other. a–anus; fd–pedal gland; l–liver; ld–hepatic intestine; m–mouth; may–stomach; o–ovary; of–aperture of the false mantle; sf–false foot; sm–false mantle; u–uterus; w–body wall of the host.

corresponds in position with the aperture above mentioned as occurring at the tip of the pathological swelling. This envelope is called the false mantle, and corresponds morphologically with the false foot of *Thyca*, much increased in size and bent back on to the shell. There occur besides a true mantle, a gill, a rudimentary foot without an operculum, eyes, auditory organs, and a typical Prosobranchiate nervous system.

The development of the remarkable false mantle no doubt signifies that, although the animal is embedded deep in the integument of the host, communication with the exterior is retained. Water for respiration can enter and flow out of the mantle cavity, and the faecal masses and genital products, and perhaps also the larvae can pass into the cavity of the false mantle and be ejected through its aperture. The sexes are separate. The snout has lengthened into a very long proboscidal tube which pierces the tissues of the integument of the Star-fish, which are rich in blood, and draws from them the necessary nourishment. Both oesophageal bulb and radula are wanting.

The two parasites just described are typical Gastropods, and are easily recognised as such when carefully examined; there are, however, two other parasitic Gastropods in which the typical organisation is so much modified that it would be difficult to recognise them as Gastropods, or even as Molluscs, were it not proved that the larvae of one of these forms at least are distinctly Gastropodan larvae. The incomplete state of our knowledge of the development of these two parasites, and the absence of any transition forms between them and the typical organisation, make them very difficult to understand.

Entocolax Ludwigii inhabits endoparasitically the body cavity of a *Holothurian* (*Myriotrochus Rinkii*), one end of its vermiform body being attached to the body wall of its host. Its organisation can be best studied with the help of some hypothetical transition forms, through which a Gastropod of the type of *Thyca* or *Stilifer* might pass in developing into an endoparasitic parasite like *Entocolax*. As shown in figure the first stage, which still much resembles *Thyca*, and is still ectoparasitic are further stages in development.

In proportion as the animal becomes endoparasitic, and gives up its relations to the external world, do the sensory organs, the shell, and the mantle cavity with the gill disappear. The stomach, as a separate section of the intestine, degenerates, the digestive gland (liver) becomes a simple unbranched diverticulum of the intestine, which loses the rectum and anus. All organs for the purpose of mastication at the anterior end of the alimentary canal are lost. The false mantle becomes larger and larger, and envelops the small visceral dome, which gradually becomes rudimentary, and finally contains merely the genital organs. At the stage, the whole animal already projects freely into the body cavity of the host, attached to its wall by a displaced portion of the false foot, and connected with the exterior only by the aperture of the false mantle.

If this last means of communication with the exterior is also abandoned, *i.e.* if the whole false mantle with its aperture becomes enclosed in the body cavity of the host, we have a form corresponding with the endoparasite *Entocolax Ludwigii*. In this form, the cavity enclosed by the false mantle, into which the ovary and its receptacula seminis open, serves as a receptacle for the fertilised eggs, which

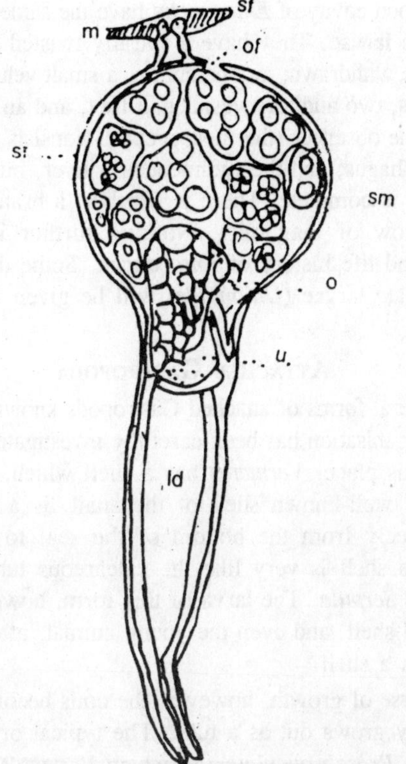

Fig. 19.2. Entocolax Ludwigii, sketch after Viogt.

were found in it in their first stage of segmentation in the one (female) specimen discovered.

Entoconcha mirabilis, an endoparasite which has been found in a *Holothurian*, *Synapta digitata*, is even more deformed than *Entocolax*. The body of this parasite is a long vermiform coiled tube, attached by one end to the intestine of the host, while the rest of the tube floats freely in the body cavity of the latter. Its organisation has as yet been imperfectly investigated. It is impossible to say how far such a comparison, which the lattering is intended to facilitate, is justifiable.

Up to the present time, no aperture leading from the ovary into the brood-chamber, which is thought to be the cavity of the false mantle, and is filled with embryos (not represented in the figure), has been observed. In a widening of the tube near its attached end, a number of free "testicular vesicles" have been found, but their real significance can only be discovered by further research. The embryos found in the brood cavity of *Entoconcha* have the same general structure as Gastropodan larvae. They have a spirally twisted shell, into which the body can be withdrawn; an operculum, a small velum, the rudiments of two tentacles, two auditory vesicles, a foot, and an intestine, which, according to one observer (the most recent), consists of only a mouth, pharynx, oesophagus, and the rudiment of a liver, but according to an older authority is complete. There is, further, a branchial cavity with a transverse row of long cilia. Nothing further is known of the development and life history of *Entoconcha*. Some details of parasite Lamellibranchiate larvae (*Unionidoe*) will be given in the section on Ontogeny.

Attached Gastropoda

Of the several forms of attached Gastropods known, only *Vermetus*, whose inner organisation has been carefully investigated, can be shortly described in this place. *Vermetus* has a shell which, instead of being coiled like the well-known shell of the snail, is a calcareous tube, which rises freely from the bottom of the sea, to which its tip is cemented. This shell is very like the calcareous tubes of tubicolous worms such as *Serpula*. The larva of this form, however, possesses a typically coiled shell, and even the young animal, after it has attached itself, has such a shell.

In the course of growth, however, the coils become loosened, and the shell finally grows out as a tube. The typical organisation of the *Monotocardian Prosobranchiates*, to which *Vermetus* belongs, is little affected by the attached manner of life. The visceral dome, like the shell, is much elongated and almost vermiform. The intestine, the circulatory system, the kidneys, the mantle, the gill, and the nervous system are typically developed. The sexes are separate, and copulatory organs, which could not be used by attached animals, are wanting. The head is well developed, and the pharynx well armed.

When the animal is slightly irritated, it is said not to withdraw at once into its shell, like other Gastropods, but to bite. The foot has the form of a truncated cylinder, and its directed anteriorly, ventrally to the head. It cannot, of course, function as a locomotory organ, but carries the operculum for closing the shell, and, by means of the pedal gland, secretes mucus. *Vermetus* is said to produce great quantities of this secretion, which it allows to float in the water for a time like a veil, and then swallows together with all that has become attached to it. In this way it fishes for the small organisms which form its food.

20

Harmful and Medical Molluscs

On the whole marine molluscs have caused relatively few deaths among man. The main dangers come from the venomous Indo-Pacific cones—they have caused only 9 or 10 deaths—from the octopus that on rare occasions inflicts a fatal bite and from bivalves that carry paralytic shellfish poisoning, responsible for more than 200 deaths. Numerous outbreaks of infectious hepatitis and typhoid have been directly attributed to clams and oysters that were collected near sewer outlets. Freshwater snails of Egypt, Venezuela, the Philippines and China have indirectly caused the deaths of hundreds of thousands of people because they are the intermediate hosts of the fatal blood fluke disease schistosomiasis, or bilharziasis.

The larval stages of the guilty trematode worm must spend several weeks developing and multiplying inside certain species of snails before they can attack and infect man. There have been no cases of this disease in Canada or mainland United States. Of more serious concern to seashore residents of North America is the possibility of becoming paralyzed and possibly dying from eating clams and mussels that are carries of paralytic shellfish poisoning. In this day of "back-to-nature" and economic food hunting caution should be exercised when eating raw or cooked clams and blue mussels taken from either of our coasts.

Fortunately the infected areas are limited, the infectious season short and the deaths not many. The bivalves become poisonous because they ingest the microscopic marine algae *Gonyaulax*. This dinoflagellate "blooms" in great quantities when the temperature of the water exceeds

10°C. The causa.ive agent on the Pacific coast is *G. acatenella*; on the Atlantic coas., it is *G. tamarensis*.

Within historic medical times, there have been 957 cases of bivalve paralytic poisoning, 222 of which have resulted in death. These figures do not include cases of severe allergic reactions to eating shellfish. Mussel poisoning, including deaths, have also been recorded from Scotland, Wales, Belgium, Ireland, France, Australia, New Zealand, South Africa and Japan. On the Atlantic coast there have been about 100 poisonings and three deaths. The areas affected were the Bay of Fundy between New Brunswick and Nova Scotia and the Gaspé Peninsula in Quebec. Bivalves carrying the position the mussels *Mytilus edulis* and *Modiolus modiolus*, the Soft-shelled Clam *Mya*, the Surf Clam *Spisula solidissima* and the razor clam *Ensis*. Poisonings occur between June 4 and October 11, with the peaks being in August and September. Ordinary cooking reduces but by no means eliminates the poison.

There is a wide range in the degree of human susceptibility—some people, especially those who do not normally eat shellfish, are very sensitive. Cats and chickens die very quickly when fed infected clams or even the "rims," or trimmings, of scallops. The muscles of scallops are not infected and therefore safe to eat. Within five minutes of chewing an infected bivalve, the victim begins to feel a stinging, tingling and numbness of the lips, tongue and throat. In the case of severe poisonings, when five to twenty bivalves have been eaten, there will be within fifteen to thirty minutes, a numbness of the fingers and toes that may spread to the knees, then hips. The face, arms and neck may become paralyzed. Death ensues when the paralysis affects the respiratory muscles. The effects usually last from eight to forty-eight hours, depending upon the dose.

The Pacific coast has had more trouble with this than the Atlantic. The earliest recorded cases occurred on June 15, 1793, when one death and four illnesses took place among Captain George Van-couver's men, who had eaten toxic mussels in British Columbia. According to historians, in 1799 the Russian Baranoff expedition lost about one hundred men near Sitka, Alaska, from eating infected mussels. The Pacific infections are recorded from eleven species of edible bivalves, including the venerid and *Mytilus* mussels. Any carnivore that eats an infected bivalve will also carry this infection. For instance, whelks and moon snails, which often feed upon clams, may also be toxic when eaten. Early man used every edible plant and animal at one time on another for medicinal purposes.

Many plants and animals were adopted as medicines for superstitious reasons or simply because they resembled the shape of some organ of the human body. A heart-shaped leaf or cockle clam was supposedly good for heart trouble. Molluscs stood high among the foods that were considered an aid to sexual activities. The belief is still prevalent that oysters are good for one's sex life. Automobile bumper stickers in the eastern United States and signs in seafood restaurants proclaim Eat Oysters, Love Longer, a slogan undoubtedly invented by the public relations firm representing the oyster industry. Lord Byron's Don Juan claimed that "Oysters are amatory food."

It is reported that some diners at Roman feasts consumed several hundred at one sitting. As if to prove this point, a Joe Garcia of Melbourne, Australia, appears to have broken the record in 1955 by eating 480 in sixty minutes! For many centuries, and even today in China, pearls ground into a powder and dispensed with milk and various herbs have been considered cures for stomach troubles. Snails were used throughout the ages as a medicine, generally for treating bad colds and for consumption. There are at least two stories about snails having been introduced to other countries by some distinguished person for the benefit of his consumptive wife. One case involved Sir Kenelm Digby, who introduced *Helix pomatia* to England; the other involved the governor of Madagascar, who in 1800 introduced the Giant African Snail to his palace gardens for the benefit of his sick wife.

In the sixteenth century, it was thought that convulsions resulting from high fever could be forestalled by "beating snales which be in shell with bay salt and mallowes, and laid to the bottomes of your feet and to the wristes of your hands." A popular Spanish cure for headache was to "Make a poultice of bruised snails. They most be broken up with their shells and put into a piece of linen folden 4 times as to make it thick, dip it in brandy, and squeeze it tolerable dry; then apply it to the forehead." The thought is enough to give moderns a headache! Pliny recommended that raw molluscs be used to cure sore throats and coughs. There seems to be some scientific basis for this cure. In the 1960s it was discovered that extracts of the Hard-shelled Clam, later called mercenine, was a strong growth-inhibitor of experimental cancers in mice.

Subsequent work by Drs. C. F. Li and Benjamin Prescott at the National Institutes of Health in Washington, D.C. found that one active fraction of raw abalone juice was very effective against penicillin-resistant strains of *Staphylococcus aureus*, *Streptococcus pyogenes*,

Salmonnella typhi and paratyphoid A and B. The new substance was called paolin. Experiments have demonstrated that the juices from oysters also have antiviral properties.

Molluscs and Pollution

Several years ago a malacologist asked an official of a large American railroad why he didn't take the precaution of protecting the wooden pilings of a train causeway that passed over a salty marsh. It was common knowledge, at least to biologists and malacologists, that the shipworm clam can thrive in such an environment and, by boring into the wood, reduce the causeway to rubble within a few years. The executive laughed and explained that there was no danger and that there was no need to spend millions of dollars impregnating the pilings with creosote. There already was a "natural" cure.

Pollution of the water the filth and oily refuse from nearby industries, was so strong in that marine marsh that no *Teredo* shipworms or boring crustacea could possibly survive. One biological was once facetiously said that if a magician could suddenly clean up our harbours and estuaries without cost, the shipping and railroad companies would go broke trying to save the wharves and bridges from being destroyed by reinvading marine organism. The absence of certain molluscs is an excellent indicator of pollution and other abnormal environmental conditions.

Despite all the expensive precautions that many enlightened manufacturers take, there is no way of predicting the effect of sewage through chemical and temperature measurements. The delicate organisms of the area are the best barometers and will soon give warnings if biological disasters is ahead. Certain sensitive species die off first. When the Florida Everglades were being drained excessively by artificial canals into the logoons and bays of east Florida, particularly Boynton Inlet, the cowries, tun shells and *Strombus* conchs were the first to succumb to the sudden drop in water salinity.

The more tolerant estuarine species, such as *Cerithidea*, *Neritina* and *Littorina*, suddenly blossomed and took over the environment. But when sulfides and oil pollutants are added, even these species disappear. Oil spills in excess of a thousand gallons are very destructive to molluscs, especially to the bivalves, whose microscopic breathing cilia are paralyzed by oil that eventually sinks to the bottom in intertidal areas. A new threat to the open expanses of the salt marshes, where there are billions of mussels and snails serving as potential nutrients for offshore fisheries, are the huge nuclear plants that give off millions of tons of heated water. A rise of only a few degrees, particularly during the higher temperatures of summer often makes deserts of marshes. Fortunately, remedial steps are well on their way toward solving this man-made imbalance. The introduction by man, either accidentally or purposely, of foreign molluscs is a very old form of pollution.

A foreign species introduced to a new environment usually has no natural enemies to keep it in check. Furthermore, it may find the new habitat particularly rich in foods not available in its homeland. If the introduced mollusc multiples rapidly and becomes a pest by destroying beneficial molluscs or crowding out useful native species, it becomes as much as factor of pollution as chemicals or excessive temperatures. sAn outstanding example of unwanted introduction is the American Slipped Shell, *Crepidula iornicata*, which was accidentally introduced to England at the end of the nineteenth century. It soon began to choke the oyster beds by its sheer numbers. Today, sea bottoms must be dredged to remove the pests so that the oysters will have room to settle. A number of snail drills were brought in with oyster seeds from Japan to the Pacific coast of the United States many years ago, and they still constitute an expensive menace. The latest, most drastic introduction was the Chinese freshwater clam *Corbicula* which has invaded and almost completely clogged the irrigation ditches of California. It is now so abundant in midwestern rivers that cement can no longer be made with the gravel contaminated by these soft shells.

21

Edible Molluscs

Edible Oyster

The oyster has been man's favourite edible mollusk since prehistoric times. The ancient kitchen middens—huge garbage dumps—found throughout the cooler parts of the world, attest to their popularity. The ancient Greeks and Romans used them as a main dish, and their taste for them encouraged the development of artificial oyster beds a century before Christ. Some enterprising Roman merchants discovered that oysters could be packed and shipped in ice and snow obtained from the north side of high mountains, and there are records of oysters being shipped for hundreds of miles to special banquets. Our earliest colonists found that the maritime Indian tribes were great oyster eaters and that when meager land crops failed, oysters supplied the necessary proteins to carry them over the hard winters. There are about a dozen species of edible oysters, belonging to the genera *Ostrea* and *Crassostrea*, that supply 90 percent of the world's harvest of oyster meats.

It would be difficult to estimate the annual world consumption, but fisheries' records show that countries having advanced shell fisheries, such as those in Europe, Japan and the United States, produce well over a billion pounds of oyster meat each year. In the eastern United States, the production has been gradually declining from an annual high of about 160 million pounds during the early 1900s to about one-third that amount today. The eastern industry is based upon the Virginia oyster, *Crassostrea, virginica*, a brackish-water species that extends from southern Canada to Mexico. In Europe, a similar species, *Crassostrea angulata*, is very popular the main species is

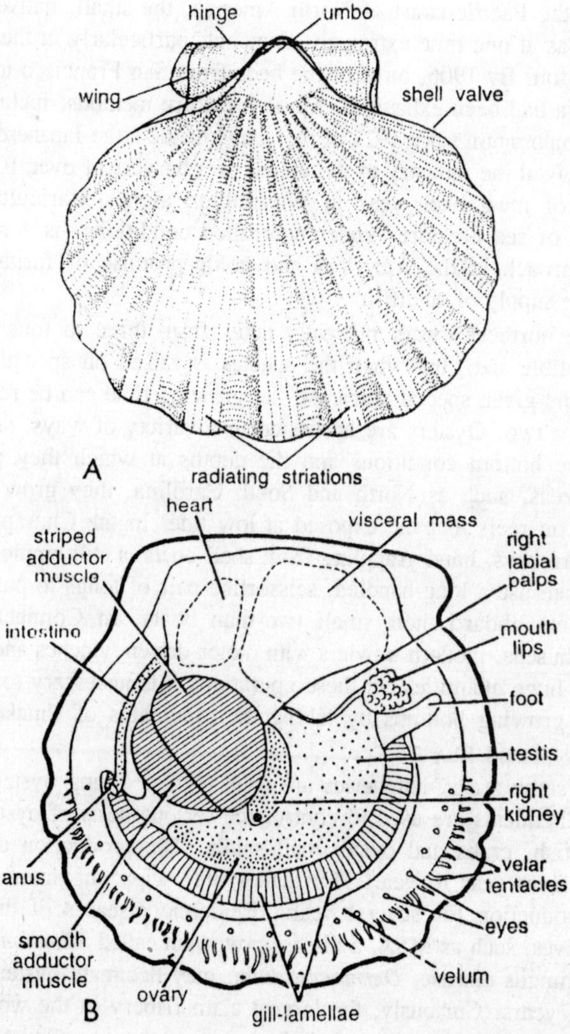

Fig. 21.1. Pecten. A—External features. B—Internal organs after removal of right shell valve.

Ostrea edulis. It differs in that it lives in clearer waters of higher salinity. It retains its fertilized eggs within its mantle cavity until they hatch into free-swimming veligers. In the *Crassostrea* forms, which live in brackish, both the eggs and sperm and shed into the open water.

On the Pacific coast of North America, the small, native *Ostrea lurida* was at one time extremely abundant, particularly in the state of Washington. By 1906, most of the beds from San Francisco to British Columbia had been exhausted. Modern culture methods, including the annual importation since 1922 of the young of the large Japanese Oyster, have revived the industry to the extent of a harvest of over 10 million pounds of meat each year in Washington alone. Mariculture—the farming of sea creatures under controlled conditions—is a relatively new approach in the Americas and holds promise of furnishing an adequate supply of shellfish.

The northern oyster normally takes from three to four years to reach edible size, but when the species is raised on special rafts in ponds and given special nutrients, the growing time can be reduced to a year or two. Oysters are harvested in a variety of ways, depending upon the bottom conditions and the depths at which they grow. In some areas, such as North and South Carolina, they grow in large clumps on reefs that the exposed at low tide. In the Chesapeake and Delaware bays, hand grappling from shell boats is still employed. The oystermen use a long-handled, scissorlike pair of tongs to pull clumps of oysters aboard their small two-man boats. In Connecticut and Massachusetts, modern trawlers with motor-driven winches and dredges harvest huge quantities. In these operations it is necessary to maintain proper growing bottoms by taking the mountains of shucked oyster shells back out to sea.

In some areas, hatcheries are used to rear young oysters, called *spat*. Chemicals are used to control the serious pests of oysters, such as starfish, crabs and oyster drill snails. The production of oysters varies from year to year. Hurricanes may wipe out natural seed or spat production for several years. Epidemic diseases of the oysters themselves, such as MSX, a disease protozoan called *Minchinia nelsoni*, and a fungus disease, *Dermocystidium*, may decimate oyster beds for several years. Curiously, the largest clam fishery in the world is the least known to the public. It is based upon the so-called Surf Clam (that rarely lives in the surf), *Spisula solidissima* a five- to seven-inch long clam living in depth from the low-tide zone to about 100 feet. The fishery is confined to a limited area off New Jersey and adjacent states. Each year about 40 million pounds of meat are landed there.

The Surf Clam was known to the Indians are early colonists but was rarely eaten because it was caught it the surf and was usually loaded with sand. Since World War II modern technological improvements in fishing gear have permitted huge offshore beds, at depths of thirty to ninety feet, to be exploited. A hydraulic dredge, equipped with water jets, a digging blade and an enormous chain bag, can deliver 400 huge bags of clams to the boat above in a matter of a few hours. The clam is then shucked, steamed, cleansed and canned in the clam factories of New Jersey and New York. It is the main ingredient in cans of chopped clams and New England clam chowder. As the supply of the Surf Clam has somewhat dwindled, another clam living in deep water has been used in the eastern clam industry. Known variously as the Arctic Clam, the Ocean Quahog and the Mahogany Clam, *Arctica islandica* has become of increasing importance to fishermen who can dredge from fifteen to eighty fathoms in depth.

Other eastern clams of lesser importance not as yet profitable commercially are the razor clam, *Ensis*, of superb culinary qualities, and the *Tagelus* clam, new being brought up and discarded during the fishing operations for the *Mya* Soft-shelled Clam. A relative newcomer to the eastern shores of the United States is the *Rangia* clam, a thick-shelled, three-inch-long, gray clam that was formerly restricted to the Gulf of Mexico. In the last thirty years it has spread northward from Florida to Delaware and is so prolific that it can be collected by the ton with little effort. It is quite edible and is now canned in combination with quahog meats. Two clams new to the commercial field in the Atlantic are the Venus Sunray Clam, *Macrocallista nimbosa*, and the Calico Clam, *Macrocallista maculata*. The former, beautiful seven-inch long clams, are being experimentally fished in the Gulf of Mexico off Florida, but the area in which they are concentrated is probably too limited to permit extensive fisheries.

Bermuda has recently been invaded by the smaller Calico Clam, thousands of which have been collected by private individuals for "cherrystone" clams on the half shell. They are sweeter and more tender than young quahogs. On the Pacific coast of the United States, there is a relatively small clam industry. The clam there is of more importance to the sportsman and private individual than it is to industry. Among the Pacific coast species are the razor clam, *Siliqua patula* (about 10 million are harvested per year); the famous Pismo Clam, *Tivela stultorum*, which has suffered a drastic decimation due to overcollecting; the *Saxidomus* butter clam (about 1 million pounds per year) and the littleneck *Protothaca* clam.

Some species are not very uncommon, such as the geoduck, *Panopea generosa*, a soft-shelled clam growing to the weight of eight pounds and living in deep, soft mud. Because its catch is limited to only a few specimens per person per week, it is of recreational rather than commercial importance. Next to the United States, Japan is the largest producer of clams. While a large proportion of its catch is locally consumed, a great many tons are smoked and canned in oil for foreign export. A unique clam industry exists in Malaya, where there is an annual catch of over 25 million pounds of ark shell meats from the *Anadara* clams. Cockles are still extensively harvested in northwest Europe and in Malaysia, with a worldwide, annual harvest of about 74 million pounds.

Edible Scallop

Although by no means as abundant as the oyster or the clam, the scallop has held a high place among the choicest of shellfish. Gourmets prefer the delicate sweet muscle of the scallop to the meat of any other mollusk when it comes to preparing a particularly tasty seafood dish. Some of the holiest of gourmet recipes have been built around the plump, pearly white body of the scallop, with the roes, shaped like scarlet cockscombs, containing the delicate, granular eggs. Scallops excel the tougher crustaceans and the coarser clams in taste. Among the practitioners of gastronomy, the cult of the scallop is revered and universal. The world production of scallops is reported to be about 300 million pounds of meat each year, with the United States producing 41 percent Japan 20 percent, Australia and Western Europe each about 10 percent. In the Mediterranean region, the shell are saved, cleaned and packed for export to foreign countries where they are used for baking dishes.

In most places, only the adductor muscle is eaten, probably because of past histories of paralytic shellfish poisoning resulting from eating the mantle and gills. The most-often-eaten scallops in the Americas are the Deep-Sea Scallop, *Placopecten magellanicus*, of the northeast United States and, to the lesser extent, the Alaskan *Pecten caurinus*, both large, flat rather smooth-shelled scallops with shells measuring from six to eleven inches across. From these come the edible portion, a round disk of white muscle about the size of a half-dollar. The eastern species lives in large colonies at depths of 60 to 150 feet, on gravel bottoms, from Nova Scotia to off Cape Hatteras, with a major industrial concentrations being on the Georges Bank and the Gulf of Marine. They are capable of swimming for short distances. Catches worth as much as $13 million have been landed in one year. Today they remain one of the most expensive seafoods, largely because of the scarcity of the product and the expense in fishing them with deep-sea trawlers.

The industry has thrived since the 1880s, off the coast of New England, but the United States is now taking second place to the efforts of the Canadians, who are working new beds off the mid-Atlantic coast and Georges Bank. Large diesel-powered vessels scrape the bottom with two large scallop dredges, usually on a twenty-four basis, with hauls being made about every half-hour. The cruises last about ten days, with the shucking of the scallops being done immediately. The muscle meats are frozen aboard in forty-pound bags. Two other kinds of scallops are fished on the east coast of the United States—the very small Bay Scallop, *Argopecten irradians*, a once very popular item in early America, and the Calico Scallop.

The Bay Scallop is readily fished in shallow water in New England and in New Jersey from grassy bottoms in two to twenty feet of water. The muscle is small but very tasty. About 2 million pounds are landed each year which means that about 20 million specimens are collected annually. Shucking is a "cottage industry," with the live scallops being taken by truck to the small scallop factories for processing. The Calico Scallop, *Argopecten gibbus*, a southern species, has recently come into its own in the Carolinas. The development of efficient shucking and eviscerating machinery has made harvesting this small two-inch-long shell an economically feasible operation. Scallops are also extensively fished in southern Australia and Japan. Minor fisheries exist in almost all of the cool-water areas throughout the world.

Other Edible Molluscs

When we consider the number of miscellaneous kinds of minor molluscan shellfish used throughout the world, we can conclude that man has eaten almost every marine species of worthy size. In areas not affected by algal poisoning, the common blue mussels of the genus *Mytilus* are cultivated in huge numbers for the table. The world annual production is about 500 million pounds, with the Netherlands and France farming about 40 percent of the supply. Chile, which has had a tradition of mussel fishing since the Spanish arrived in the 1600s, leads among the South American countries.

Mytilus edulis is the most abundant shellfish in New England, but it has never gained the popularity that it has in France, where tons are eaten. A small fishery in New England produces about a million pounds of meat and shells, but they are distributed mainly to large cities where there are people of French and Italian extraction. Mussels are not generally eaten to any great extent on the Pacific coast of the United States, mainly because of the dangers of paralytic mussel poisoning at certain seasons. The leading gastropod fishery in the United States involves several large species of *Haliotis* abalones found in commercial sizes and quantity only on the Pacific coast.

Each year, about a million pounds of meat are marketed. It wholesale value exceeds that of the 23 million pounds of squid caught annually in United States waters. The abalone fishery goes back into Indian days, long before the advent of white man. Overfishing and the return of the sea otter have reduced this fishery in recent years. Formerly, helmet divers, and now, scuba divers, obtained the abalones in commercial quantities. The most important species is the large Red Abalone, which reaches a size of about ten inches. It customarily lives attached to rocks, where it grazes on algae at a depth of twenty to fifty feet, and is particularly abundant along the coast of central California. The other ten species of Pacific coast abalones are rarely exploited commercially because of their smaller size and increasing rarity. Sportsmen obtain large quantities, but strict California fisheries laws reduce to catch by size restrictions and limitation of the manner in which they may be collected.

Eventually, these species may come back to their former abundance. Abalones are also extensively fished in Japan, South Africa and South Australia. The foot, when pounded to make it tender tastes not unlike scallop. The remainder of the soft parts is discarded. Throughout the West Indies, and particularly in the Bahamas, the nearly foot-long Pink Conch, *Strombus gigas*, has been a mainstay meat for the natives for many centuries. At its height, the stromb fishery in the Bahamas was worth nearly a quarter of a million dollars a year. The huge piles of dead shells seen almost everywhere throughout the Bahamas attest to their use as a food. Supplies are dwindling, and in the Lower Florida Keys adult specimens are now rarely found. The large muscular foot is used in chowders, fritters and, when pounded, sliced and diced, is used raw in conch cocktails and seafood salads.

In the eastern United States, the *Busycon* whelks, or conchs, were popular with the Indians, and in recent times, largely because of Italian tastes, a large fishery is developing from Massachusetts to Virginia. The conchs, of which there are three main species, reach a shell length of six to ten inches. They are gathered by hand or dredged in shallow water.

Many are sent alive to fish markets; others are sent to large canning operations in New Jersey and New York. From one-half to one million pounds are landed each year. Some are ground up and dumped into the water to serve as bait for sport fishing. In England, Ireland and to a limited extent in New England, the Common European Periwinkle is gathered along the rocky shores by hand and sent alive to market in baskets. Before being eaten, the small snails are boiled and sometimes soaked in vinegar or dipped in hot garlic butter. About 50,000 pounds were harvested in New England in 1967, but an amount many hundred times that is annually consumed in northwest Europe. To a lesser extent, and mostly by picnickers visiting the seashore, the *Nucella* dogwinkles are eaten, either roasted over open or boiled.

Edible Cephalopods

The thought of eating squids and octopus is not particularly appealing to Americans and northern Europeans, despite the fact that they are fond of fish and many other shellfish. Yet the cephalopods from a worldwide standpoint, and second only to the oysters in annual catch. The Japanese land and eat over a billion pounds of squid a year, and the Spanish annually catch over 17 million pounds of *Sepia* squid, which they can in the squid's black ink and export to foreign countries. A mere 200,000 pounds of *Octopus* are caught and eaten each year by the Japanese, and throughout the East Indies and South Seas, women make nightly trips to the reef to catch the next day's meal of octopus. However, over 8,000 tons of squid are annually sold in the fish markets of California, Oregon, Washington and New York City. Modern supermarkets in most part of the United States now shell five-pound packages of frozen squid.

In the west, they are eaten mainly by people of Chinese and Japanese extraction and in the east by Italians, Spaniards and Puerto Ricans. The canning of huge quantities of squid for export has been thriving in California for more than fifty years. In 1946, some 30 million cans of squid were sold to Mediterranean and Far Eastern countries. For many years, in Monterey, squid were slit open, hung on open racks and sun-cured for future shipment to the Orient. Squid are caught by jigging, a procedure used to attract and ensnare the squid. One fisherman may sometimes land squid at the rate of 1,200 an hour. Nets, traps and creels are also used in many parts of the world. Octopuses are sometimes, caught by baiting them into jars lowered to the bottom of the sea. Off Naples, Italy, a basket similar to a lobster pot is used, from which the octopus cannot escape.

Squid are sometimes caught in large numbers, occasionally as many as 16,000 bushels in a single evening off the coast of Maine. The main squid fishery of eastern America is centered about Newfoundland, where *illex illecebrosus* is a seasonal migrant from July to September. The annual catch fluctuates radically, but in good years as many as 20 million pounds may be landed, frozen and sold in Norway and Portugal for use as codfish bait. Near Newfoundland and to Cape Cod and Cape Hatteras, North Carolina, the bait squid, *Loligo pealei*, is commonly trawled in large numbers, usually about 1,500 metric tons per year. Many of them are sold to Spanish canners. The Philippines produce, for internal consumption, meat from the chambered nautilus. The shells are saved and sold to a ready overseas market. About a hundred thousand *Nautilus* are fished each year in this erea.

The *Sepia* squids of the Mediterranean produce an internal, calcareous cuttlefish bone which is commercially harvested by the ton and exported for use in the manufacture of toothpaste and as "bones" for pet cage-birds. In Roman times, the soft, lightweight cuttlebones were ground to a powder, tinted with purple dye and used as a cosmetic rouge. Yearly, several hundred tons are collected along the southern shores of the Mediterranean and sold to wholesalers. For many centuries, the ink from the sacs of the *Sepia* cuttlefish of the Mediterranean was a major source of brown coloring matter. It was used also as a writing ink. Its use has now been replaced by aniline dyes because sepia fades in light and will not mix well in oils.

22

PEARL AND PEARL INDUSTRY

A pearl is a consecretion produced by a mollusc. Its culture is paying industry these days. The oldest record of using pearls go back to the Chinese people. They are said to have used them as for back as 2300 B. C. Royal families and nobles of every country in age have adorned their persons with pearls.

What is a Pearl?

A pearl is a concretion formed by mollusc. It consists of *nacre* or mother of pearl as the mollu's shell. It is secreted by the mantle, the inner muscular lining of the shell, as a means of protection against foreign body. A faultless pearl in formed by the deposition of nacreous

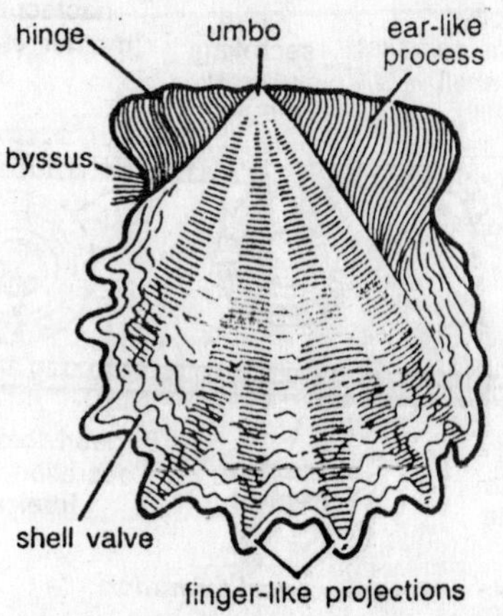

Fig. 22.1. Pearl oyster (Pinctada vulgaris).

substances arranged one after the other in a concentric layer and laid down by the epithelium of mantle.

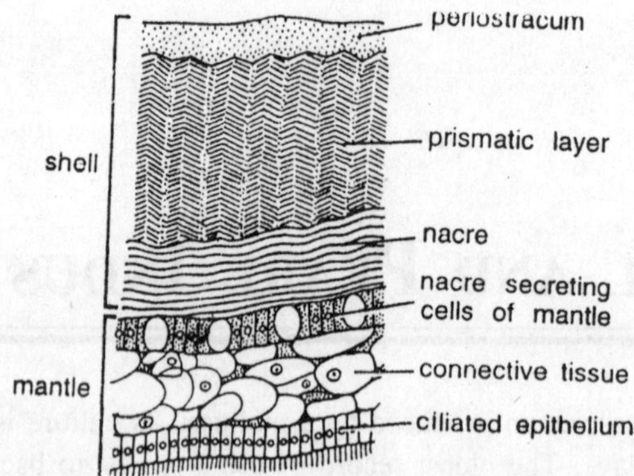

Fig. 22.2. Unio. A part of shell and mantle in T.S.

Pearl Formation

The pearl formation is an adaptation against outside materials. When a foreign material such as a sand-grain or a prarsite happens to enter the body it adheres with the mantle. The mantle epithelium immediately grows over the material in the form of a sac and enclose it. This mantle epithelium start secreting concentric layers of nacre around the foreign material. The complete structure is called pearl.

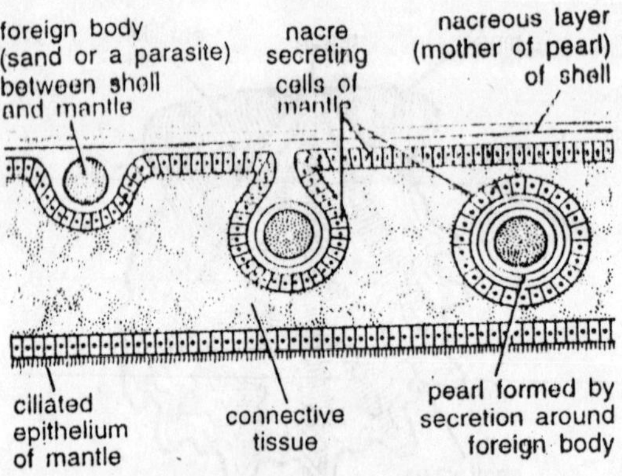

Fig. 22.3. Stages in pearl formation

Composition of Pearl

The pearl is formed by nacre. The nacre is formed of two substances, the calcium carbonate which is in the form of *argonite* or *calcite* and albuminoid substance called *conchiolin* ($C_{30}H_{42}N_2O_{11}$). The chemical analysis of a typical pearl is:

Calcium Carbonate	88.90%
Organic matter	3.8–5.9%
Water	2–3%
Residue	0.1–0.8%

Types of Pearls

Following are the four main types of pearls:

1. *Cultured pearls.* These are the pearls obtained from cultured species of pearls oysters.
2. *Hem pearls.* These are formed near the margin of mantle where the dark pigments usually produce brown or blackish pearls.
3. *Blister pearls.* When some foreign matter such as and, worms, crabs, fish or pieces of shell becomes lodged between the fleshy mantle and the shell blister pearls are formed, such pearls are attached to shell.
4. *Seed pearls*. The small pearls are called seed pearls.
5. *Baroque pearls.* There pearls are most likely to be perfectly spherical, for they are formed inside the flesh of mantle where succeeding layers of nacre may be added on all sides.

Pearl Producing Animals

Pearls are produced by bivalve molluscs. These are marine as well as fresh water:

1. The fresh water mollusc is: Unio margaritifera.
2. The marine molluscs are: *Pinctada Vulgaris*, *P. fucta*, *P. chemnitzi*, *P. margartifera*, *P. anomioides*, *P. atropurpurea*, *Haliotis*, *Mytilus*, *Placuna blacenta*, *P. maxima*. The species of mollusc which is mainly is *Pinctada vulgaris*.

Culture of Pearls

The culture of pearls is a complex but sensitive process. It is involves the following steps:

Collection of Oysters

1. *Japanese Method of Oyster Culture.* The larvae of the oyster freely swim in the shallow waters of the sea they are called *spats*. They are maintained in nurseries upto a certain time. For collection of spats different methods are employed. They are: (a) Semi-cylindrical tiles, (b) Bamboo poles with branches, (c) Shells of various molluscs, (d) Plastic nests.

2. The oysters are collected from the bottom of the sea.
3. In the laboratories the eggs of pearl oysters are fertilized and young ones are obtained.

Preparation of Graft Tissue

The piece of tissue which is inserted into the oyster is called graft-tissue. The graft must be in the form of square of 2 × 2 mm in size and usually it is cut off form the mantle of another oyster.

Preparation of Nucleus

The nucleus is a foreign material which is inserted into the oyster and is in the form of a bead 2 mm in diameter.

Implantation

The oyster's foot is exposed and a small incision is made on the foot. On this incision the graft tissue in placed. The nucleus is placed on the graft tissue. Then the oyster is released in the cage.

Rearing of Oysters

The operated oysters are placed in cages and are suspended in the sea from rafts. This type of *roft culture*. Rafts are rectangular wooden frames floating at the water surface. They are made to float by empty diesel drums. The rafts is kept in place by anchors. Besides raft other methods are there, these are long-line method, rack culture, pole culture and bottom-culture method.

Harvesting

Pearls attain their maximum size in three years and after the oysters are removed from cages and the pearl is taken out.

INDEX

R